STEWART'S CALCULUS
SECOND EDITION

Lecture Guide
and Student Notes
Volume I, Chapters 0 – 3

STEWART'S CALCULUS
SECOND EDITION
Lecture Guide
and Student Notes
Volume I, Chapters 0 – 3

Allen R. Strand
Colgate University

Stephen M. Kokoska
Bloomsburg University

Brooks/Cole Publishing Company
Pacific Grove, California

Brooks/Cole Publishing Company
A Division of Wadsworth, Inc.

Printed in the United States of America
10 9 8 7 6 5 4 3 2 1

Sponsoring Editor: *Jeremy Hayhurst*
Marketing Representative: *Mark DeWeese, Nathalie Cunningham*
Editorial Associate: *Nancy Champlin*
Production Editor: *Nancy L. Shammas*
Cover Design: *Katherine Minerva*
Cover Photo: *Lee Hocker Photography*
Typesetting: *Allen R. Strand, Stephen M. Kokoska*
Cover Printing: *Malloy Lithographing, Inc.*
Printing and Binding: *Malloy Lithographing, Inc.*

To the User

≡≡≡≡≡

   This manual is a companion to the Review and Preview Chapter and Chapters 1 – 3 from Stewart's *Calculus*, Second Edition. It forms a comprehensive basis for lecture notes for the teacher and student. First and foremost, this manual considers the needs of students learning calculus and removes the tedious task of note taking during lecture sessions. The intent is to remove the burden that often accompanies note taking during formal lectures and, consequently, hinders the student's understanding of the underlying notions of the "whats," "whys," and "hows" of calculus. Lectures can also be based upon the use of transparencies made from manual pages.

   The contents are organized to correspond directly with the textbook and include most definitions, theorems (some with proofs), and formulas appearing in the text. Concepts and solutions are explained in an informal, step-by-step fashion replete with supporting justifications to insure a better understanding. Note that the justifications are enclosed in parentheses and are placed in the text in a twofold manner. Justification for a relationship is indicated by an arrow when more than one relationship occurs per line or justification is placed on the right side of a page when a single relationship occurs per line. Examples are carefully written with detailed explanations and, when appropriate, are supported with accurate graphs. The "Remarks" sections offer comments about common errors, frequent misinterpretations, helpful hints, and alternative methods of solution.

   Students of calculus thrive with practice in problem-solving. By reducing the amount of note taking and increasing discussion during class sessions, the student gains immediate feedback related to difficulties. Using this manual helps develop student confidence and, ultimately, enhances understanding of calculus by fostering active exchanges between teacher and students.

Allen R. Strand
Stephen M. Kokoska

Contents

Chapter 3   The Mean Value Theorem and Curve Sketching

STEWART'S CALCULUS
SECOND EDITION

Lecture Guide
and Student Notes

Volume I, Chapters 0 – 3

0

Review and Preview

0.1   Numbers, Inequalities, and Absolute Values

Definition: A **set** is a collection of objects, and the objects are called **elements** of the set.

Notation:

(N1) Generally, use upper-case letters to denote sets and lower-case letters to represent elements.

(N2) To describe sets use braces { } within which specific elements are characterized.

(N3) If S is a set, the notation $x \in S$ indicates the object is an element of S, whereas $x \notin S$ means x is not an element of S.

(N4) A set having no elements, denoted $\emptyset$, is called the **null set** or **empty set**.

The set of real numbers, usually denoted by **R**, consists of the **rational numbers** and the **irrational numbers**.

Remarks:

(R1) The elements of the rational numbers are the **integers**

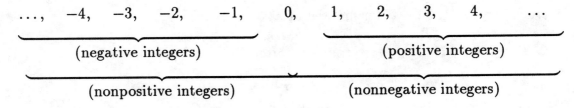

$$\ldots, \quad -4, \quad -3, \quad -2, \quad -1, \quad 0, \quad 1, \quad 2, \quad 3, \quad 4, \quad \ldots$$

(negative integers)   (positive integers)

(nonpositive integers)   (nonnegative integers)

and any number $r = m/n$, where m and n are integers and $n \neq 0$.

Note!

(N1) $r = m/n$ expressed as a ratio of integers.

(N2) If $n = 1$, the integers are represented as a ratio of integers.

(N3) The decimal representation of a rational number is a repeating decimal.

Example 1:

(A1) $\dfrac{7}{11} = 0.636363\ldots = 0.\,\underbrace{\overline{63}}$

(repeating sequence of digits)

(A2) $\dfrac{11}{7} = 1.571428571428\ldots = 1.\underbrace{\overline{571428}}$

(repeating sequence of digits)

(A3) $\dfrac{11}{8} = 1.375\,\underbrace{0000\ldots} = 1.375$ (a terminating decimal)

(repeating sequence of digits)

Note! A terminating decimal representation occurs when the repeating sequence of digits consists of zeros only.

(R2) The elements of the irrational numbers cannot be expressed as a ratio of integers.

Note!

(N1) The decimal representation of an irrational number is a nonrepeating decimal.

Example 2:

(A1) $\pi = 3.14159265358979323846264338327 9\ldots$

(A2) $\sqrt{3} = 1.7320508075688772935274463415 05\ldots$

(N2) Real numbers whose decimal representations do not terminate can be approximated by terminating its decimal expansion at a chosen place.

Example 3:

(A1) $\pi \approx 3.14159$:   terminated after 5 decimal places

(A2) $\dfrac{11}{7} \approx 1.571428$:   terminated after 6 decimal places

(R3) The real numbers can be represented by a **number scale** which associates the real numbers with points on a straight line. To set up a number scale:

(S1) select any point on the line as the **origin**, a reference point designated by the letter O and assigned the real number 0;

(S2) choose a sense of positive direction to the line and with a convenient unit of measure, locate the number 1 a unit's distance in the positive direction from the origin;

(S3) if x is a positive real number, represent x by a point x units in the positive direction from the origin whereas, if $-x$ is a negative number, represent $-x$ by a point x units in the negative direction from the origin.

Terminology: The number associated with a point P on the line is called the **coordinate** of P and the line is called a **coordinate line**.

Illustration: On a coordinate line, indicate the unit by marking the points associated with the integers.

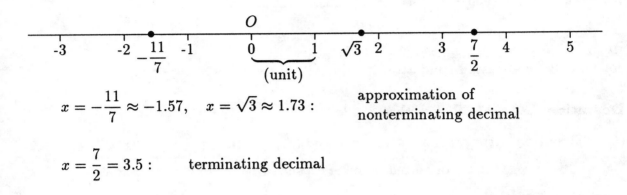

$$x = -\frac{11}{7} \approx -1.57, \quad x = \sqrt{3} \approx 1.73 : \qquad \text{approximation of nonterminating decimal}$$

$$x = \frac{7}{2} = 3.5 : \qquad \text{terminating decimal}$$

Definition: Order Relations for real numbers. Let $a, b \in \mathbf{R}$.

(D1) $a < b$ means $b - a$ is positive.   (a is less than b)

Note! On a coordinate line, the point associated with b lies more in the positive direction than the point associated with a.

(D2) $a \leq b$ means $a < b$ or $a = b$.   (a is less than or equal to b)

(D3) $a > b$ means $b < a$.   (a is greater than b)

(D4) $a \geq b$ means $a > b$ or $a = b$.   (a is greater than or equal to b)

Remarks:

(R1) $b > 0$ or $0 < b$: b is a positive number.

(R2) $b < 0$ or $0 > b$: b is a negative number.

Definition: Let $a, b \in \mathbf{R}$ such that $a \neq b$. The real number x is **between** a and b provided either

$$a < x < b \qquad \text{or} \qquad b < x < a$$

Illustration: Using to-the-right as the positive direction,

(I1) $a < x < b$ (hence, $a < b$)

(I2) $b < x < a$ (hence, $b < a$)

Remark: Between a and b indicates it is immaterial whether $a < b$ or $a > b$.

Definition: Let $a, b \in \mathbf{R}$ such that $a < b$.

(D1) The **open interval** from a to b, denoted (a, b), is the set $\underbrace{\{x \mid a < x < b\}}$.

(the set of all real numbers x such that x is between a and b)

Note! The **endpoints** a and b are not included in an open interval.

(D2) The **closed interval** from a to b, denoted $[a, b]$, is the set $\{x \mid a \leq x \leq b\}$.

Note!

(N1) $\{x \mid a \leq x \leq b\} = \{x \mid a < x < b\} \underset{\uparrow}{\cup} \{a, b\}$

(**union** operation: includes all elements in one set or the other (or in both sets))

(N2) The endpoints a and b are included in a closed interval.

(D2) The **half-open intervals** (or **half-closed intervals**) from a to b, denoted $(a, b]$ or $[a, b)$, are sets $\{x \mid a < x \leq b\}$ or $\{x \mid a \leq x < b\}$, respectively.

Note! Exactly one endpoint is not included in a half-open interval.

Remark: The use of a parenthesis in interval notation indicates an endpoint is excluded from the interval, whereas the use of a bracket indicates an endpoint is included in the interval.

Illustration:

(I1) Interval (a, b):

(I2) Interval $[a, b]$:

(I3) Interval $(a, b]$:

(I4) Interval $[a, b)$:

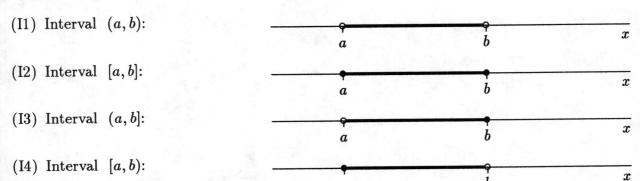

Remarks:

(R1) Geometrically, a **finite interval** (a, b), $[a, b]$, $(a, b]$, or $[a, b)$, when graphed on a coordinate line, corresponds to a *bounded* line segment.

(R2) Other intervals, which have at most one finite endpoint, are called **infinite intervals**.

Interval (a, ∞): $\{x \mid x > a\}$

Interval $[a, \infty)$: $\{x \mid x \geq a\}$

Interval $(-\infty, b)$: $\{x \mid x < b\}$

Interval $(-\infty, b]$: $\{x \mid x \leq b\}$

Interval $(-\infty, \infty)$: $\{x \mid x \text{ is a real number}\} = \mathbf{R}$

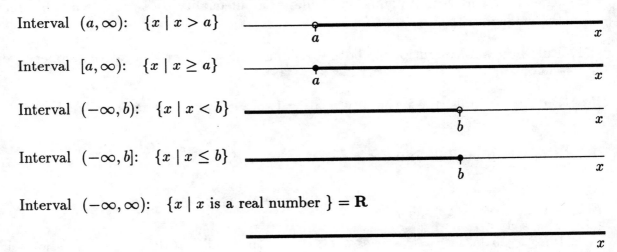

Note! The presence of a parenthesis when the symbols ∞ or $-\infty$ appear indicates that the interval, when graphed on a coordinate line, corresponds to an *unbounded* line segment which extends indefinitely in the direction indicated by these symbols.

Rules for Inequalities (2):

1. If $a < b$, then $a + c < b + c$.

2. If $a < b$ and $c < d$, then $a + c < b + d$.

3. If $a < b$ and $c > 0$, then $ac < bc$.

4. If $a < b$ and $c < 0$, then $ac > bc$.

5. If $0 < a < b$, then $1/a > 1/b$.

Remarks:

(R1) Rule 1, in words: adding the same number to both sides of an inequality retains the inequality.

(R2) Rule 2, in words: adding two inequalities of the same sense retains that inequality.

(R3) Rule 3, in words: multiplying both sides of an inequality by a positive number retains the inequality.

(R4) Rule 4, in words: multiplying both sides of an inequality by a negative number reverses the inequality.

(R5) Rule 5, in words: the inequality relating two positive numbers is reversed when their reciprocals are taken.

(R6) It follows from Rule 3: if $0 < a < b$, then $a^2 < b^2$.

Example 4: Solve $1 - 5x < 5 + 3x$.

$$1 - 5x < 5 + 3x \implies -5x < 4 + 3x \qquad \text{(Inequality Rule 1: add } -1 \text{ to both sides)}$$

$$\implies -8x < 4 \qquad \text{(Inequality Rule 1: add } -3x \text{ to both sides)}$$

$$\implies x > -\frac{1}{2} \quad \text{or} \quad \left(-\frac{1}{2}, \infty\right) \qquad \text{(Inequality Rule 4: multiply by } -1/8 < 0)$$

Check in original inequality:

$$x = 0 \in \left[-\frac{1}{2}, \infty\right): \qquad 1 - 5(0) < 5 + 3(0) ? \qquad 1 < 5: \text{ yes}$$

$$x = -1 \notin \left[-\frac{1}{2}, \infty\right): \qquad 1 - 5(-1) < 5 + 3(-1) \ ? \qquad\qquad 6 < 2: \ \text{no}$$

Example 5: Solve $x + 3 \le 2x - 1 < 3x$.

Solve each inequality separately.

$$x + 3 \le 2x - 1 \implies 4 \le x \qquad\qquad\qquad \text{(IR1: add } -x + 1 \text{ to both sides)}$$

$$2x - 1 < 3x \implies -1 < x \qquad\qquad\qquad \text{(IR1: add } -2x \text{ to both sides)}$$

Both inequalities must hold, so determine the **intersection** of the two sets (the elements common to both sets).

Overlap:

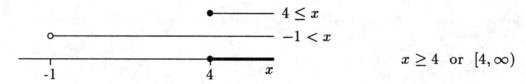

$$x \ge 4 \quad \text{or} \quad [4, \infty)$$

Check:

$$x = 5 \in [4, \infty): \quad 5 + 3 \le 2(5) - 1 < 3(5) \ ? \quad 8 \le 9 < 15: \ \text{yes}$$

$$x = 3 \notin [4, \infty): \quad 3 + 3 \le 2(3) - 1 < 3(3) \ ? \quad 6 \le 5 < 9: \ \text{no}$$

Example 6: Solve $2x - 1 \le 3x < x + 3$.

$$2x - 1 \le 3x \implies -1 \le x \qquad\qquad\qquad \text{(IR1: add } -2x \text{ to both sides)}$$

$$3x < x + 3 \implies 2x < 3 \qquad\qquad\qquad \text{(IR1: add } -x \text{ to both sides)}$$

$$\implies x < \frac{3}{2} \qquad\qquad\qquad \text{(IR3: multiplied both sides by } 1/2 > 0)$$

Overlap:

$$-1 \le x < \frac{3}{2} \quad \text{or} \quad \left[-1, \frac{3}{2}\right)$$

Check:

$$x = 0 \in \left[-1, \frac{3}{2}\right): \quad 2(0) - 1 \le 3(0) < 0 + 3? \quad -1 \le 0 < 3: \text{ yes}$$

$$x = 2 \notin \left[-1, \frac{3}{2}\right): \quad 2(2) - 1 \le 3(2) < 2 + 3? \quad 3 \le 6 < 5: \text{ no}$$

$$x = -2 \notin \left[-1, \frac{3}{2}\right): \quad 2(-2) - 1 \le 3(-2) < -2 + 3? \quad -5 \le -6 < 1: \text{ no}$$

Example 7: Solve $x^2 \ge x$.

Note! Can not multiply both sides by $1/x$ unless the cases for $x \ne 0$, namely, $x > 0$ and $x < 0$ are considered separately.

Another method is to use IR1 to obtain zero on one side of the inequality.

$$x^2 \ge x \;\;\Longrightarrow\;\; x^2 - x \ge 0 \;\;\Longrightarrow\;\; x(x-1) \ge 0$$
$$\uparrow$$

(IR1: add $-x$ to both sides)

Set each factor equal to zero: $\quad x = 0; \qquad x - 1 = 0 \;\;\Longrightarrow\;\; x = 1$

$$
\begin{array}{ccccccc}
+ & & 0 & & - & & 0 & & + & & x(x-1)
\end{array}
$$

$$
\begin{array}{ccccc}
 & 0 & & 1 & \\
x = -1 & & x = 1/2 & & x = 2 \\
\uparrow & & \uparrow & & \uparrow
\end{array} \quad x
$$

$$\left(\begin{array}{l} \text{use a test value of } x \text{ in each open interval to} \\ \text{determine the sign of } x(x-1) \text{ in that interval} \end{array} \right)$$

$$x(x-1) \ge 0: \qquad x \le 0 \quad \text{or} \quad x \ge 1$$

Example 8: Solve $x^3 - 2x^2 - 3x < 0$.

$$x^3 - 2x^2 - 3x < 0 \;\;\Longrightarrow\;\; x(x^2 - 2x - 3) < 0 \;\;\Longrightarrow\;\; x(x-3)(x+1) < 0$$

$$x = 0; \qquad x - 3 = 0 \implies x = 3; \qquad x + 1 = 0 \implies x = -1$$

$$
\begin{array}{ccccccc}
- & 0 & + & 0 & - & 0 & + \qquad x(x-3)(x+1)
\end{array}
$$

$$
\begin{array}{cccc}
& -1 & 0 & 3 \qquad x
\end{array}
$$

$$x(x-3)(x+1) < 0 \implies x < -1 \quad \text{or} \quad 0 < x < 3$$

Example 9: Solve $x - 2 > \dfrac{4}{x+1}$.

Note! Can not multiply both sides by $x + 1$ unless the cases for $x + 1 \neq 0$, namely, $x + 1 > 0$ and $x + 1 < 0$ are considered separately.

Another method is to use IR1 to obtain zero on one side of the inequality.

$$x - 2 > \frac{4}{x+1} \implies x - 2 - \frac{4}{x+1} > 0$$

$$\implies \frac{(x-1)(x+1) - 4}{x+1} > 0 \qquad\qquad (x+1: \text{ common denominator})$$

$$\implies \frac{x^2 - x - 6}{x+1} > 0 \implies \frac{(x-3)(x+2)}{x+1} > 0$$

Set each factor in numerator and denominator equal to zero:

$$x - 3 = 0 \implies x = 3; \qquad x + 2 = 0 \implies x = -2; \qquad x + 1 = 0 \implies x = -1$$

Note!

(N1) numerator $= 0$, denominator $\neq 0 \implies \dfrac{\text{numerator}}{\text{denominator}} = 0$

(N2) denominator $= 0 \implies \dfrac{\text{numerator}}{\text{denominator}}$ does not exist: DNE

$$
\begin{array}{ccccccc}
- & 0 & + & \text{DNE} & - & 0 & + \qquad \dfrac{(x-3)(x+2)}{x+1}
\end{array}
$$

$$
\begin{array}{cccc}
& -2 & -1 & 3 \qquad x
\end{array}
$$

$$\frac{(x-3)(x+2)}{x+1} > 0: \qquad -2 < x < -1 \quad \text{or} \quad x > 3$$

Example 10: Solve $\dfrac{x^2 + 2x}{x^2 + 1} \geq 0$.

$x^2 + 1 > 0$ for all $x \implies x^2 + 2x \geq 0$ need only be solved

↑

(IR3: multiplied both sides by $x^2 + 1 > 0$ for all x)

$\implies x(x+2) \geq 0$

$x = 0; \qquad x + 2 = 0 \implies x = -2$

```
        +      0        −        0    +      x(x + 2)
   ─────────────────┬──────────────────┬──────────
                   -2                   0          x
```

$x(x+2) \geq 0: \qquad x \leq -2 \quad \text{or} \quad x \geq 0$

Definition: The **absolute value** of a real number a, denoted by $|a|$, is defined by

$$|a| = \begin{cases} a & \text{if } a \geq 0 \\ -a & \text{if } a < 0 \end{cases}$$

Remarks:

(R1) Sometimes stated as: $|a| = \max\{a, -a\}$ \qquad (choose the greater between a and $-a$)

(R2) $\begin{cases} a \geq 0 \implies |a| = a \geq 0 \\ a < 0 \implies -a > 0 \implies |a| = -a > 0 \end{cases} \implies |a| \geq 0$

Note! $|a| = 0 \iff a = 0$

(R3) $\sqrt{a^2} = |a|$ \qquad (≥ 0: positive square root)

Example 11:

(A1) $\sqrt{3^2} = |3| = 3$ \hfill $(3 > 0)$

(A2) $\sqrt{(-3)^2} = |-3| = -(-3) = 3$ $\qquad\qquad\qquad\qquad\qquad\qquad\qquad$ $(-3 < 0)$

$\qquad\qquad$ *Note!* $\sqrt{(-3)^2} = \sqrt{9} = 3$

(R4) $|a|$: the **distance** between a and O on a coordinate line.

$\qquad$ *Note!*

$\qquad$ (N1) $|a| \geq 0$: distance is nonnegative.

$\qquad$ (N2)

(R5) Property: $|a| = |-a|$.

$\qquad$ Proof:

$$|-a| = \begin{cases} -a & \text{if } -a \geq 0 \\ -(-a) & \text{if } -a < 0 \end{cases} = \begin{cases} -a & \text{if } a \leq 0 \\ a & \text{if } a > 0 \end{cases} \underset{\uparrow}{=} \begin{cases} -a & \text{if } a < 0 \\ a & \text{if } a \geq 0 \end{cases} = |a|$$

$$(a = 0: -a = a)$$

Properties of Absolute Values (5): Suppose a and b are any real numbers and n is an integer. Then

1. $|ab| = |a||b|$

2. $\left|\dfrac{a}{b}\right| = \dfrac{|a|}{|b|}$ $\qquad b \neq 0$

3. $|a^n| = |a|^n$

Properties of Absolute Values (6): Suppose $a > 0$. Then

4. $|x| = a$ if and only if $x = \pm a$.

5. $|x| < a$ if and only if $-a < x < a$.

6. $|x| > a$ if and only if $x > a$ or $x < -a$.

Remarks:

(R1) Properties 4-6 are used to solve equations and inequalities involving absolute values.

(R2) Properties 4 and 5 remain true if the *strict* inequalities $<$ and $>$ are replaced by $\leq$ and $\geq$, respectively.

(R3) For $a > 0$: $|a| = a$

$$|x| < a \quad \Longleftrightarrow \quad -a < x < a$$

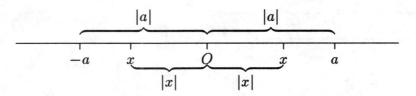

$|x| < a$ represents the set of all points x on a coordinate line such that the distance between x and O is less than the distance between a and O and $-a$ and O.

Note! $x \leq 0$ is possible.

(R4) For $a > 0$: $|a| = a$

$$|x| > a \quad \Longleftrightarrow \quad x > a \quad \text{or} \quad x < a$$

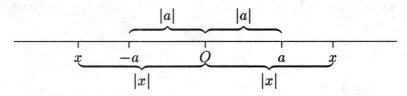

$|x| > a$ represents the set of all points x on a coordinate line such that the distance between x and O is greater than the distance between a and O and $-a$ and O.

Example 12: Rewrite $|3 - 2x|$ without using the absolute value symbol.

$$|3 - 2x| = \begin{cases} 3 - 2x & \text{if } 3 - 2x \geq 0 \\ -(3 - 2x) & \text{if } 3 - 2x < 0 \end{cases} = \begin{cases} 3 - 2x & \text{if } x \leq 3/2 \\ 2x - 3 & \text{if } x > 3/2 \end{cases}$$

Example 13: Solve.

(A1) $|3 - 2x| = 4$

$$|3 - 2x| = 4 \implies 3 - 2x = 4 \quad \text{or} \quad 3 - 2x = -4 \qquad (|\pm 4| = 4)$$

$$\implies \quad x = -\frac{1}{2} \quad \text{or} \quad x = \frac{7}{2} \implies x = -\frac{1}{2}, \frac{7}{2}$$

Check: $x = -\frac{1}{2}$: $\left| 3 - 2\left(-\frac{1}{2}\right) \right| = |4| = 4$: yes

$$x = \frac{7}{2}: \quad \left| 3 - 2\left(-\frac{7}{2}\right) \right| = |-4| = 4: \quad \text{yes}$$

(A2) $|3 - 2x| < 4$

$$|3 - 2x| < 4 \implies -4 < 3 - 2x < 4 \implies -7 < -2x < 1$$
$$\uparrow$$
(Absolute Value Property 5: $a = 4 > 0$)

$$\implies \quad \frac{7}{2} > x > -\frac{1}{2} \quad \text{or, rewritten,} \quad -\frac{1}{2} < x < \frac{7}{2} \implies x \in \left(-\frac{1}{2}, \frac{7}{2}\right)$$

(A3) $|3 - 2x| \geq 4$

$$|3 - 2x| \geq 4 \implies \left\{ \begin{array}{l} 3 - 2x \geq 4 \\ 3 - 2x \leq -4 \end{array} \right\} \implies \left\{ \begin{array}{l} -2x \geq 1 \\ -2x \leq -7 \end{array} \right\}$$
$$\uparrow$$
(Absolute Value Property 6: $a = 4 > 0$)

$$\implies \left\{ \begin{array}{l} x \leq -1/2 \\ x \geq 7/2 \end{array} \right\} \implies x \in \left(-\infty, -\frac{1}{2}\right] \cup \left[\frac{7}{2}, \infty\right)$$

(A4) $|3 - 2x| = -4$

$|3 - 2x| \geq 0$ for all $x \in (-\infty, \infty) \implies$ no x satisfy $|3 - 2x| = -4 < 0$

(A5) $|3 - 2x| < -4$

$|3 - 2x| \geq 0$ for all $x \in (-\infty, \infty) \implies$ no x satisfy $|3 - 2x| < -4 < 0$

(A6) $|3 - 2x| > -4$

$$|3 - 2x| \geq 0 \text{ for all } x \in (-\infty, \infty) \implies |3 - 2x| > -4 \text{ for all } x \in (-\infty, \infty)$$

$$\uparrow$$

$$(0 > -4)$$

(A7) $|3 - 2x| = 0$

$$|3 - 2x| = 0 \implies 3 - 2x = 0 \implies x = \frac{3}{2}$$

(A8) $1 < |3 - 2x| \leq 5$

(Case 1) Assume $3 - 2x \geq 0$: $x \leq \dfrac{3}{2}$

$$1 < |3 - 2x| \leq 5 \implies 1 < 3 - 2x \leq 5 \implies -2 < -2x \leq 2$$

$$\implies 1 > x \geq -1 \quad \text{or, rewritten,} \quad -1 \leq x < 1$$

Overlap (for assumption and inequality):

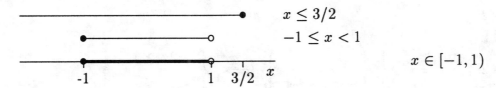

$x \leq 3/2$

$-1 \leq x < 1$

$x \in [-1, 1)$

(Case 2) Assume $3 - 2x < 0$: $x > \dfrac{3}{2}$

$$1 < |3 - 2x| \leq 5 \implies 1 < -(3 - 2x) \leq 5 \implies 4 < 2x \leq 8 \implies 2 < x \leq 4$$

Overlap (for assumption and inequality):

$x > 3/2$

$2 < x \leq 4$

$x \in (2, 4]$

Combine cases: $x \in [-1, 1) \cup (2, 4]$

(A9) $0 < |3 - 2x| \leq 5$

$$|3 - 2x| > 0 \implies 3 - 2x \neq 0 \implies x \neq \frac{3}{2}$$

$$\uparrow$$

$$(3 - 2x = 0 \implies x = 3/2)$$

$$|3 - 2x| \leq 5 \implies -5 \leq 3 - 2x \leq 5 \implies -8 \leq -2x \leq 2$$

$$\implies 4 \geq x \geq -1 \quad \text{or, rewritten,} \quad -1 \leq x \leq 4$$

Hence, $-1 \leq x \leq 4, \ x \neq \dfrac{3}{2} \implies x \in \left[-1, \dfrac{3}{2}\right) \cup \left(\dfrac{3}{2}, 4\right]$

(A10) $|9 - x^2| < 5$

    (C1) Assume $9 - x^2 \geq 0$

$$9 - x^2 \geq 0 \implies x^2 - 9 \leq 0 \implies (x - 3)(x + 3) \leq 0$$

$$x - 3 = 0 \implies x = 3; \quad x + 3 = 0 \implies x = -3$$

$$
\begin{array}{ccccccc}
+ & & 0 & & - & & 0 & & + & & (x-3)(x+3) \\
\hline
& & & -3 & & & & 3 & & & x
\end{array}
$$

$$(x - 3)(x + 3) \leq 0 \implies 9 - x^2 \geq 0: \ -3 \leq x \leq 3$$

$$9 - x^2 < 5 \implies 4 - x^2 < 0 \implies x^2 - 4 > 0 \implies (x - 2)(x + 2) > 0$$

$$\implies x - 2 = 0 \implies x = 2; \quad x + 2 = 0 \implies x = -2$$

$$
\begin{array}{ccccccc}
+ & & 0 & & - & & 0 & & + & & (x-2)(x+2) \\
\hline
& & & -2 & & & & 2 & & & x
\end{array}
$$

$$(x - 2)(x + 2) > 0 \implies 9 - x^2 < 5: \ x < -2 \text{ or } x > 2$$

Overlap (for assumption and inequality)

$-3 \leq x \leq 3$

$x < -2 \text{ or } x > 2$

$$\implies \quad x \in [-3, -2) \cup (2, 3]$$

(C2) Assume $9 - x^2 < 0$: $\underbrace{x < -3 \ \text{ or } \ x > 3}_{\text{(from (C1) above)}}$

$$-(9 - x^2) < 5 \quad \implies \quad x^2 - 14 < 0 \quad \implies \quad (x - \sqrt{14})(x + \sqrt{14}) < 0$$

$$x - \sqrt{14} = 0 \quad \implies \quad x = \sqrt{14} \approx 3.74; \quad x + \sqrt{14} = 0 \quad \implies \quad x = -\sqrt{14} \approx -3.74$$

$$(x - \sqrt{14})(x + \sqrt{14}) < 0 \quad \implies \quad -(9 - x^2) < 5: \quad -\sqrt{14} < x < \sqrt{14}$$

Overlap:

$$\implies \quad x \in (-\sqrt{14}, -3) \cup (3, \sqrt{14})$$

Combine cases: $x \in (-\sqrt{14}, -2) \cup (2, \sqrt{14})$

(A11) $x \le |3 - 2x|$

(C1) Assume $3 - 2x \ge 0$: $x \le \dfrac{3}{2}$

$$x \le |3 - 2x| \quad \implies \quad x \le 3 - 2x \quad \implies \quad 3x \le 3 \quad \implies \quad x \le 1$$

Overlap:

(C2) Assume $3 - 2x < 0$: $x > \dfrac{3}{2}$

$$x \leq |3 - 2x| \implies x \leq -(3 - 2x) \implies 3 \leq x$$

Overlap:

$$x \in [3, \infty)$$

Combine cases: $x \in (-\infty, 1] \cup [3, \infty)$

(A12) $|x| \leq |3 - 2x|$

Note! The method of solution using cases would require four cases: two for $|x|$ and, for each of those, two for $|3 - 2x|$.

$$|x| \leq |3 - 2x| \implies |x|^2 \leq |3 - 2x|^2$$
$$\uparrow$$
$$(0 \leq |x| \leq |3 - 2x|)$$

$$\implies |x^2| \leq |(3 - 2x)^2| \qquad\qquad \text{(Absolute Value Property 3: } n = 2)$$

$$\implies x^2 \leq (3 - 2x)^2 \implies x^2 \leq 9 - 12x + 4x^2 \implies 3x^2 - 12x + 9 \geq 0$$
$$\uparrow$$
$$\left(x^2 \geq 0: \ |x^2| = x^2; \ (3 - 2x)^2 \geq 0: \ |(3 - 2x)^2| = (3 - 2x)^2\right)$$

$$\implies x^2 - 4x + 3 \geq 0 \implies (x - 3)(x - 1) \geq 0$$

$$x - 3 = 0 \implies x = 3; \quad x - 1 = 0 \implies x = 1$$

$|x| \leq |3 - 2x|$: $x \leq 1$ or $x \geq 3$

$$\implies x \in (-\infty, 1] \cup [3, \infty)$$

(A13) $\left|\dfrac{3 - 2x}{x + 1}\right| \geq 1$

Denominator: $x + 1 \neq 0 \implies x \neq -1$

$\left| \dfrac{3 - 2x}{x + 1} \right| \geq 1 \implies \dfrac{|3 - 2x|}{|x + 1|} \geq 1$ (absolute Value Property 2)

$\implies |3 - 2x| \geq |x + 1|$ ($|x + 1| > 0$)

$\implies |3 - 2x|^2 \geq |x + 1|^2$ ($|3 - 2x| \geq |x + 1| \geq 0$)

$\implies |(3 - 2x)^2| \geq |(x + 1)^2|$ (Absolute Value Property 3: $n = 2$)

$\implies (3 - 2x)^2 \geq (x + 1)^2$ (($3 - 2x)^2 \geq 0$; $(x + 1)^2 \geq 0$)

$\implies 9 - 12x + 4x^2 \geq x^2 + 2x + 1 \implies 3x^2 - 14x + 8 \geq 0$

$\implies (3x - 2)(x - 4) \geq 0$

$3x - 2 = 0 \implies x = \dfrac{2}{3}$; $x - 4 = 0 \implies x = 4$

$\left| \dfrac{3 - 2x}{x + 1} \right| \geq 1 :\ x \neq -1$ with $x \leq 2/3$ or $x \geq 4$

$\implies x \in (-\infty, -1) \cup (-1, 2/3] \cup [4, \infty)$

Definition: If a and b are real numbers, then the **distance** between a and b on a coordinate line is $|a - b|$.

Remarks:

(R1) $|a - b| = |(-1)(b - a)| = |-1||b - a| = |b - a|$
                            ↑
                (Absolute Value Property 1)

$\implies$ the distance between a and b is also given by $|b - a|$

(R2) $a > b$

$$|a - b| = |b - a|$$

(R3) $a < b$

$$|a - b| = |b - a|$$

(R4) In geometric terms: if P_1 and P_2 are the points on a coordinate line associated with a and b, respectively, then the distance between P_1 and P_2, denoted by $|P_1 P_2|$, is $|P_1 P_2| = |a - b|$.

Note!

(N1) $|P_1 P_2|$ denotes the length of a line segment connecting P_1 and P_2

↑   ↑

(not the absolute value symbol)

(N2) $|P_1 P_2| = |P_2 P_1|$

(N3)

$$|P_1 P_2| = |a - b| = |b - a| \geq 0$$

with $|P_1 P_2| = 0 \iff P_1$ and P_2 are the same point

(R5) Let a and b be real numbers such that $b > 0$. $|x - a| < b$ represents the set of all points x on a coordinate line such that the distance between x and a is less than the value b.

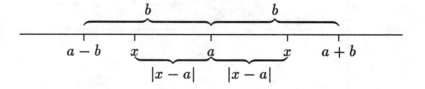

Note!

(N1) Distance between $a - b$ and a:

$$|(a - b) - a| = |-b| = -(-b) = b$$

↑

$$(b > 0 \implies -b < 0)$$

(N2) Distance between a and $a + b$: $|a - (a + b)| = |-b| = b$

(N3) $|x - a| < b \implies -b < x - a < b$ (Absolute Value Property 5)

$\implies a - b < x < a + b \implies x \in (a - b, a + b)$

Property: If a is a real number, then $-|a| \le a \le |a|$.

Proof:

(C1) Assume $a \ge 0$.

$$a \ge 0 \implies \left\{ \begin{array}{l} -a \le 0 \\ |a| = a \end{array} \right\} \implies -a \le a = |a| \implies -|a| \le a = |a|$$

$$\underset{(-a = -|a|)}{\uparrow}$$

(C2) Assume $a < 0$

$$a < 0 \implies \left\{ \begin{array}{l} -a > 0 \\ |a| = -a \end{array} \right\} \implies -a = |a| > a$$

$$\implies \underset{\substack{\uparrow \\ \text{(multiply by } -1 \ (< 0))}}{a = -|a| < -a} \quad \implies \underset{\substack{\uparrow \\ (-a = |a|)}}{-|a| = a < |a|}$$

Combine cases: $-|a| \le a \le |a|$

The Triangle Inequality (7): If a and b are any real numbers, then

$$|a + b| \le |a| + |b|$$

Proof:

Preceding property: $-|a| \le a \le |a|$ and $-|b| \le b \le |b|$

$\implies -|a| + (-|b|) \le a + b \le |a| + |b|$ (Inequality Rule 2)

$\implies -(|a| + |b|) \le a + b \le |a| + |b|$

$$\Longrightarrow \quad |a + b| \le \Big||a| + |b|\Big| \qquad\qquad \text{(preceding property: } |a| + |b| \text{ replaces } a)$$

$$\Longrightarrow \quad |a + b| \le |a| + |b| \qquad\qquad\qquad\qquad (|a| + |b| \ge 0)$$

Remarks:

(R1) a, b: same sign $|a + b| = |a| + |b|$

(R2) a, b: opposite signs $|a + b| < |a| + |b|$

Example 14: If $|x+2| < 1$ and $|y-3| < 4$, use the Triangle Inequality to estimate $|x+2y-4|$.

$$|x + 2y - 4| = |(x + 2 - 2) + 2(y - 3 + 3) - 4| \qquad \left(\begin{array}{l} \text{given information in terms of} \\ x + 2 \text{ and } y - 3; \ \ \text{add zero} \end{array}\right)$$

$$= |(x + 2) + 2(y - 3) + (-2 + 6 - 4)| = |(x + 2) + 2(y - 3)|$$

$$\le |x + 2| + |2(y - 3)| \qquad\qquad \left(\begin{array}{l} \text{Triangle Inequality} \\ a = x + 2, \ \ b = 2(y - 3) \end{array}\right)$$

$$= |x + 2| + 2|y - 3| \qquad\qquad\qquad \text{(Absolute Value Property 1: } |2| = 2)$$

$$< 1 + 2(4) = 9 \Longrightarrow \ \ |x + 2y - 4| < 9$$

Remark: The Triangle Inequality can be extended to the sum of more than two real numbers.

The Triangle Inequality: Let n be a positive integer. If $a_1, a_2, \ldots, a_n$ are real numbers, then

$$|a_1 + a_2 + \cdots + a_n| \le |a_1| + |a_2| + \cdots + |a_n|$$

Remark: In words: the absolute value of a sum is less than or equal to the sum of the absolute values.

Example 15: If $|x+2| < 1$ and $|y-3| < 4$, use the Triangle Inequality to estimate $|x+2y-5|$.

$$|x + 2y - 5| = |(x + 2 - 2) + 2(y - 3 + 3) - 5| \qquad\qquad \text{(add zero)}$$

$$= |(x + 2) + 2(y - 3) + (-2 + 6 - 5)| = |(x + 2) + 2(y - 3) + (-1)|$$

$$\leq |x+2| + |2(y-3)| + |-1| \qquad \text{(Triangle Inequality)}$$

$$= |x+2| + 2|y-3| + 1 \qquad (|2(y-3)| = |2||y-3| = 2|y-3|)$$

$$< 1 + 2(4) + 1 = 10 \implies |x+2y-5| < 10$$

Example 16: If $|x-1| < 2$, show that $|8-3x| < 11$.

$$|8-3x| = |8-3(x-1+1)| \qquad \text{(given information in terms of } x-1; \text{ add zero)}$$

$$= |8-3-3(x-1)| = |5-3(x-1)|$$

$$\leq |5| + |-3(x-1)| \qquad \text{(Triangle Inequality: } 5-3(x-1) = 5 + [-3(x-1)])$$

$$= 5 + 3|x-1| < 5 + 3(2) = 11 \implies |8-3x| < 11$$

0.2   Coordinate Geometry and Lines

A **rectangular**, or **Cartesian**, **coordinate system** is constructed in a plane by drawing two coordinate lines *perpendicular* to one another such that they intersect at the zero points of each line. These coordinate lines, one of which is usually drawn horizontally while the other is drawn vertically, are called **coordinate axes**. The horizontal axis, usually referred to as the **x axis**, has its positive direction to the right while the vertical axis, usually called the **y axis**, has its positive direction upwards.

A point P in the plane is associated with an *order pair* (a, b), where a and b are real numbers. To plot a point $P(a, b)$ locate the value a, called the **abscissa** or **x coordinate**, on the x axis and draw a line through a which is parallel to the y axis and then locate the value b, called the **ordinate** or **y coordinate**, on the y axis and draw a line through b which is parallel to the x axis. The point of intersection of these two lines is the location of the point $P(a, b)$ in the plane, called the **Cartesian plane**.

Remarks:

(R1) All points of the form $(x, 0)$ lie on the x axis.

(R2) All points of the form $(0, y)$ lie on the y axis.

(R3) $(0, 0)$, the point of intersection of the axes, is called the **origin** and designated by the letter O.

(R4) The coordinate axes separate the plane into four **quadrants** and the quadrant in which a point lies is determined by the sign of its coordinates:

Quadrant	Coordinate Signs
I	$(+,+)$
II	$(-,+)$
III	$(-,-)$
IV	$(+,-)$

(R5) The Cartesian Plane:

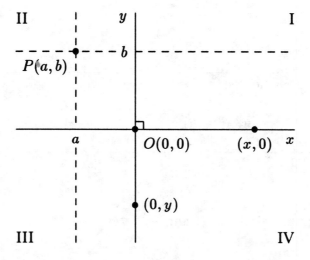

(R6) The unit measure may differ between the axes.

Example 1: Locate the points $(4,5)$, $(-1,3)$, $(0,-2)$, $(-3,-1)$, $(-1,-3)$, and $(3,0)$ in a Cartesian coordinate system.

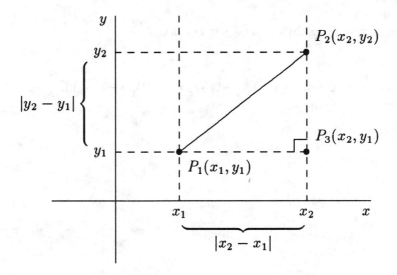

Distance Formula (8): The distance between the points $P_1(x_1, y_1)$ and $P_2(x_2, y_2)$ is

$$|P_1 P_2| = \sqrt{(x_2 - x_1)^2 + (y_2 - y_1)^2}$$

Remarks:

(R1) Derivation of formula via right triangles:

Let P_3 be the point with coordinates (x_2, y_1)

$\implies$ P_1, P_3, and P_2 determine a right triangle

$\implies$ $|P_1P_2|^2 = |P_1P_3|^2 + |P_3P_2|^2$ (Pythagorean Theorem)

$\implies$ $|P_1P_2| = +\sqrt{|P_1P_3|^2 + |P_3P_2|^2}$ (distance ≥ 0)

$$= \sqrt{|x_2 - x_1|^2 + |y_2 - y_1|^2} = \sqrt{(x_2 - x_1)^2 + (y_2 - y_1)^2}$$
$\uparrow$
$(|P_1P_3| = |x_2 - x_1|,\ |P_3P_2| = |y_2 - y_1|$: distance along coordinate line)

(R2) $|P_2P_1| = \sqrt{(x_1 - x_2)^2 + (y_1 - y_2)^2} = \sqrt{(x_2 - x_1)^2 + (y_2 - y_1)^2} = |P_1P_2|$
$\uparrow$
$(x_1 - x_2 = -(x_2 - x_1);\ \ y_1 - y_2 = -(y_2 - y_1))$

$\implies$ the choice of coordinates to associate with P_1 and P_2 is immaterial.

(R3) $|P_1P_2| \geq 0$ with $|P_1P_2| = 0 \iff x_2 = x_1$ and $y_2 = y_1$: P_1 and P_2 are the same point.

(R4) $|OP_1| = \sqrt{(x_1 - 0)^2 + (y_1 - 0)^2} = \sqrt{x_1^2 + y_1^2}$

Example 2: Find the distance between the points:

(A1) $(-1, 3)$ and $(4, 5)$.

A choice: $P_1(-1, 3)$, $P_2(4, 5)$

$\implies$ $|P_1P_2| = \sqrt{[4 - (-1)]^2 + (5 - 3)^2} = \sqrt{5^2 + 2^2} = \sqrt{29} \approx 5.39$

or, another choice: $P_1(4, 5)$, $P_2(-1, 3)$

$\implies$ $|P_1P_2| = \sqrt{(-1 - 4)^2 + (3 - 5)^2} = \sqrt{(-5)^2 + (-2)^2} = \sqrt{29} \approx 5.39$

(A2) $(-1, 3)$ and $(0, -2)$.

A choice: $P_1(-1, 3)$, $P_2(0, -2)$

$$\implies \quad |P_1P_2| = \sqrt{[0-(-1)]^2 + (-2-3)^2} = \sqrt{26} \approx 5.10$$

Definition: Let L be a line in the x, y plane passing through the distinct points (x_1, y_1) and (x_2, y_2)

(D1) If $x_1 \neq x_2$, then L is a **nonvertical line.**

(D2) If $x_1 = x_2$, then L is a **vertical line.**

Remarks:

(R1) A nonvertical line is not parallel to the y axis.

(R2) A vertical line is parallel to the y axis.

(R3) A nonvertical line which is parallel to the x axis is called a **horizontal line.**

Definition (9): The **slope** of a nonvertical line that passes through the points $P_1(x_1, y_1)$ and $P_2(x_2, y_2)$ is

$$m = \frac{y_2 - y_1}{x_2 - x_1}$$

The slope of a vertical line is not defined.

Remarks:

(R1) Let L be a nonvertical line.

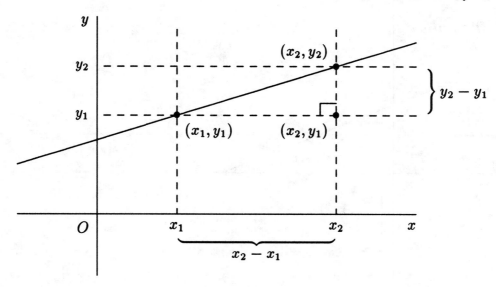

$$m = \frac{\text{change in } y}{\text{change in } x} = \frac{y_2 - y_1}{x_2 - x_1} = \frac{-(y_2 - y_1)}{-(x_2 - x_1)} = \frac{y_1 - y_2}{x_1 - x_2}$$

$\implies$ the choice of coordinates to associate with P_1 and P_2 is immaterial.

Note!

(N1) In the formula for m, the x coordinate of a point is "directly below" its corresponding y coordinate at the point.

(N2) If L is a horizontal line, then $y_2 = y_1$ for the distinct points P_1 and P_2

$$m = \frac{y_2 - y_1}{x_2 - x_1} \underset{\uparrow}{=} \frac{0}{x_2 - x_1} \underset{\uparrow}{=} 0$$

$$(y_2 = y_1) \qquad (x_2 \neq x_1 \implies x_2 - x_1 \neq 0)$$

(R2) Consider the points $P_1(x_1, y_1)$ and $P_2(x_2, y_2)$ on L such that $x_2 > x_1$.

Note! $x_1 \neq x_2 \implies L$ is a nonvertical line.

(C1) $y_2 > y_1 \implies m > 0$ and L rises as x increases from x_1 to x_2.

(C2) $y_2 < y_1 \implies m < 0$ and L falls as x increases from x_1 to x_2.

(C3) $y_2 = y_1 \implies m = 0$ and L is a horizontal line.

Illustration:

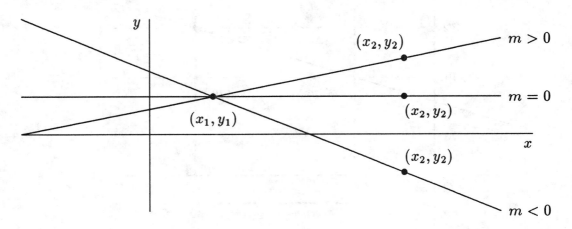

(R3) A vertical line is said to have *no slope* in the sense that the slope is not defined.

Note! Be careful not to use no slope (which has no change in x along L) and slope of 0 (which has no change in y along L) interchangeably.

Example 3: Find the slope of the line through the points:

(A1) $(-1, 3)$ and $(4, 5)$.

$$m = \frac{5 - 3}{4 - (-1)} = \frac{2}{5}$$

Note! $m > 0$ and, for a change in x of 5 units, y increases by 2 units.

(A2) $(-1, 3)$ and $(0, -2)$.

$$m = \frac{-2 - 3}{0 - (-1)} = \frac{-5}{1} = -5$$

Note! $m < 0$ and, for a change in x of 1 unit, y decreases by 5 units.

(A3) $(-1, 3)$ and $(0, 3)$.

$$m = \frac{3 - 3}{0 - (-1)} = \frac{0}{1} = 0$$

Note! $m = 0$ and, for any change in x, y does not change $\implies$ points lie on a horizontal line.

(A4) $(-1, 3)$ and $(-1, -2)$

$$m = \frac{-2 - 3}{-1 - (-1)} = \frac{-5}{0}: \quad \text{does not exist}$$

Note! m does not exist $\implies$ points lie on a vertical line.
$$\uparrow$$
(x values at the points are the same)

Problem: Find an equation of a given line L.

Remark: An equation of L must be satisfied by the coordinates of all points on L and by no other points.

A Solution: Consider a nonvertical line L
$$\left\{ \begin{array}{l} \text{passing through the point } P_1(x_1, y_1) \\ \text{and} \\ \text{having slope } m \end{array} \right\}.$$

Let $P(x, y)$ be any point on L. If $x \neq x_1$, then the slope of L (determined by the distinct points P_1 and P) is

$$m = \frac{y - y_1}{x - x_1}$$

$\implies y - y_1 = m(x - x_1)$: an equation of L which holds for all points on L except P_1.

Insert $x = x_1$ into the equation:

$$y - y_1 = m(x_1 - x_1) = 0 \implies y = y_1$$

$\implies$ the coordinates of P_1 also satisfy the equation.

Point-Slope Form of the Equation of a Line (10): An equation of the line passing through the point $P_1(x_1, y_1)$ and having slope m is

$$y - y_1 = m(x - x_1)$$

Example 4: Find an equation of the line through $(-1, 3)$ with slope -2.

$$\left\{ \begin{array}{l} P_1(-1,3) \\ m = -2 \end{array} \right\} \implies \text{use point-slope form: } y - 3 = -2[x - (-1)] \implies y = -2x + 1$$

Note! Sketch graph of the line.

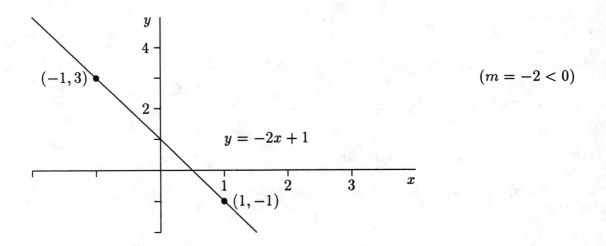

$(m = -2 < 0)$

Need two points: given $(-1,3)$; choose $x = arb$1a choice: $y = -2(1) + 1 = -1$

Example 5: Find an equation of the line through $(0,3)$ with slope -2.

$$\left\{ \begin{array}{l} P_1(0,3) \\ m = -2 \end{array} \right\} \implies \text{use point-slope form: } y - 3 = -2[x - 0] \implies y = -2x + 3$$

Note! Sketch graph of the line.

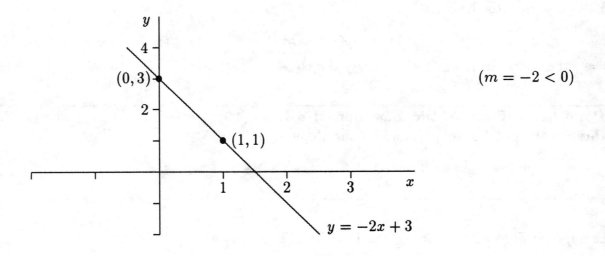

$(m = -2 < 0)$

Need two points: given $(0,3)$; choose $x = 1$: $y = -2(1) + 3 = 1$

Remarks:

(R1) The point $(0, b)$, where b is a real number, lies on the y axis; the value b is called the **y intercept** of the line.

(R2) Consider a nonvertical line L having

$$\left\{ \begin{array}{l} y \text{ intercept } b \\ \text{and} \\ \text{slope } m \end{array} \right\}.$$

$\implies$ L passes through the point $(0, b)$ and has slope m

$\implies$ use point-slope form: $y - b = m(x - 0)$ $\implies$ $y = mx + b$

Slope-Intercept Form of the Equation of a Line (11): An equation of the line with slope m and y-intercept b is
$$y = mx + b$$

Example 6: Find an equation of the line with y intercept -1 and slope 2.

$$\left\{ \begin{array}{l} b = -1 \\ m = 2 \end{array} \right\} \implies \text{use slope-intercept form: } y = 2x + (-1) \implies y = 2x - 1$$

Note! Sketch graph of the line.

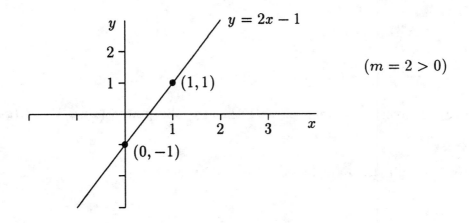

$(m = 2 > 0)$

Need two points: given $(0, -1)$; choose $x = 1:$ $y = 2(1) - 1 = 1$

Example 7: Find an equation of the line through $(-1, -3)$ and $(1, 1)$.

Use the two points to determine slope: $m = \dfrac{1 - (-3)}{1 - (-1)} = 2$

Know a point (choose one of the given points) and the slope

$\Longrightarrow$ use point-slope form with $P_1(-1, -3)$

$y - (-3) = 2[x - (-1)] \quad \Longrightarrow \quad y = 2x - 1$

Note!

(N1) Another choice: $P_1(1, 1)$; $y - 1 = 2(x - 1) \quad \Longrightarrow \quad y = 2x - 1$ (as above)

(N2) From slope-intercept form: $y = 2x - 1 \quad \Longrightarrow \quad m = 2, \quad b = -1$

(N3) Sketch graph of the line

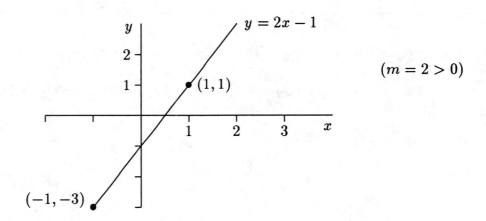

$(m = 2 > 0)$

Given two points: $(-1, -3), \quad (1, 1)$

(R3) Consider a nonvertical line L passing through the points (x_1, y_1) and (x_2, y_2), where $x_1 \neq x_2$.

L passes through the point (x_1, y_1) and has slope $m = \dfrac{y_2 - y_1}{x_2 - x_1}$

$\Longrightarrow$ use point-slope form: $y - y_1 = \dfrac{y_2 - y_1}{x_2 - x_1}(x - x_1),$

the **two-point** form of an equation of a line.

Comment: Given two points, the choice of points to associate with (x_1, y_1) and (x_2, y_2) is immaterial.

Example 8: Find an equation of the line through $(-1, -2)$ and $(1, -2)$.

Use two-point form: $y - (-2) = \dfrac{-2 - (-2)}{1 - (-1)}[x - (-1)]$

$\underbrace{\qquad\qquad\qquad\qquad\qquad\qquad}$

(a choice: $(x_1, y_1) = (-1, -2),\quad (x_2, y_2) = (1, -2)$)

$\Longrightarrow\quad y + 2 = 0(x + 1)\quad \Longrightarrow\quad y = -2$

Note!

(N1) $m = 0\ \Longrightarrow\ $ horizontal line

(N2) Sketch of the line.

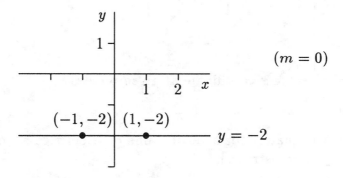

$y_1 = -2 = y_2$: same y coordinate at given points

(N3) $\left\{ \begin{array}{l} y = mx + b \\ m = 0 \end{array} \right\}\ \Longrightarrow\ $ an equation of a horizontal line is $y = b$

Comment: An equation for a horizontal line restricts y to a single value (no change in y) while x is unrestricted.

Example 9: Find an equation of the line through $(-2, -1)$ and $(-2, 1)$.

Use two-point form with choices: $(x_1, y_1) = (-2, -1)$, $(x_2, y_2) = (-2, 1)$

$$y - (-1) = \frac{1 - (-1)}{-2 - (-2)}[x - (-2)] \implies y + 1 = \frac{2}{0}(x + 2) \implies m \text{ does not exist}$$

$\implies$ the line is a vertical line $\implies$ an equation is $x = -2$

↑

$$\left(\begin{array}{l} x_1 = -2 = x_2 : \text{ can not use two-point form;} \\ x \text{ coordinate does not change on a vertical line} \end{array} \right)$$

Note!

(N1) Sketch graph of the line.

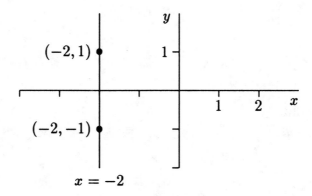

$x_1 = -2 = x_2$: same x coordinate at given points

(N2) An equation for a vertical line is $x = a$

Comment: An equation for a vertical line restricts x to a single value (no change in x) while y is unrestricted.

(N3) The equation $x = 0$ represents the y axis (a vertical line).
The equation $y = 0$ represents the x axis (a horizontal line).

Every line can be associated with an equation of the form

$$Ax + By + C = 0$$

where A, B, and C are constants such that $\underbrace{A^2 + B^2 \neq 0}$.

(at least one of A and B is nonzero)

This equation is called a **linear equation** or the **general equation of a line**.

Remarks:

(R1) $B \neq 0$

$$y = -\frac{A}{B}x - \frac{C}{B}: \text{line is a nonvertical with slope} -\frac{A}{B} \text{ and } y \text{ intercept } -\frac{C}{B}$$

Note! $A = 0 \implies m = 0$: line is horizontal with equation $y = -\underbrace{\frac{C}{B}}$

$(y$ is a constant)

(R2) $B = 0$

Hence, $A \neq 0 \implies x = -\underbrace{\frac{C}{A}}$: line is vertical

$(x$ is a constant)

Example 10: Sketch the graph of the equation $3x - 4y + 12 = 0$.

$3x - 4y + 12 = 0$: linear equation with both variables present

$\implies$ the line is neither vertical nor horizontal

Determine two points:

y intercept (set $x = 0$): $y = 3 \implies (0,3)$

$\underline{x \text{ intercept}}$ (set $y = 0$): $x = -4 \implies (-4,0)$

(the x value where the line intersects the x axis)

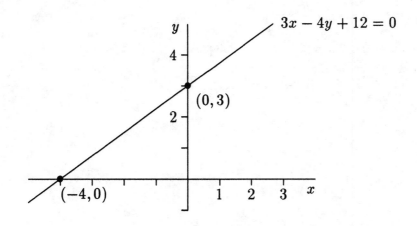

Note! Any two distinct points can be used to sketch the graph: assign a value to $\left\{ \begin{array}{c} x \\ y \end{array} \right\}$ and solve for $\left\{ \begin{array}{c} y \\ x \end{array} \right\}$.

Problem: In the x, y plane, sketch the region

$$\{(x,y) \mid Ax + By + C \leq 0, \quad \text{where} \quad A^2 + B^2 \neq 0\}$$

A Procedure:

(S1) The region, which has the line $Ax + By + C = 0$ as a boundary, includes all points on the line (due to the $=$ sign in the inequality) and all points to one side of the line. Sketch the line.

(S2) Check the inequality $Ax + By + C < 0$ at a point not on the line $Ax + By + C = 0$.

(C1) If the inequality is satisfied, the region includes all points on the side of the line where the point lies.

(C2) If the inequality is not satisfied, the region includes all points on the side of the line opposite from where the point lies.

Remark: If the given inequality is strict, then the line in (S1) is not included in the region: sketch line via a dashed line.

Example 11: Sketch the region $\{(x,y) \mid 3x - 2y + 6 \leq 0\}$.

    Bounding line: $3x - 2y + 6 = 0$

    y intercept $(x = 0)$: $y = 3 \implies (0,3)$

    x intercept $(y = 0)$: $x = -2 \implies (-2,0)$

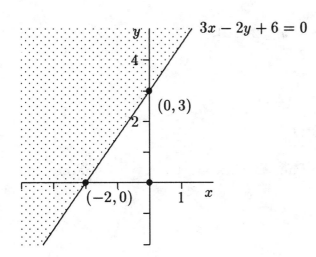

$(0,0)$, a choice not on line:

$3(0) - 2(0) + 6 < 0$?   no

$\implies$ opposite side of line
from $(0,0)$

Example 12: Sketch the region $\{(x, y) \mid |x + 2y| < 2\}$.

$|x + 2y| < 2 \implies -2 < x + 2y < 2$

Solve each inequality separately:

$x + 2y = 2$:   two points are $(0, 1)$ and $(2, 0)$

$x + 2y = -2$:   two points are $(0, -1)$ and $(-2, 0)$

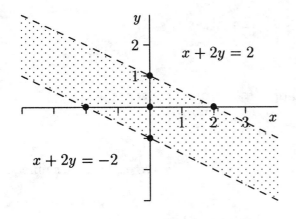

$(0,0)$ not on either line:

$0 + 2(0) < 2$?   yes

$-2 < 0 + 2(0)$?   yes

Example 13: Sketch the region $\{(x,y) \mid x < y \le 2 - x\}$.

Solve each inequality separately:

$y = 2 - x$:   two points are $(0, 2)$ and $(2, 0)$

$y = x$:   two points are $(0, 0)$ and $(1, 1)$

Note! $y = x$ passes through the origin, so a second point with nonzero coordinates must be found.

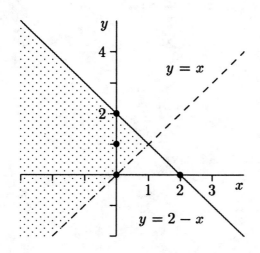

$(0, 1)$ not on either line:

$1 \le 2 - 0$?   yes

$0 < 1$?   yes

Example 14: Sketch the region $\{(x,y) \mid |x| \le 3 \quad \text{and} \quad |y - 1| < 2\}$

Solve each inequality separately:

$|x| \le 3 \implies -3 \le x \le 3$: solve each of these inequalities separately

$x = 3$:   vertical line; $x = -3$:   vertical line

$|y - 1| < 2 \implies -2 < y - 1 < 2 \implies -1 < y < 3$:   solve each of these inequalities separately

$y = 3$:   horizontal line; $y = -1$:   horizontal line

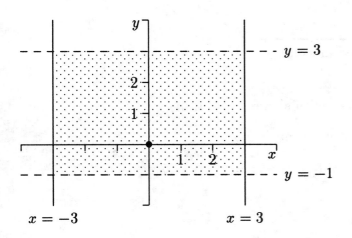

$(0,0)$ not on the four lines:

$0 \le 3$?   yes

$-3 \le 0$?   yes

$0 < 3$?   yes

$-1 < 0$?   yes

Example 15: Sketch the region $\{(x,y) \mid x(y-1) > 0\}$.

Consider the equation $x(y-1) = 0$

$$\implies \begin{cases} x = 0: & \text{vertical line } (y \text{ axis}) \\ y - 1 = 0 \implies & y = 1: \quad \text{horizontal line} \end{cases}$$

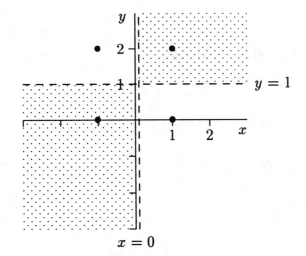

$(1,0)$ not on either line:

$(1)(0-1) > 0$?   no

$(-1,0)$ not on either line:

$(-1)(0-1) > 0$?   yes

$(1,2)$ not on either line:

$(1)(2-1) > 0$?   yes

$(-1,2)$ not on either line:

$(-1)(2-1) > 0$?   no

Note!

(N1) Since a product of variables occur in the given inequality, check a point in each of the four regions determined by $x(y-1) = 0$.

(N2) The given region can be written as

$$\{(x,y) \mid \underbrace{x > 0 \quad \text{and} \quad y - 1 > 0}_{x(y-1) > 0} \quad \text{or} \quad \underbrace{x < 0 \quad \text{and} \quad y - 1 < 0}_{x(y-1) > 0}\}$$

$$= \underbrace{\{(x,y) \mid x > 0,\ y > 1 \quad \text{or} \quad x < 0,\ y < 1\}}$$
(the two regions shaded in sketch)

Theorem: Consider two nonvertical lines with slopes m_1 and m_2.

(T1) The lines are parallel if and only if $m_1 = m_2$.

(T2) The lines are perpendicular if and only if $m_1 m_2 = -1$.

Remarks:

(R1) In words: two nonvertical lines are parallel iff they have the same slope.

(R2) In words: two lines are perpendicular iff the product of their slopes is a negative one.

(R3) If the slopes are nonzero, the slopes of perpendicular lines are negative reciprocals: $m_2 = -1/m_1$.

(R4) If a line has slope zero (hence, a horizontal line), a line perpendicular to it must be a vertical line.

Example 16: Find an equation of the line through $(-1, 3)$ and:

(A1) parallel to the line $2x - y + 4 = 0$.

$$2x - y + 4 = 0 \implies y = 2x + 4 \implies m_1 = 2$$

Know point $(-1, 3)$ and slope $m_2 = 2$ (parallel: $m_2 = m_1$)

$\implies$ use point-slope form: $y - 3 = 2[x - (-1)] \implies y = 2x + 5$

Check: $y = 2x + 5$

$m = 2$: yes; $3 = 2(-1) + 5$? yes $\implies$ $(-1, 3)$ on line

(A2) perpendicular to the line $2x - y + 4 = 0$

From (A1): $m_1 = 2$

Know point $(-1, 3)$ and slope $m_2 = -\dfrac{1}{2}$ (perpendicular: $m_2 = -1/m_1$)

$\implies$ use point-slope form: $y - 3 = -\dfrac{1}{2}[x - (-1)] \implies y = -\dfrac{1}{2}x + \dfrac{5}{2}$

Check: $y = -\dfrac{1}{2}x + \dfrac{5}{2}$

$m = -\dfrac{1}{2}$: yes; $3 = -\dfrac{1}{2}(-1) + \dfrac{5}{2}$? yes $\implies$ $(-1, 3)$ on line

(A3) Sketch the graph of the given line and the lines found in (A1) and (A2) on the same set of axes.

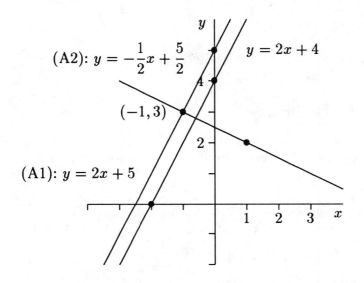

Example 17: Consider the lines $2x + y - 4 = 0$ and $x - y + 1 = 0$.

(A1) Show that the lines are not parallel.

$2x + y - 4 = 0 \implies y = -2x + 4 \implies m_1 = -2$

$x - y + 1 = 0 \implies y = x + 1 \implies m_2 = 1$

$m_1 \neq m_2 \implies$ lines are not parallel.

Note! $m_1 m_2 = (-2)(1) = -2 \neq -1 \implies$ lines are not perpendicular.

(A2) Find the point of intersection of the lines

$$2x + y - 4 = 0 \quad \Longrightarrow \quad 2x + y = 4$$
$$x - y + 1 = 0 \quad \Longrightarrow \quad x - y = -1$$

$\left.\right\}$ system of two linear equations in two unknowns; add (to eliminate y)

$$3x = 3 \quad \Longrightarrow \quad x = 1 \quad \Longrightarrow \quad 2(1) + y = 4 \quad \Longrightarrow \quad y = 2$$

$\Longrightarrow$ point of intersection: $(1, 2)$

(A3) Find an equation of the line through the point of intersection found in (A2) and perpendicular to the line $2x + y - 4 = 0$.

$$2x + y - 4 = 0 \quad \Longrightarrow \quad y = -2x + 4 \quad \Longrightarrow \quad m_1 = -2$$

Know point $(1, 2)$ from (A2) and slope $m_2 = -\dfrac{1}{m_1} = -\dfrac{1}{-2} = \dfrac{1}{2}$

$\Longrightarrow$ use point-slope form: $y - 2 = \dfrac{1}{2}(x - 1) \quad \Longrightarrow \quad y = \dfrac{1}{2}x + \dfrac{3}{2}$

(A4) Sketch the graph of the given lines and the line found in (A3).

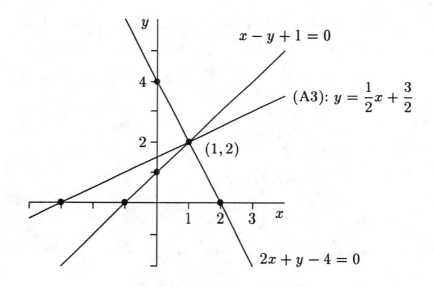

0.3   Graphs of Second-Degree Equations

Definition: A circle is the set of points $P(x, y)$ whose distance from the center $C(h, k)$ is r.

Remarks:

(R1) r: radius of circle (≥ 0)

(R2) $r = 0$: a *degenerate case* where the circle consists solely of the center (h, k)

(R3) $r = 1$: the circle is called a **unit circle**

(R4) $P(x, y)$ is on the circle if and only if $|CP| = r$

$$\implies \sqrt{(x - h)^2 + (y - k)^2} = r \implies (x - h)^2 + (y - k)^2 = r^2$$

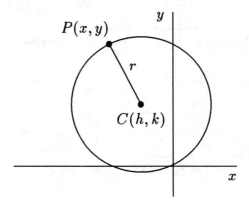

Equation of a Circle (14): The equation of a circle with center (h, k) and radius r is

$$(x - h)^2 + (y - k)^2 = r^2$$

In particular, if the center is the origin $(0, 0)$, the equation is

$$x^2 + y^2 = r^2$$

Example 1: Find an equation of a circle with radius $\sqrt{2}$ and center $(-1, 3)$.

$$\left\{ \begin{array}{l} r = \sqrt{2} \\ (h, k) = (-1, 3) \end{array} \right\} \implies [x - (-1)]^2 + (y - 3)^2 = (\sqrt{2})^2$$

$$\implies (x + 1)^2 + (y - 3)^2 = 2$$

Note! The equation can be written as $x^2 + 2x + 1 + y^2 - 6y + 9 = 2$

$$\implies x^2 + y^2 + 2x - 6y + 8 = 0$$

Example 2: Consider the equation $x^2 + y^2 - 6x + 2y + 6 = 0$.

(A1) Show that the given equation represents a circle and find the center and the radius.

$$x^2 + y^2 - 6x + 2y + 6 = 0 \implies \underbrace{(x^2 - 6x) + (y^2 + 2y)}_{} = -6$$

$$\text{(group terms involving same variable)}$$

$$\implies \underbrace{\left[x^2 - 6x + \left(\frac{-6}{2}\right)^2 - \left(\frac{-6}{2}\right)^2 \right]}_{\text{(complete the square)}} + \underbrace{\left[y^2 + 2y + \left(\frac{2}{2}\right)^2 - \left(\frac{2}{2}\right)^2 \right]}_{\text{(complete the square)}} = -6$$

Recall: When coefficient of x^2 is 1:

$$x^2 + bx + \underbrace{\left(\frac{b}{2}\right)^2 - \left(\frac{b}{2}\right)^2}_{\text{(add zero)}} = \underbrace{\left(x + \frac{b}{2}\right)^2 - \left(\frac{b}{2}\right)^2}_{\text{(completed the square)}}$$

$$\implies [x + (-3)]^2 - (-3)^2 + (y + 1)^2 - (1)^2 = -6$$

$$\implies (x - 3)^2 + (y + 1)^2 = -6 + 9 + 1 = 4 = 2^2 \quad (\geq 0)$$

$$\implies \text{circle with center } (3, -1) \text{ and radius } 2$$

Note! To find coordinates of center:

$$\left. \begin{cases} x - 3 = 0 \implies x = 3 \\ y + 1 = 0 \implies y = -1 \end{cases} \right\} \implies \text{center } (3, -1)$$

(A2) Sketch the graph of the given equation.

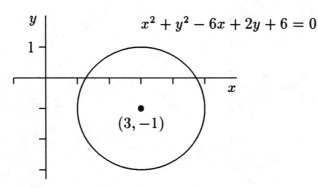

$$x^2 + y^2 - 6x + 2y + 6 = 0$$

$(3, -1)$

Example 3: Consider the equation $2x^2 + 2y^2 + 4x - 8y + 7 = 0$.

(A1) Show that the given equation represents a circle and find the center and the radius.

$$2x^2 + 2y^2 + 4x - 8y + 7 = 0 \implies (x^2 + 2x) + (y^2 - 4y) = -\frac{7}{2}$$

$\Big($ x^2 and y^2 have same coefficient: divide by coefficient so x^2 and y^2 have a coefficient of 1; group terms according to variable $\Big)$

$$\implies \left[x^2 + 2x + \left(\frac{2}{2}\right)^2 - \left(\frac{2}{2}\right)^2 \right] + \left[y^2 - 4y + \left(\frac{-4}{2}\right)^2 - \left(\frac{-4}{2}\right)^2 \right] = -\frac{7}{2}$$

(completing the square: adding zero)

$$\implies (x + 1)^2 + [y + (-2)]^2 = -\frac{7}{2} + 1 + 4$$

$$\implies (x + 1)^2 + (y - 2)^2 = \frac{3}{2} = \left(\sqrt{\frac{3}{2}}\right)^2 \quad (\geq 0)$$

$$\implies \quad \text{circle with center } (-1, 2) \text{ and radius } \sqrt{\frac{3}{2}} \approx 1.22$$

(A2) Sketch the graph of the given equation.

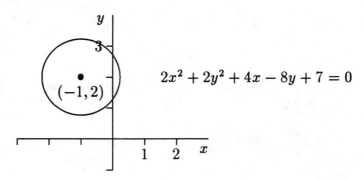

$$2x^2 + 2y^2 + 4x - 8y + 7 = 0$$

The curve with equation

$$y = ax^2 + bx + c$$

where a, b, and c are real numbers such that $a \neq 0$, is called a **parabola**.

Remarks:

(R1) To obtain the x intercepts of the parabola $y = ax^2 + bx + c$, set $y = 0$

$$\Longrightarrow \quad \underbrace{ax^2 + bx + c = 0}_{\text{(quadratic equation)}} \quad \Longrightarrow \quad \underbrace{x = \frac{-b \pm \sqrt{b^2 - 4ac}}{2a}}_{\text{(quadratic formula: } a \neq 0\text{)}}$$

Note!

(N1) The number of x intercepts depends upon the sign of the **discriminant** $b^2 - 4ac$:

$b^2 - 4ac$	Number of x intercepts
> 0	two
$= 0$	one
< 0	zero

(N2) May also solve $ax^2 + bx + c = 0$ by factoring.

(R2) The vertical line $x = -\dfrac{b}{2a}$ is the **axis (line of symmetry)** of the parabola.

Note!

(N1) The equation $x = \dfrac{-b}{2a}$ of the axis occurs in the quadratic formula: ignore the term $\sqrt{b^2 - 4ac}$

(N2) If $b^2 - 4ac = 0$, the single x intercept occurs at $x = \dfrac{-b}{2a}$.

(N3) If $b^2 - 4ac > 0$, each x intercept occurs a distance

$$\frac{\sqrt{b^2 - 4ac}}{2|a|} \qquad (> 0)$$

to the left and to the right of the axis.

(R3) The point of intersection of the parabola with its axis is called the **vertex** of the parabola.

Note!

(N1) $x = -\dfrac{b}{2a} \implies y = a\left(-\dfrac{b}{2a}\right)^2 + b\left(-\dfrac{b}{2a}\right) + c = -\dfrac{b^2 - 4ac}{4a}$

$\implies$ vertex : $\left(-\dfrac{b}{2a}, -\dfrac{b^2 - 4ac}{4a}\right)$

(N2) $y = ax^2 + bx + c = a\left(x^2 + \dfrac{b}{a}x\right) + c = a\left[x^2 + \dfrac{b}{a}x + \left(\dfrac{b}{2a}\right)^2 - \left(\dfrac{b}{2a}\right)^2\right] + c$

$\uparrow$ $\qquad\qquad\qquad\qquad$ $\uparrow$

$\left(\begin{array}{l}\text{coefficient of } x^2\text{: 1;} \\ \text{group } x \text{ terms}\end{array}\right)$ $\qquad$ $\left(\begin{array}{l}\text{complete the square:} \\ \text{add zero}\end{array}\right)$

$= a\left(x + \dfrac{b}{2a}\right)^2 - a\left(\dfrac{b}{2a}\right)^2 + c = a\left(x + \dfrac{b}{2a}\right)^2 - \dfrac{b^2 - 4ac}{4a}$

$\left(x + \dfrac{b}{2a}\right)^2 \geq 0 \implies \left\{\begin{array}{ll} y \geq -\dfrac{b^2 - 4ac}{4a} & \text{if } a > 0 \\[2mm] y \leq -\dfrac{b^2 - 4ac}{4a} & \text{if } a < 0 \end{array}\right\}$

$\implies \left\{\begin{array}{ll} a > 0 : & \text{parabola opens upward} \\ a < 0 : & \text{parabola opens downward} \end{array}\right\}$ from vertex

Illustrations:

(I1) $a > 0, \quad b^2 - 4ac > 0$

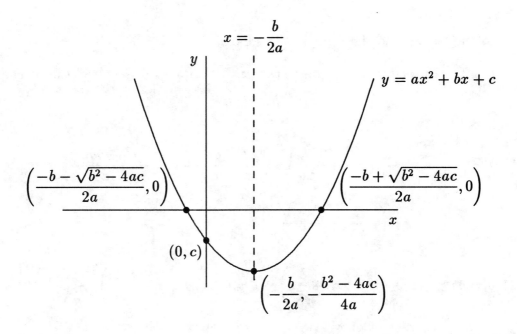

(I2) $a < 0, \quad b^2 - 4ac > 0$

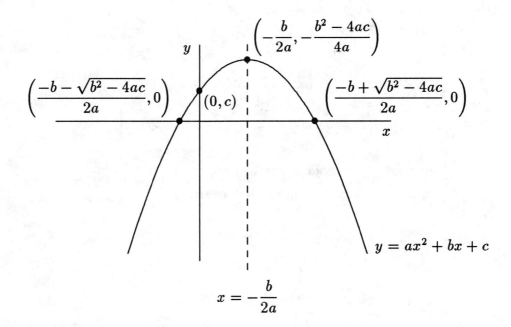

(N3) $b^2 - 4ac = 0 \implies$ vertex: $\left(-\dfrac{b}{2a}, 0\right)$, which is the x intercept

$$\implies \left\{ \begin{array}{ll} a > 0: & \text{parabola opens upward} \\ a < 0: & \text{parabola opens downward} \end{array} \right\} \text{ from it } x \text{ intercept}$$

Example 4: Sketch the graph of the parabola $x^2 + 2x + y - 3 = 0$.

$x^2 + 2x + y - 3 = 0 \implies y = -x^2 - 2x + 3$

Note! $a = -1 \ (\neq 0), \quad b = -2, \quad c = 3; \quad a = -1 < 0:$ opens downward

x intercepts $(y = 0)$: $-x^2 - 2x + 3 = 0 \implies x^2 + 2x - 3 = 0$

$\implies (x + 3)(x - 1) = 0 \implies x = -3, \ 1$

vertex: x value is midpoint of x intercepts $\implies x = \dfrac{-3 + 1}{2} = -1$

$\implies y = -(-1)^2 - 2(-1) + 3 = 4 \implies (-1, 4)$

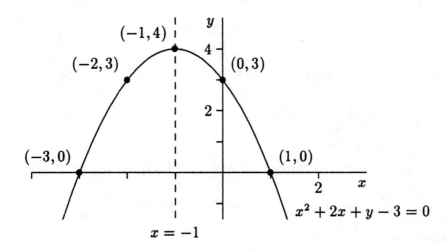

Note! y intercept: $x = 0 \implies y = 3 \ (= c) \implies (-2, 3)$ is also on graph

$\uparrow$

$\left(\begin{array}{l} \text{symmetry with respect to } x = -1: \\ x = 0 \text{ is 1 unit to the right and } x = -2 \text{ is 1 unit to the left} \end{array} \right)$

Example 5: Sketch the graph of the parabola $y = x^2 + 2x + 3$.

    Note! $a = 1,\quad b = 2,\quad c = 3;\quad a = 1 > 0$: opens upward

    x intercepts: $x^2 + 2x + 3 = 0 \implies x = \dfrac{-2 \pm \sqrt{2^2 - 4(1)(3)}}{2(1)}$: $\underbrace{\text{none}}$

$$\left(b^2 - 4ac = -8 < 0\right)$$

    vertex: $x = \underbrace{\dfrac{-2}{2(1)}}_{} = -1 \implies y = (-1)^2 + 2(-1) + 3 = 2 \implies (-1, 2)$

(part of quadratic formula above)

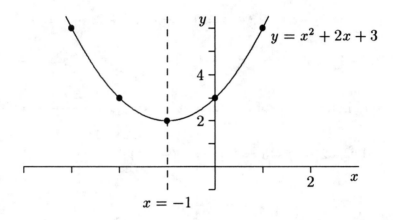

    Note! Use symmetry with respect to $x = -1$ to obtain more points:

    $(0, 3)$, hence $(-2, 3)$; $(1, 6)$, hence $(-3, 6)$

If the variables x and y are interchanged in $y = ax^2 + bx + c$, where $a \neq 0$, to obtain

$$x = ay^2 + by + c,$$

the curve is also a parabola.

Remarks:

(R1) The interchange of x and y reflects the graph of $y = ax^2 + bx + c$ about the diagonal line $y = x$ to obtain the graph of $x = ay^2 + by + c$.

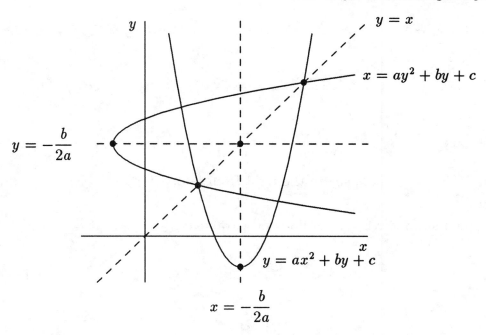

Note!

(N1) Vertical lines reflect into horizontal lines (and vice versa)

(N2) Points of the graph on the line $y = x$ remain on that line after the reflection.

(R2) To sketch the graph of $x = ay^2 + by + c$, proceed in a manner similar to that given above for $y = ax^2 + bx + c$.

$$\begin{cases} a > 0 : & \text{parabola opens to the right} \\ a < 0 : & \text{parabola opens to the left} \end{cases} \text{ from the vertex } \left(-\frac{b^2 - 4ac}{4a}, -\frac{b}{2a}\right),$$

which is located on the axis $y = -\dfrac{b}{2a}$, a horizontal line of symmetry.

The y intercepts depend upon the sign of the discriminant $b^2 - 4ac$: $y = \dfrac{-b \pm \sqrt{b^2 - 4ac}}{2a}$

Note! If $b^2 - 4ac > 0$, the axis is midway between the two y intercepts.

Example 6: Sketch the region bounded by the line $x - y - 1 = 0$ and the parabola $y^2 - 2x - 6 = 0$.

$$x - y - 1 = 0 \implies y = x - 1$$

$$y^2 - 2x - 6 = 0 \implies x = \frac{1}{2}y^2 - 3 \implies a = \frac{1}{2}, \quad b = 0, \quad c = -3$$

y intercepts $(x = 0)$: $\dfrac{1}{2}y^2 - 3 = 0 \implies y^2 = 6 \implies y = \pm\sqrt{6} \approx \pm\sqrt{2.45}$

$\implies$ at vertex: $y = \dfrac{+\sqrt{6} + (-\sqrt{6})}{2} = 0 \implies x = -3 : \ (-3, 0)$
$\uparrow$
(midpoint of y intercepts)

$a = \dfrac{1}{2} > 0$: parabola opens to the right from $(-3, 0)$

Find points of intersection of line and parabola:

$\left\{ \begin{array}{l} y = x - 1 \\ x = (1/2)y^2 - 3 \end{array} \right\} \implies x = \dfrac{1}{2}(x-1)^2 - 3 \implies 2x = x^2 - 2x + 1 - 6$
$\uparrow$
(substitution)

$\implies x^2 - 4x - 5 = 0 \implies (x-5)(x+1) = 0 \implies x = 5, \ -1$

$\implies (5, 4), \ (-1, -2)$ $\hfill (y = x - 1)$

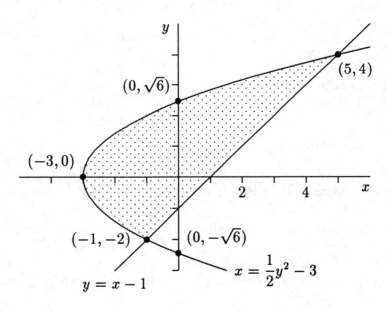

Example 7: Sketch the region bounded by the parabolas $x = y^2$ and $x = 2y - y^2$.

$x = y^2 \implies a = 1, \ b = 0 = c$

y intercepts: $y^2 = 0 \implies y = 0 \implies$ at vertex: $y = 0 \implies x = 0$

$\uparrow$

(the single y intercept)

$a = 1 > 0$: parabola opens to the right from $(0,0)$

$x = 2y - y^2 \implies a = -1, \ b = 2, \ c = 0$

y intercepts: $2y - y^2 = 0 \implies y(2 - y) = 0 \implies y = 0, \ 2$

$\implies$ at vertex: $y = \dfrac{0+2}{2} = 1 \implies x = 2 - 1 = 1$

$\uparrow$

(midpoint of y intercepts)

$a = -1 < 0$: parabola opens to the left from $(1,1)$

Find points of intersection of the parabolas:

$\left\{ \begin{array}{l} x = y^2 \\ x = 2y - y^2 \end{array} \right\} \implies y^2 = 2y - y^2 \implies 2y^2 - 2y = 0 \implies 2y(y-1) = 0$

$\uparrow$

(substitution)

$\implies y = 0, \ 1 \implies (0,0), \ (1,1)$

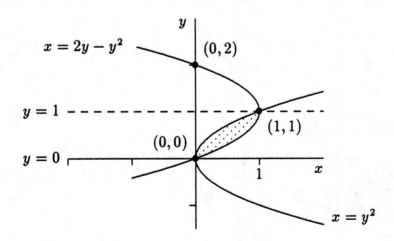

The curve with equation

$$\frac{(x-h)^2}{a^2} + \frac{(y-k)^2}{b^2} = 1$$

where a, b, h, and k are real numbers such that $a > 0$ and $b > 0$, is called an **ellipse**.

Remarks:

(R1) The point $C(h, k)$ is the center of the ellipse.

(R2) If $a = b$, the ellipse is a circle with center (h, k) and radius a.

(R3) $y = k \implies \dfrac{(x-h)^2}{a^2} = 1 \implies (x-h)^2 = a^2 \implies \sqrt{(x-h)^2} = \sqrt{a^2}$

$\implies |x - h| = |a| \implies \pm(x - h) = a \implies x - h = \pm a \implies x = h \pm a$

$\underset{\uparrow}{}$

$(a > 0 : |a| = a)$

$\implies$ the ellipse intersects the horizontal line $y = k$ (passing through the center) at the points $(h + a, k)$ and $(h - a, k)$.

Note! The center is the midpoint of these points of intersection.

(R4) Similarly (as in (R3)), the ellipse intersects the vertical line $x = h$ (passing through the center) at the points $(h, k + b)$ and $(h, k - b)$.

Note! The center is the midpoint of these points of intersection.

(R5) $\dfrac{(x-h)^2}{a^2} = 1 - \underbrace{\dfrac{(y-k)^2}{b^2}}_{\geq 0} \leq 1 \implies (x-h)^2 \leq a^2 \implies \sqrt{(x-h)^2} \leq \sqrt{a^2}$

$\implies |x - h| \leq |a| = a \implies -a \leq x - h \leq a \implies h - a \leq x \leq h + a$

$\underset{\uparrow}{}$

$(a > 0)$

Similarly, $k - b \leq y \leq k + b$

Hence the ellipse lies inside the rectangle bounded by the lines $x = h - a$, $x = h + a$, $y = k - b$, and $y = k + b$.

Illustration: $a > b > 0$

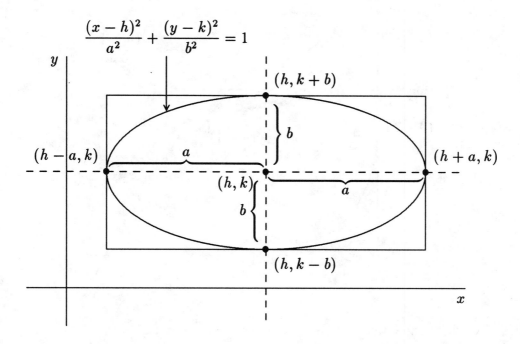

Example 8: Sketch the graph of the ellipse $\dfrac{(x-2)^2}{4} + \dfrac{(y+1)^2}{9} = 1$

$a^2 = 4 \implies a = 2 \ (> 0); \qquad b^2 = 9 \implies b = 3 \ (> 0)$

$$\left\{ \begin{array}{l} x - 2 = 0 \implies x = 2 = h \\ y + 1 = 0 \implies y = -1 = k \end{array} \right\} \implies \text{center: } (2, -1)$$

Note! First plot the center $(h, k) = (2, -1)$; then plot the four "turning" points

$(h + a, k) = (4, -1), \ (h - a, k) = (0, -1), \ (h, k + b) = (2, 2), \ (h, k - b) = (2, -4)$

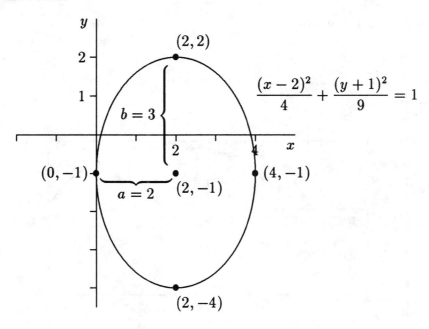

$$\frac{(x-2)^2}{4} + \frac{(y+1)^2}{9} = 1$$

Example 9: Consider the equation $4x^2 + y^2 + 8x - 4y + 4 = 0$.

(A1) Show that the given equation represents an ellipse and find its center.

$$4x^2 + y^2 + 8x - 4y + 4 = 0 \implies \underbrace{(4x^2 + 8x) + (y^2 - 4y)}_{} = -4$$

(group terms involving same variable)

$$\implies \underbrace{4(x^2 + 2x)}_{} + (y^2 - 4y) = -4$$

$\left(\begin{array}{l} \text{factor a constant from each grouping (if necessary)} \\ \text{to obtain a coefficient of 1 for each variable squared} \end{array} \right)$

$$\implies 4\left[x^2 + 2x + (1)^2 - (1)^2\right] + \left[y^2 - 4y + (-2)^2 - (-2)^2\right] = -4$$
↑

(completing the square; add zero)

$$\implies 4(x+1)^2 + [y + (-2)]^2 = -4 + 4(1)^2 + (-2)^2$$

$$\implies 4(x+1)^2 + (y-2)^2 = 4 \qquad (> 0)$$

$$\implies (x+1)^2 + \frac{(y-1)^2}{4} = 1: \quad \text{an ellipse}$$

$$\begin{cases} x + 1 = 0 & \Longrightarrow & x = -1 = h \\ y - 2 = 0 & \Longrightarrow & y = 2 = k \end{cases} \Longrightarrow \quad \text{center } (-1, 2)$$

(A2) Sketch the graph of the given equation.

$$(x + 1)^2 + \frac{(y - 1)^2}{4} = 1 \quad \Longrightarrow \quad \begin{cases} a = 1 \ (> 0); \quad b = 2 \ (> 0) \\ \text{center } (-1, 2) \end{cases}$$

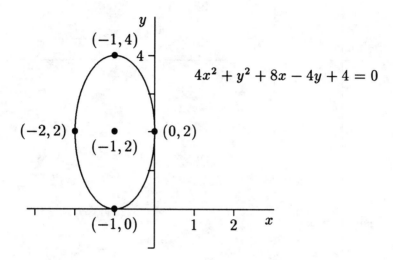

$$4x^2 + y^2 + 8x - 4y + 4 = 0$$

Example 10: Sketch the region $\{(x, y) \mid 4x^2 + 9y^2 - 16x - 18y < 11\}$.

Determine the curve associated with the equality:

$$4x^2 + 9y^2 - 16x - 18y = 11 \quad \Longrightarrow \quad 4(x^2 - 4x) + 9(y^2 - 2y) = 11$$

$$\Longrightarrow \quad 4[x^2 - 4x + (-2)^2 - (-2)^2] + 9[y^2 - 2y + (-1)^2 - (-1)^2] = 11$$
↑

(complete the square: add zero)

$$\Longrightarrow \quad 4[x + (-2)]^2 + 9[y + (-1)]^2 = 11 + 4(-2)^2 + 9(-1)^2$$

$$\Longrightarrow \quad 4(x - 2)^2 + 9(y - 1)^2 = 36 \quad \Longrightarrow \quad \frac{(x - 2)^2}{9} + \frac{(y - 1)^2}{4} = 1:$$

an ellipse with $\begin{cases} a = 3 \ (> 0); \quad b = 2 \ (> 0) \\ \text{center } (2, 1) \end{cases}$

Note! Given inequality is a strict inequality, so the ellipse is not included in the region.

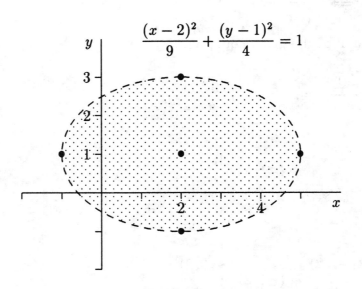

$$\frac{(x-2)^2}{9} + \frac{(y-1)^2}{4} = 1$$

center $(2,1)$ not on ellipse: $4(2)^2 + 9(1)^2 - 16(2) - 18(1) < 11$?   yes

$\Longrightarrow$ region inside ellipse

The curve with equation

$$\frac{(x-h)^2}{a^2} - \frac{(y-k)^2}{b^2} = 1,$$

where a, b, h, and k are real numbers such that $a > 0$ and $b > 0$, is called a **hyperbola**.

Remarks:

(R1) The squared terms are joined by a $\left\{ \begin{array}{l} \text{plus sign for an ellipse} \\ \text{minus sign for a hyperbola} \end{array} \right\}$

(R2) $\dfrac{(x-h)^2}{a^2} - \dfrac{(y-k)^2}{b^2} = 1 \implies \dfrac{(x-h)^2}{a^2} = 1 + \underbrace{\dfrac{(y-k)^2}{b^2}}_{\geq 0} \geq 1$

$\implies (x-h)^2 \geq a^2 \implies \underset{\underset{(a>0)}{\uparrow}}{|x-h| \geq |a| = a} \implies x - h \geq a$ or $x - h \leq -a$

$\implies x \geq h + a$ or $x \leq h - a$

$\Longrightarrow$ the hyperbola lies to the right of the vertical line $x = h + a$ and to the left of the vertical line $x = h - a$

$\Longrightarrow$ the hyperbola consists of two parts, called **branches**

Note! $h + a > h - a$ $\hfill (a > 0)$

(R3) $x = h + a$ $\Longrightarrow$ $\dfrac{(h-k)^2}{b^2} = \dfrac{a^2}{a^2} - 1 = 0$ $\Longrightarrow$ $y = k$

Similarly, $x = h - a$ $\Longrightarrow$ $y = k$

The points $(h \pm a, k)$ are the "turning" points of the hyperbola

(R4) Consider the equation $\dfrac{(x-h)^2}{a^2} - \dfrac{(y-k)^2}{b^2} = 0$

$\dfrac{(x-h)^2}{a^2} = \dfrac{(y-k)^2}{b^2}$ $\Longrightarrow$ $\dfrac{y-k}{b} = \pm \dfrac{x-h}{a}$

$\Longrightarrow$ $y - k = \pm\dfrac{b}{a}(x-h)$:

nonvertical and nonhorizontal lines which are called the **asymptotes** of the hyperbola.
$\left(m = \pm\dfrac{b}{a} \right)$ $(b \neq 0 \ \Longrightarrow\ m \neq 0)$

Note!

(N1) The point (h, k) lies on both asymptotes.

(N2) The points $(h+a, k+b)$ and $(h-a, k-b)$ lie on the asymptote $y - k = \dfrac{b}{a}(x-h)$

(N3) The points $(h+a, k-b)$ and $(h-a, k+b)$ lie on the asymptote $y - k = -\dfrac{b}{a}(x-h)$

(N4) The asymptotes pass through the corners of the rectangle determined by the four corners $(h \pm a, k \pm b)$.

Illustration:

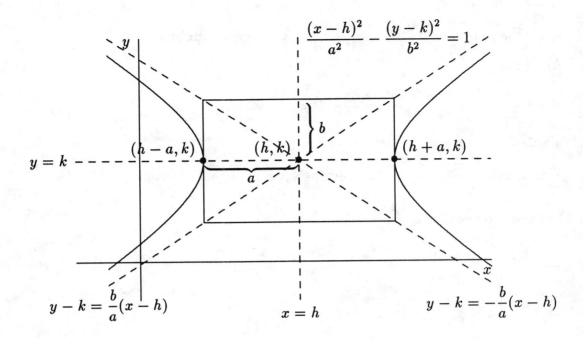

$$\frac{(x - h)^2}{a^2} - \frac{(y - k)^2}{b^2} = 1$$

Example 11: Sketch the graph of the hyperbola $\dfrac{(x - 2)^2}{4} - \dfrac{(y + 1)^2}{9} = 1$

$$a^2 = 4 \implies a = 2\ (> 0); \qquad b^2 = 9 \implies b = 3\ (> 0)$$

$$\left\{ \begin{array}{l} x - 2 = 0 \implies x = 2 = h \\ y + 1 = 0 \implies y = -1 = k \end{array} \right\} \implies (2, -1) \text{ is center of ``reference'' rectangle}$$

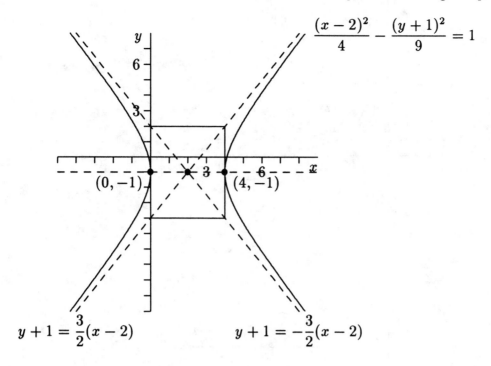

$$\frac{(x-2)^2}{4} - \frac{(y+1)^2}{9} = 1$$

$$y + 1 = \frac{3}{2}(x-2) \qquad\qquad y + 1 = -\frac{3}{2}(x-2)$$

Note! First construct the reference rectangle and draw the two asymptotes; then plot the two turning points.

Example 12: Consider the equation $4x^2 - y^2 + 16x + 2y + 11 = 0$.

(A1) Show that the given equation represents a hyperbola.

$$4x^2 - y^2 + 16x + 2y + 11 = 0 \implies 4(x^2 + 4x) - (y^2 - 2y) = -11$$

$$\implies 4[x^2 + 4x + (2)^2 - (2)^2] - [y^2 - 2y + (-1)^2 - (-1)^2] = -11$$

$$\implies 4(x+2)^2 - [y + (-1)]^2 = -11 + 4(2)^2 - (-1)^2$$

$$\implies 4(x+2)^2 - (y-1)^2 = 4 \implies (x+2)^2 - \frac{(y-1)^2}{4} = 1: \text{ a hyperbola}$$

(A2) Sketch the graph of the given equation.

$$(x+2)^2 - \frac{(y-1)^2}{4} = 1 \implies \begin{cases} a = 1; \quad b = 2 \\ (h,k) = (-2,1) \end{cases}$$

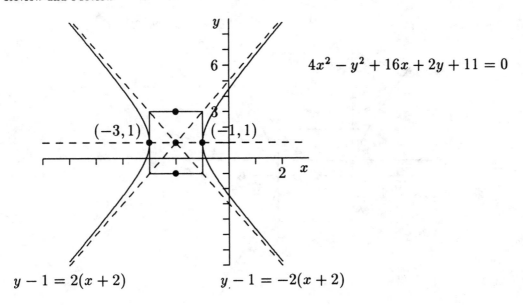

$$4x^2 - y^2 + 16x + 2y + 11 = 0$$

$(-3, 1)$ $(-1, 1)$

$y - 1 = 2(x + 2)$ $y - 1 = -2(x + 2)$

The curve with equation

$$\frac{(y-k)^2}{b^2} - \frac{(x-h)^2}{a^2} = 1$$

is a hyperbola which has a branch turning upward and a branch turning downward at the points $(h, k + b)$ and $(h, k - b)$, respectively, on the boundary of the reference rectangle.

Example 13: Sketch the graph of the hyperbola $4(x + 2)^2 - (y - 1)^2 + 4 = 0$.

$$4(x + 2)^2 - (y - 1)^2 + 4 = 0 \implies 4(x + 2)^2 - (y - 1)^2 = -4$$

$$\implies \frac{(y-1)^2}{4} - (x + 2)^2 = 1 \text{: a hyperbola with } \begin{cases} a = 1; \quad b = 2 \\ (h, k) = (-2, 1) \end{cases}$$

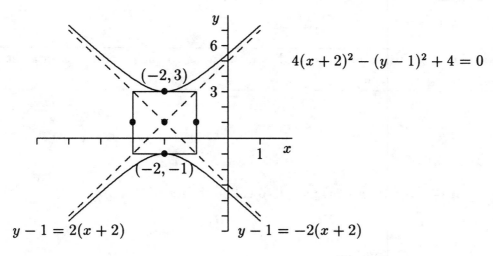

Consider the equation
$$Ax^2 + By^2 + Cx + Dy + E = 0,$$
where A, B, C, D, and E are real numbers.

(C1) $A = B = 0$, $C^2 + D^2 \neq 0$ (i.e., $C \neq 0$ or $D \neq 0$):   line

(C2) $A = 0$, $B \neq 0$, $C \neq 0$:   parabola

   Note! Parabola opens to the right or to the left.

(C3) $A \neq 0$, $B = 0$, $D \neq 0$:   parabola

   Note! Parabola opens upward or downward.

(C4) $A \neq 0$, $B \neq 0$ such that $A = B$:   circle

   Note! Complete the square; the sum of squares must equal a nonnegative constant.

(C5) $A \neq B$ such that $AB > 0$ (i.e., A and B have same sign):   ellipse

   Note! Complete the square; the sum of squares must equal a positive constant.

(C6) $AB < 0$ (i.e., A and B have opposite signs):   hyperbola

   Note! Complete the square.

   (N1) To obtain a hyperbola the difference of squares must equal a nonzero constant.

   (N2) If the difference of squares equals zero, the graph of the equation is a pair of intersecting lines (see (R4) regarding the asymptotes of a hyperbola).

Remark: If $E = 0$, the graph of the curve passes through the origin.

0.4   Functions and Their Graphs

Definition (19): A **function** f is a rule that assigns to each element x in a set A exactly one element, called $f(x)$, in a set B.

Remarks:

(R1) A: **domain of f**

(R2) $\underbrace{f(x)}_{(f \text{ of } x)}$:   **image of x under f**;   also called **value of f at x**

(R3) $\underbrace{\{f(x) \mid x \in A\}}$:   **range of f**

   (the set of all images of x under f)

   Note! $\{f(x) \mid x \in A\} \subset B$

Illustration: **Arrow Diagram**

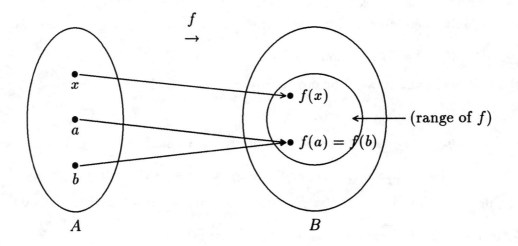

Note! A rule associating elements of A and B such that

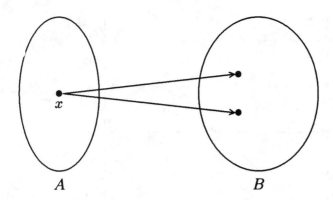

is not a function from the set of pre-images in A into B since $x \in A$ is assigned to two different elements in B.

If A and B are subsets of the real numbers, a function f is often represented by

(S1) specifying the domain of A, and

(S2) giving an explicit rule whereby the value $f(x)$ for each $x \in A$ may be determined.

Example 1: Consider f defined on $\underset{\text{(domain)}}{\underline{A = \mathbf{R}}}$ such that $\underset{\text{(rule given by a formula)}}{\underline{f(x) = |x|}}$ for each $x \in A$.

Note!

(N1) $x \in \mathbf{R} \implies |x| \in \mathbf{R} \implies$ range of f is a subset of $\mathbf{R}$

(N2) $|x| \geq 0 \implies$ range of f is the set of nonnegative real numbers

(N3) *Illustration*:

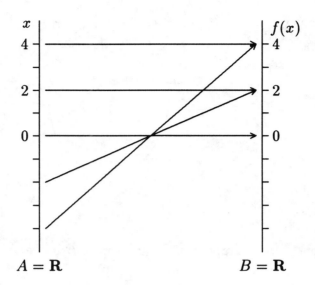

$$A = \mathbf{R} \qquad\qquad B = \mathbf{R}$$

Convention: Let A and B be subsets of the real numbers. If a function is defined via a formula and the domain A is not stated explicitly, then A is assumed to be the set of all numbers for which the formula defines a real number.

Example 2: Find the domain of the function $f(x) = \dfrac{x-1}{x^2 - 2x - 3}$.

The quotient exists if the denominator is nonzero:

$$x^2 - 2x - 3 = 0 \implies (x-3)(x+1) = 0 \implies x = 3, \ -1$$

Hence, $\quad x^2 - 2x - 3 \neq 0 \iff x \neq -1, 3$

$\implies \quad$ domain of f: $\underbrace{(-\infty, -1) \cup (-1, 3) \cup (3, \infty)}_{\{x \mid x \neq -1, \ 3\}}$

Example 3: Find the domain of the function $f(x) = \sqrt{x^2 - 2x - 3}$.

$\underbrace{x^2 - 2x - 3 \geq 0}_{\text{(square root of nonnegative numbers only)}} \implies (x-3)(x+1) \geq 0$

$$x - 3 = 0 \implies x = 3; \qquad x + 1 = 0 \implies x = -1$$

$$\begin{array}{ccccccc} + & 0 & - & 0 & + & & (x-3)(x+1) \\ \hline & -1 & & 3 & & x & \end{array}$$

$(x-3)(x+1) = x^2 - 2x - 3 \geq 0$: $x \in (-\infty, -1] \cup [3, \infty)$

$\implies$ domain of f: $x \in (-\infty, -1] \cup [3, \infty)$

Note!

(N1) Consider $y = x^2 - 2x - 3$: parabola

   $a = 1 > 0$: opens upward

   x intercepts: $x^2 - 2x - 3 = 0 \implies x = 3, -1$ (Example 2 above)

   Sketch graph (interested in sign of $y = x^2 - 2x - 3$, so need not exhibit units on the y axis):

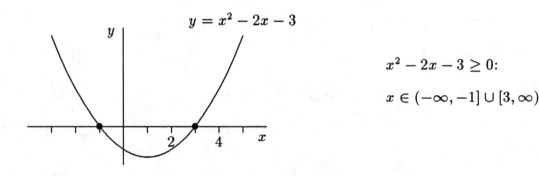

$x^2 - 2x - 3 \geq 0$:

$x \in (-\infty, -1] \cup [3, \infty)$

(N2) Range of f: nonnegative real numbers

Example 4: Find the domain of the function $f(x) = \dfrac{1}{\sqrt{|x| - 1}}$.

$\left\{ \begin{array}{l} \text{square root:} \quad |x| - 1 \geq 0 \\ \text{denominator:} \quad |x| - 1 \neq 0 \end{array} \right\} \implies |x| - 1 > 0 \implies |x| > 1$

$\implies x > 1$ or $x < -1 \implies$ domain of f: $x \in (-\infty, -1) \cup (1, \infty)$

Definition: Let A and B be subsets of the real numbers. If f is a function with domain A and range contained in B, then the **graph of f** is the set of ordered pairs $\{(x, f(x)) \mid x \in A\}$.

Remarks:

(R1) The graph of f is a method for visualizing the function f.

(R2) In words: the graph of f consists of all points (x, y) in the coordinate plane such that $y = f(x)$, where x is in the domain of f.

(R3) The graph of f is also referred to as the graph of $y = f(x)$.

(R4) Domain of f: subset of real numbers on the x axis

   Note! The domain is the set of permitted values of the first coordinate of the ordered pair (x, y), where $y = f(x)$.

(R5) Range of f: subset of real numbers on the y axis

   Note! The range is the set of assigned values of the second coordinate of the ordered pair (x, y), where $y = f(x)$.

Example 5: Sketch the graph of $y = x + 3$.

   $y = x + 3$:   line

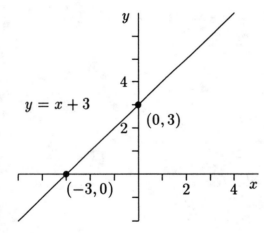

Note!

(N1) Plotted the intercepts to determine the graph:

$$x = 0 : \quad y = 0 + 3 = 3 \Longrightarrow (0, 3)$$
$$y = 0 : \quad 0 = x + 3 \Longrightarrow x = -3 \Longrightarrow (-3, 0)$$

(N2) Domain: $(-\infty, \infty)$

(N3) Range: $\underbrace{(-\infty, \infty)}$

   (from graph: $y \in (-\infty, \infty)$)

Example 6: Consider $f(x) = x^2 - 2x$.

(A1) Sketch the graph of $y = f(x)$.

   $y = x^2 - 2x$:   parabola

   $a = 1 > 0$:   opens upward

x intercepts: $x^2 - 2x = 0 \implies x(x-2) = 0 \implies x = 0, 2$

$\implies$ at vertex: $x = \dfrac{0+2}{2} = 1 \implies y = -1$

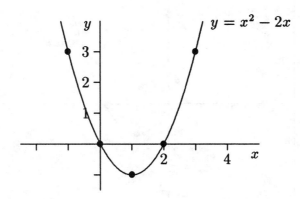

(A2) State the domain and range of f.

domain of f: $x \in (-\infty, \infty)$;   range of f: $y \in [-1, \infty)$

Example 7: Consider $f(x) = \begin{cases} x + 3 & \text{if } x < -1 \\ x^2 - 2x & \text{if } -1 \le x < 3 \\ 4 & \text{if } x \ge 3 \end{cases}$

Note! f is defined by *cases*.

(N1) More than one formula is given, but the formula used to calculate $f(x)$ for a given x depends upon which restriction x satisfies.

(N2) To insure f is a function, each x in the domain of f satisfies exactly one restriction.

(A1) Sketch the graph of $y = f(x)$

$y = x + 3$:   line (see Example 5 above), $x < -1$

$y = x^2 - 2x$:   parabola (see Example 6 above), $-1 \le x < 3$

$y = 4$:   horizontal line, $x \ge 3$

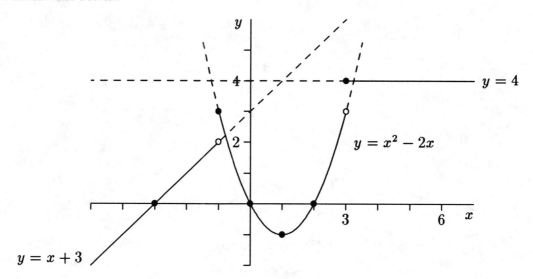

(A2) State the domain and range of f.

domain of f: $x \in (-\infty, \infty)$;   range of f: $y \in (-\infty, 3] \cup \{4\}$

Example 8: Consider $f(x) = \dfrac{x^2 + |x - 1| - 1}{|x - 1|}$.

(A1) State the domain of f.

nonzero denominator $\Longrightarrow$ $|x - 1| \neq 0$: $|x - 1| = 0$ $\Longrightarrow$ $x - 1 = 0$ $\Longrightarrow$ $x = 1$

$\Longrightarrow$ domain of f: $x \in (-\infty, 1) \cup (1, \infty)$ (i.e., $\{x \mid x \neq 1\}$)

(A2) Sketch the graph of $y = f(x)$.

Remove the absolute value symbol:

$$|x - 1| = \begin{cases} x - 1 & \text{if } x - 1 \geq 0 \\ -(x - 1) & \text{if } x - 1 < 0 \end{cases} = \begin{cases} x - 1 & \text{if } x \geq 1 \\ 1 - x & \text{if } x < 1 \end{cases}$$

For $x \neq 1$:

$$x > 1 : \quad f(x) = \frac{x^2 + (x - 1) - 1}{x - 1} = \frac{x^2 + x - 2}{x - 1} = \frac{(x + 2)(x - 1)}{x - 1} = x + 2$$
$$\uparrow$$
$$(x \neq 1)$$

$$x < 1: \quad f(x) = \frac{x^2 + (1-x) - 1}{1-x} = \frac{x^2 - x}{1-x} = \frac{x(x-1)}{1-x} = -x$$

$$\uparrow$$
$$(x \neq 1)$$

Hence, f can be written as cases: $f(x) = \begin{cases} -x & \text{if } x < 1 \\ x + 2 & \text{if } x > 1 \end{cases}$

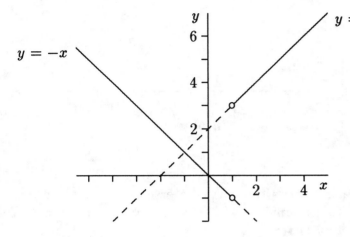

$y = x + 2$

$y = -x$

$y = -x$: line, $x < 1$

$y = x + 2$: line, $x > 1$

(A3) State the range of f

range of f: $y \in (-1, \infty)$

The Vertical Line Test (20): A curve in the plane is the graph of a function if and only if no vertical line intersects the curve more than once.

Remarks:

(R1) If the vertical line $x = a$ intersects the curve at more than one point, say at the two distinct points (a, b) and (a, c) where $b \neq c$, the curve associates two different values with $x = a$; hence the curve can not represent a function of the variable x.

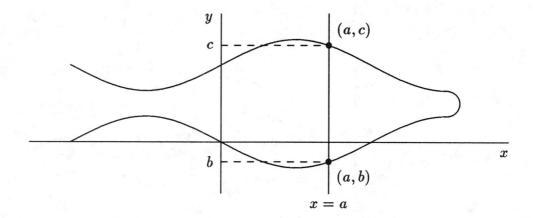

(R2) If each vertical line $x = a$, which intersects the curve, intersects exactly once, say at the point (a, b), the curve associates the unique value b with a and a function f of the variable x can be defined at $x = a$ by letting $f(a) = b$.

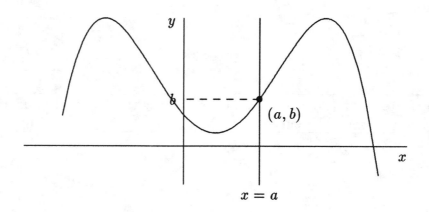

Example 9: Consider the function $f(x) = \sqrt{4 - x}$.

(A1) State the domain of f.

square root: $4 - x \geq 0 \implies x \leq 4 \implies$ domain of f: $x \in (-\infty, 4]$

(A2) Sketch the graph of $y = f(x)$.

$$y = \sqrt{4-x} \implies y^2 = \left(\sqrt{4-x}\right)^2 = 4 - x \implies x = -y^2 + 4: \text{ parabola}$$

$a = -1 < 0$: opens to the left

y intercepts: $-y^2 + 4 = 0 \implies y^2 = 4 \implies y = \pm 2$

$\implies$ at vertex: $y = \dfrac{2 + (-2)}{2} = 0 \implies x = -(0)^2 + 4 = 4$

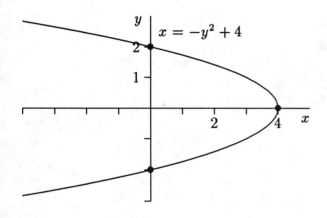

Note! Vertical lines $x = a$, where $a < 4$, intersect the parabola twice
$\implies$ the parabola is not a function of x.

Given equation: $y = \sqrt{4-x} \implies y \geq 0$:

the graph of $y = \sqrt{4-x}$ is the part of the parabola for which $y \geq 0$

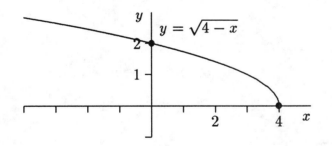

Note! From the graph,

domain of f: $x \in (-\infty, 4]$

(A3) State the range of f.

range of f: $y \in [0, \infty)$

Example 10: Consider the function $f(x) = 1 - \sqrt{4 - x^2}$.

(A1) State the domain of f.

square root: $4 - x^2 \geq 0 \implies x^2 \leq 4 \implies |x| \leq 2 \implies -2 \leq x \leq 2$

domain of f: $x \in [-2, 2]$

(A2) Sketch the graph of $y = f(x)$.

$$y = 1 - \sqrt{4 - x^2} \implies y - 1 = -\sqrt{4 - x^2} \implies (y - 1)^2 = 4 - x^2$$

$$\implies x^2 + (y - 1)^2 = 4: \quad \text{circle with center } (0, 1) \text{ and radius } 2$$

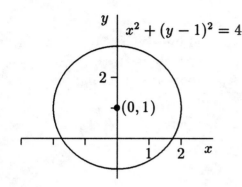

Note! Vertical lines $x = a$, where $-2 < a < 2$,
intersect the circle twice
$\implies$ the circle is not a function of x

Given equation: $y = 1 - \sqrt{4 - x^2} \implies y \leq 1$:
$$\uparrow$$
$$\left(\sqrt{4 - x^2} \geq 0\right)$$

the graph of $y = 1 - \sqrt{4 - x^2}$ is the part of the circle for which $y \leq 1$

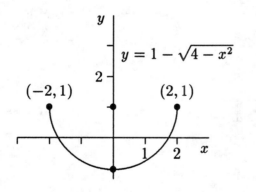

Note! From the graph,
domain of f: $x \in [-2, 2]$

(A3) State the range of f.

range of f: $y \in [-1, 1]$

Definition: Let f be a function with the property that if $x \in A$, then $-x \in A$, where A is the domain of f.

(D1) f is an **even function** if $f(-x) = f(x)$ for each $x \in A$.

(D2) f is an **odd function** if $f(-x) = -f(x)$ for each $x \in A$.

Remarks:

(R1) f: an even function $\implies$ graph of $y = f(x)$ is symmetric with respect to the y axis.

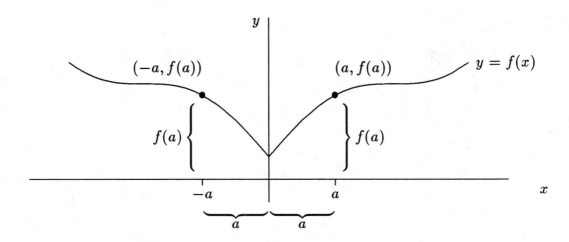

Note! Plot the graph of $y = f(x)$ for $x \geq 0$; then reflect this portion of the graph about the y axis to obtain the graph of $y = f(x)$ for $x < 0$.

(R2) f: an odd function $\implies$ the graph of $y = f(x)$ is symmetric about the origin.

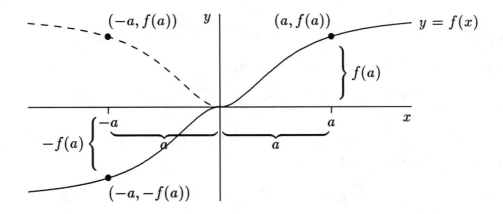

Note! Plot the graph of $y = f(x)$ for $x \geq 0$; reflect this portion of the graph about the y axis (as done for f an even function) and then reflect this later part about the x axis to obtain the graph of $y = f(x)$ for $x < 0$.

Example 11: Consider the function $f(x) = \begin{cases} 1 - |x| & \text{if } |x| \leq 1 \\ 2 & \text{if } |x| > 1 \end{cases}$

(A1) Show f is an even function.

$$|x| \leq 1: \; f(-x) = 1 - |-x| = 1 - |x| = f(x)$$
$$\uparrow$$
$$(|-x| = |(-1)x| = |-1||x| = |x|)$$

$$|x| > 1: \; f(-x) = 2 = f(x)$$

Hence, $f(-x) = f(x)$ for $x \in (-\infty, \infty) \implies f$ is an even function.

(A2) Sketch the graph of $y = f(x)$.

$$x \geq 0: \quad f(x) = \begin{cases} 1 - x & \text{if } 0 \leq x \leq 1 \\ 2 & \text{if } x > 1 \end{cases}$$
$$\uparrow$$
$$(|x| = x)$$

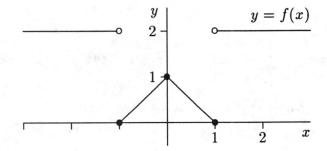

Note!

$$\left\{ \begin{array}{ll} y = 1 - x & : \text{line} \\ y = 2 & : \text{line} \end{array} \right\};$$

then reflect about y axis

Example 12: Consider the function $f(x) = x|x|$.

(A1) State the domain of f.

domain of f: $x \in (-\infty, \infty)$

(A2) Show f is an odd function.

$$f(-x) = (-x)|-x| = -x|x| = -f(x) \implies f \text{ is an odd function.}$$
$$\uparrow$$
$$(|-x| = |x|)$$

(A3) Sketch the graph of $y = f(x)$.

$x \geq 0$: $f(x) = x(x) = x^2$
$\uparrow$
$(|x| = x)$

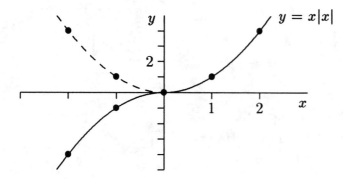

Note! $y = x^2$: parabola;

reflect about the y axis,

then about the x axis

Example 13: Consider the function $f(x) = |x| + 2x$.

(A1) State the domain of f.

domain of f: $x \in (-\infty, \infty)$

(A2) Show f is neither an even nor an odd function.

$$f(-x) = |-x| + 2(-x) = |x| - 2x \neq \begin{cases} f(x) = |x| + 2x & \Longrightarrow \quad \text{not even} \\ -f(x) = -(|x| + 2x) & \Longrightarrow \quad \text{not odd} \end{cases}$$
$$\uparrow$$
$$(|-x| = |x|)$$

(A3) Sketch the graph of $y = f(x)$.

$$y = |x| + 2x = \begin{cases} x + 2x = 3x & \text{if } x \geq 0 \\ -x + 2x = x & \text{if } x < 0 \end{cases}$$

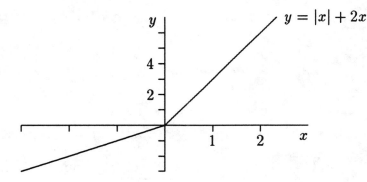

Note!

$y = 3x$: line

$y = x$: line

Let g be a function. Sketch the graph of the function $f(x) = |g(x)|$.

Note! $f(x) = \begin{cases} g(x) & \text{if } g(x) \geq 0 \\ -g(x) & \text{if } g(x) < 0 \end{cases}$

A Procedure:

(S1) Sketch the graph of $y = g(x)$.
  Note! Ignore the absolute value symbol: sketch the graph of the "inside" function.

(S2) Reflect the graph of $y = g(x)$ about the x axis if $g(x) < 0$.

(S3) The graph of $y = f(x)$ is $\left\{ \begin{array}{c} \text{the graph of } y = g(x) \text{ if } g(x) \geq 0 \\ \text{and} \\ \text{the reflected portion from (S2)} \end{array} \right\}$

  Illustration:

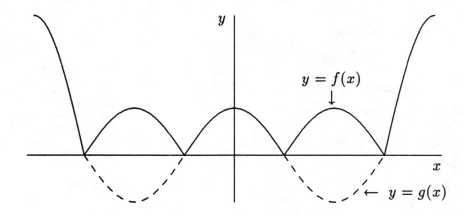

Note! $y = |g(x)| \geq 0$

Example 14: Sketch the graph of $f(x) = |x + 3|$.

Sketch graph of $y = x + 3$: line   (inside function $g(x) = x + 3$ in procedure)

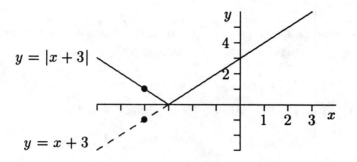

Note! Could have sketched via cases:

$$y = |x + 3| = \begin{cases} x + 3 & \text{if } x + 3 \geq 0 \\ -(x + 3) & \text{if } x + 3 < 0 \end{cases} = \begin{cases} x + 3 & \text{if } x \geq -3 \\ -x - 3 & \text{if } x < -3 \end{cases}$$

Example 15: Sketch the graph of $f(x) = |x^2 - 2x|$.

Sketch the graph of $\underbrace{y = x^2 - 2x: \text{parabola}}_{\text{(see Example 6 above)}}$ (inside function $g(x) = x^2 - 2x$)

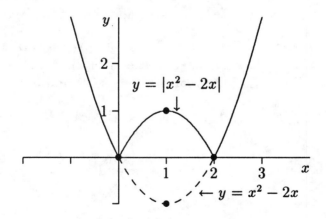

0.5   Combinations of Functions

This section demonstrates how functions can be combined in order to obtain another function.

Algebra of Functions (22): Let f and g be functions with domains A and B. Then the functions $f + g$, $f - g$, fg, and f/g are defined as follows:

$$(f + g)(x) = f(x) + g(x) \qquad \text{domain} = A \cap B$$

$$(f - g)(x) = f(x) - g(x) \qquad \text{domain} = A \cap B$$

$$(fg)(x) = f(x)g(x) \qquad \text{domain} = A \cap B$$

$$\left(\frac{f}{g}\right)(x) = \frac{f(x)}{g(x)} \qquad \text{domain} = \{x \in A \cap B \mid g(x) \neq 0\}$$

Remarks:

(R1) The **sum of functions** f and g, denoted $f + g$, is defined to be a function h whose domain is $A \cap B$, which is the common domain of f and g, such that, for each $x \in A \cap B$ ($\neq \emptyset$),

$$h(x) = (f + g)(x) = f(x) + g(x).$$

Note! The $+$ sign in $(f+g)(x)$ designates the operation of addition of functions, whereas the $+$ sign in $f(x) + g(x)$ denotes the usual addition of the real numbers $f(x) + g(x)$.

(R2) Similarly, the **difference of functions** $f - g$, the **product of functions** fg, and the **quotient of functions** f/g are defined, respectively, in terms of the usual subtraction, multiplication, and division of the real numbers $f(x)$ and $g(x)$ for each $x \in A \cap B$.

Note! The function f/g is not defined at those $x \in A \cap B$ such that $g(x) = 0$.

Example 1: Consider the functions $f(x) = \sqrt{4 - x^2}$ and $g(x) = \sqrt{1 - x}$.

(A1) Find the domain of f.

square root: $4 - x^2 \geq 0 \implies x^2 \leq 4 \implies |x| \leq 2 \implies -2 \leq x \leq 2$

$\implies$ domain of f: $[-2, 2]$

(A2) Find the domain of g.

square root: $1 - x \geq 0 \implies x \leq 1 \implies$ domain of g: $(-\infty, 1]$

(A3) Find the function $f + g$.

domain of $f + g$: $\underbrace{[-2, 2]}_{\text{(A1)}} \cap \underbrace{(-\infty, 1]}_{\text{(A2)}} = [-2, 1]$

$\implies (f + g)(x) = \underbrace{\sqrt{4 - x^2}}_{f(x)} + \underbrace{\sqrt{1 - x}}_{g(x)}, \qquad x \in [-2, 1]$

(A4) Find the function $f - g$.

$(f - g)(x) = \underbrace{\sqrt{4 - x^2}}_{f(x)} - \underbrace{\sqrt{1 - x}}_{g(x)}, \qquad \underbrace{x \in [-2, 1]}_{\text{(from (A3))}}$

(A5) Find the function fg.

$(fg)(x) = \underbrace{\sqrt{4 - x^2}}_{f(x)} \; \underbrace{\sqrt{1 - x}}_{g(x)} = \sqrt{(4 - x^2)(1 - x)}$

$= \sqrt{x^3 - x^2 - 4x + 4}, \qquad \underbrace{x \in [-2, 1]}_{\text{(from (A3))}}$

(A6) Find the function $\dfrac{f}{g}$.

$\left(\dfrac{f}{g}\right)(x) = \underset{\underset{\left(\frac{f(x)}{g(x)}\right)}{\uparrow}}{\dfrac{\sqrt{4 - x^2}}{\sqrt{1 - x}}} = \sqrt{\dfrac{4 - x^2}{1 - x}}, \qquad \underbrace{x \in [-2, 1)}_{\text{(from (A3): } g(1) = 0)}$

(A7) Find the function $\dfrac{g}{f}$.

$\left(\dfrac{g}{f}\right)(x) = \underset{\underset{\left(\frac{g(x)}{f(x)}\right)}{\uparrow}}{\dfrac{\sqrt{1 - x}}{\sqrt{4 - x^2}}} = \sqrt{\dfrac{1 - x}{4 - x^2}}, \qquad \underbrace{x \in (-2, 1]}_{\text{(from (A3): } f(-2) = 0)}$

The graph of $f + g$ can be obtained from the graphs of f and g via a procedure called **graphical addition**, which uses the sum of the corresponding y coordinates: for each a in the common domain of f and g, the point $(a, (f + g)(a))$ on the graph of $f + g$ can be obtained by proceeding along the line $x = a$ from the point $(a, f(a))$

$$\left\{ \begin{array}{l} \text{upwards the distance } g(a) \text{ if } g(a) \geq 0 \\ \text{downwards the distance } |g(a)| \text{ if } g(a) < 0 \end{array} \right\}$$

Note! $g(a) = 0 \implies (f+g)(a) = f(a)$: need not move from $(a, f(a))$ to obtain $(a, (f+g)(a))$.

Illustration:

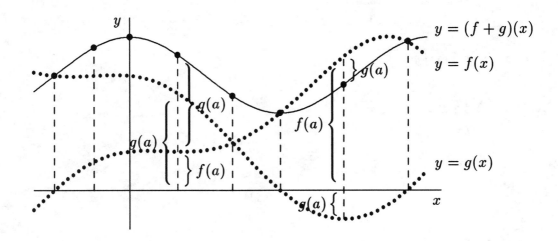

Example 2: Use graphical addition to obtain the graph of $y = (f + g)(x)$ where $f(x) = \sqrt{4 - x^2}$ and $g(x) = \sqrt{1 - x}$ (from Example 1 above).

Note!

(N1) $y = \sqrt{4 - x^2}$: upper semicircle of circle $x^2 + y^2 = 4$

(N2) $y = \sqrt{1 - x}$: "upper half" of parabola $x = -y^2 + 1$

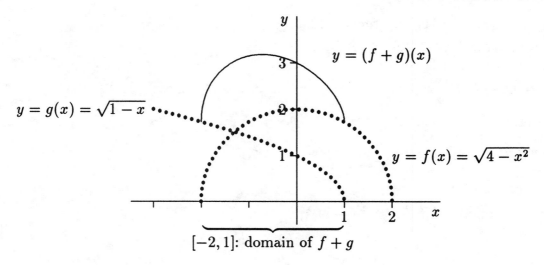

$$y = (f + g)(x)$$

$$y = g(x) = \sqrt{1 - x}$$

$$y = f(x) = \sqrt{4 - x^2}$$

$[-2, 1]$: domain of $f + g$

The following way of combining functions treats a function as an "operator": the value $f(x)$ results from f operating on x according to some rule.

Definition (23): Given two functions f and g, the **composite function** $f \circ g$ (also called the **composition** of f and g) is defined by

$$(f \circ g)(x) = f(g(x))$$

Remarks:

(R1) To obtain the value $(f \circ g)(x)$, first g operates on x to obtain the value $g(x)$ and then f operates on the value $g(x)$ to obtain the value $f(g(x))$.

(R2) The domain of $f \circ g$ are those x in the domain of g (hence, $g(x)$ is defined) such that $g(x)$ is in the domain of f (hence, $f(g(x))$ is defined).

(R3) Arrow diagram:

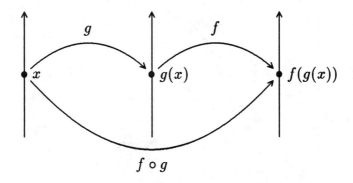

Example 3: Consider the functions $f(x) = \sqrt{x}$ and $g(x) = 1 - x$.

(A1) Find the function $f \circ g$.

domain of f: $x \in [0, \infty)$;   domain of g: $x \in (-\infty, \infty)$

Determine domain of $f \circ g$:

$$\underbrace{g(x) \in [0, \infty)}_{\text{(domain of } f)} \implies 0 \leq \underbrace{1 - x}_{(g(x))} \implies x \leq 1$$

$$\implies \text{ domain of } f \circ g: \ x \in (-\infty, 1] \subset \underbrace{(-\infty, \infty)}_{\text{(domain of } g)}$$

$$(f \circ g)(x) = f(g(x)) = f(1 - x) = \sqrt{1 - x}, \quad x \in (-\infty, 1]$$

(A2) Find the function $g \circ f$.

Determine domain of $g \circ f$:

$$\underbrace{f(x) \in (-\infty, \infty)}_{\text{(domain of } g)} \implies \underbrace{-\infty < \sqrt{x}}_{(f(x))} < \infty, \text{ which holds for all } x \in \underbrace{[0, \infty)}_{\text{(domain of } f)}$$

$$\implies \text{ domain of } g \circ f: \ x \in [0, \infty)$$

$$(g \circ f)(x) = g(f(x)) = g(\sqrt{x}) = 1 - \sqrt{x}, \quad x \in [0, \infty)$$

(A3) Using the results of (A1) and (A2), sketch the graphs of $f \circ g$ and $g \circ f$ on the same set of axes.

$$f \circ g: \ y = \sqrt{1 - x} \ (\geq 0) \qquad\qquad y^2 = 1 - x \implies x = -y^2 + 1: \text{ parabola}$$

$$g \circ f: \ y = 1 - \sqrt{x} \ (\leq 1) \qquad\qquad \sqrt{x} = 1 - y \implies x = (1 - y)^2: \text{ parabola}$$

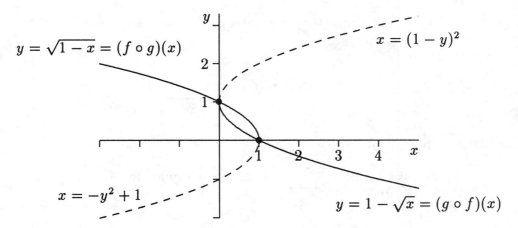

$$y = \sqrt{1-x} = (f \circ g)(x)$$

$$x = (1-y)^2$$

$$x = -y^2 + 1$$

$$y = 1 - \sqrt{x} = (g \circ f)(x)$$

Note!

(N1) $\left\{ \begin{array}{l} \text{domain of } f \circ g: \ (-\infty, 1] \\ \text{domain of } g \circ f: \ [0, \infty) \end{array} \right\} \implies f \circ g \neq g \circ f$ \hfill (different domains)

(N2) On common domain: $(-\infty, 1] \cap [0, \infty = [0, 1]$. From graph above:

$$(f \circ g)(x) \neq (g \circ f)(x), \quad x \in (0, 1)$$

$$(f \circ g)(x) = (g \circ f)(x), \quad x = 0, 1$$

(A4) Find the function $f \circ f$.

Determine domain of $f \circ f$:

$$\underbrace{f(x) \in [0, \infty)}_{\text{(domain of } f)} \implies 0 \leq \underbrace{\sqrt{x}}_{(f(x))} < \infty, \text{ which holds for all } x \in \underbrace{[0, \infty)}_{\text{(domain of } f)}$$

$\implies$ domain of $f \circ f$: $x \in [0, \infty)$

$$(f \circ f)(x) = f(f(x)) = f(\sqrt{x}) = \sqrt{\sqrt{x}} = (x^{1/2})^{1/2} = x^{1/4} = \sqrt[4]{x}, \quad x \in [0, \infty)$$

(A5) Find the function $g \circ g$.

Determine domain of $g \circ g$:

$$\underbrace{g(x) \in (-\infty, \infty)}_{\text{(domain of } g)} \implies -\infty < \underbrace{1 - x}_{(g(x))} < \infty, \text{ which holds for all } x \in \underbrace{(-\infty, \infty)}_{\text{(domain of } g)}$$

$\implies$ domain of $g \circ g$: $x \in (-\infty, \infty)$

$$(g \circ g)(x) = g(g(x)) = g(1 - x) = 1 - (1 - x) = x, \quad x \in (-\infty, \infty)$$

Example 4: Consider the functions $f(x) = x^2$ and $g(x) = \sqrt{1-x}$.

(A1) Find the function $f \circ g$.

domain of f: $x \in (-\infty, \infty)$;   domain of g: $\underbrace{x \in (-\infty, 1]}$

$$(1 - x \geq 0 \Longrightarrow x \leq 1)$$

Determine domain of $f \circ g$:

$g(x) \in \underbrace{(-\infty, \infty)}_{\text{(domain of } f)} \Longrightarrow -\infty < \underbrace{\sqrt{1-x}}_{g(x)} < \infty$, which holds for all $x \in \underbrace{(-\infty, 1]}_{\text{(domain of } g)}$

$\Longrightarrow$ domain of $f \circ g$: $x \in (-\infty, 1]$

$$(f \circ g)(x) = f(g(x)) = f\left(\sqrt{x}\right) = \left(\sqrt{1-x}\right)^2 = 1 - x, \quad x \in (-\infty, 1]$$

Note! If the final expression $(f \circ g)(x) = 1 - x$ were used to determine the domain of $f \circ g$, then $x \in (-\infty, \infty)$ are permitted values; however, for $x > 1$, $g(x) = \sqrt{1-x}$ is not defined.

Reminder: The domain of $f \circ g$ is always a subset of the domain of g.

(A2) Find the function $g \circ f$.

$f(x) \in \underbrace{(-\infty, 1]}_{\text{(domain of } g)} \Longrightarrow \underbrace{x^2}_{(f(x))} \leq 1 \Longrightarrow |x| \leq 1 \Longrightarrow -1 \leq x \leq 1$

$\Longrightarrow$ domain of $g \circ f$: $x \in [-1, 1] \subset \underbrace{(-\infty, \infty)}_{\text{(domain of } f)}$

$$(g \circ f)(x) = g(f(x)) = g(x^2) = \sqrt{1 - x^2}, \quad x \in [-1, 1]$$

(A3) Sketch the graphs of f, g, $f \circ g$, and $g \circ f$ on the same set of axes.

$y = f(x) = x^2$: parabola

$y = g(x) = \sqrt{1-x}$:   part of parabola $x = 1 - y^2$

$y = (f \circ g)(x) = 1 - x, \quad x \in (-\infty, 1]$:   part of line $1 - x$

$y = (g \circ f)(x) = \sqrt{1 - x^2}$:   upper semicircle of $x^2 + y^2 = 1$

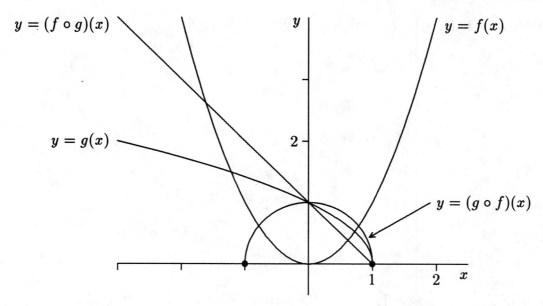

Example 5: Consider the functions $f(x) = x^2$, $g(x) = \sqrt{1-x}$, and $h(x) = x+1$.

(A1) Determine the expression for $(f \circ g \circ h)(x)$.

$$(f \circ g \circ h)(x) = f(g(h(x))) = f(g(x+1)) = f\left(\sqrt{1-(x+1)}\right)$$

$$= f\left(\sqrt{-x}\right) = \left(\sqrt{-x}\right)^2 = -x$$

(A2) Determine the domain of $f \circ g \circ h$.

domain of f: $x \in (-\infty, \infty)$;   domain of g: $\underbrace{x \in (-\infty, 1]}_{(1-x \geq 0)}$;

domain of h: $x \in (-\infty, \infty)$

domain of $g \circ h$: $h(x) \in \underbrace{(-\infty, 1]}_{\text{(domain of } g)} \implies \underset{h(x)}{\underbrace{x+1}} \leq 1 \implies x \leq 0$

$\implies$ domain of $g \circ h$: $x \in (-\infty, 0] \subset \underbrace{(-\infty, \infty)}_{\text{(domain of } h)}$

domain of $f \circ g \circ h$: $(g \circ h)(x) \in \underbrace{(-\infty, \infty)}_{\text{(domain of } f)} \implies -\infty < \sqrt{-x} < \infty,$

$\uparrow$

(from (A1): $(g \circ h)(x) = g(h(x))$)

which holds for all $x \in (-\infty, 0]$, the domain of $g \circ h$

(A3) Sketch the graph of $f \circ g \circ h$.

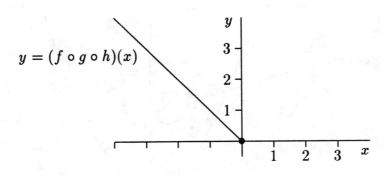

$$y = (f \circ g \circ h)(x)$$

Example 6: Express $h(x) = |x - 1|$ in the form $(f \circ g)(x)$ such that $f(x) \neq x$ or $g(x) \neq x$.

A choice: $\left\{ \begin{array}{l} f(x) = |x| \\ g(x) = x - 1 \end{array} \right\} \implies (f \circ g)(x) = f(g(x)) = f(x - 1) = |x - 1| = h(x)$

Note! $g(x) \in \underbrace{(-\infty, \infty)}_{\text{(domain of } f)} \implies -\infty < \underbrace{x - 1}_{g(x)} < \infty,$

which holds for all $x \in \underbrace{(-\infty, \infty)}_{\text{(domain of } g)} \implies$ domain of $f \circ g$: $x \in (-\infty, \infty)$

$\implies h(x) = (f \circ g)(x)$ for all $x \in (-\infty, \infty)$, the domain of h

Example 7: Express $h(x) = \dfrac{1}{1 - \sqrt{x}}$ in the form $(f \circ g)(x)$ such that $f(x) \neq x$ or $g(x) \neq x$.

A choice: $\left\{ \begin{array}{l} f(x) = \dfrac{1}{x} \\ g(x) = 1 - \sqrt{x} \end{array} \right\} \implies (f \circ g)(x) = f(g(x)) = f(1 - \sqrt{x}) = \dfrac{1}{1 - \sqrt{x}}$

Note! Another choice: $\left\{ \begin{array}{l} f(x) = \dfrac{1}{1 - x} \\ g(x) = \sqrt{x} \end{array} \right\}$

$\implies (f \circ g)(x) = f(g(x)) = f(\sqrt{x}) = \dfrac{1}{1 - \sqrt{x}}$

Example 8: If $f(x) = \dfrac{1}{2x-1}$ and $h(x) = \dfrac{1}{4x-3}$, find a function g such that $f \circ g = h$.

$$\left\{ \begin{array}{l} (f \circ g)(x) = f(g(x)) = \dfrac{1}{2g(x)-1} \\[4mm] (f \circ g)(x) = h(x) = \dfrac{1}{4x-3} \end{array} \right\}$$

$$\implies \dfrac{1}{2g(x)-1} = \dfrac{1}{4x-3} \implies 2g(x) - 1 = 4x - 3 \implies 2g(x) = 4x - 2$$

$$\implies g(x) = 2x - 1 \qquad\qquad\qquad\qquad\qquad \text{(domain of } g: \ (-\infty, \infty))$$

Check: $(f \circ g)(x) = f(g(x)) = f(2x - 1) = \dfrac{1}{2(2x-1)-1} = \dfrac{1}{4x-3} = h(x)$

Example 9: If $g(x) = x^2$ and $h(x) = \dfrac{|x|}{x^2+4}$, find a function f such that $f \circ g = h$.

$$\left\{ \begin{array}{l} (f \circ g)(x) = f(g(x)) = f(x^2) \\[4mm] (f \circ g)(x) = h(x) = \dfrac{|x|}{x^2+4} \end{array} \right\} \implies f(x^2) = \underset{\uparrow}{\dfrac{|x|}{x^2+4}} = \dfrac{\sqrt{x^2}}{x^2+4}$$

$$\left(\text{express } \dfrac{|x|}{x^2+4} \text{ in terms of } x^2 \right)$$

$$\implies f(x) = \dfrac{\sqrt{x}}{x+4}$$

Example 10: If $g(x) = 2x - 1$ and $h(x) = \dfrac{1}{4x+5}$, find a function f such that $f \circ g = h$.

$$\left\{ \begin{array}{l} (f \circ g)(x) = f(g(x)) = f(2x-1) \\[4mm] (f \circ g)(x) = h(x) = \dfrac{1}{4x+5} \end{array} \right\} \implies f(2x-1) = \underset{\uparrow}{\dfrac{1}{4x+5}} = \dfrac{1}{2(2x)+5}$$

$$\left(\text{express } \dfrac{1}{4x+5} \text{ in terms of } 2x-1 \right)$$

$$\underset{\uparrow}{=} \dfrac{1}{2(2x-1+1)+5} = \dfrac{1}{2(2x-1)+7} \implies f(x) = \dfrac{1}{2x+7}$$

(added zero)

0.6 Types of Functions; Shifting and Scaling

This section presents some basic functions which commonly occur in calculus and certain transformations which enable the graphs of many functions to be obtained from those of the basic functions.

Definition: A function of the form $f(x) = c$, where c is a constant, is called a **constant function**.

Remarks:

(R1) Domain of a constant function: $(-\infty, \infty)$

(R2) $f(-x) = c = f(x) \implies f$ is an even function.

(R3) Range of a constant function: $\{c\}$

(R4) Graph of $y = c$:

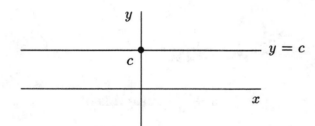

Note! Graph is a horizontal line with y intercept c

Definition: A function of the form $f(x) = x^a$, where a is a constant, is called a **power function**.

Consider special cases for the real number a:

(C1) $a = n$, where n is a positive integer: $f(x) = x^n$

Note!

(N1) Domain of $f(x) = x^n$: $(-\infty, \infty)$

(N2) $n = 1 \implies f(x) = x$: **identity function** (since the rule maps the element in the domain to itself)

(N3) $f(-x) = (-x)^n = (-1)^n x^n = \left\{ \begin{array}{ll} x^n = f(x) & \text{if } n \text{ even} \\ -x^n = -f(x) & \text{if } n \text{ odd} \end{array} \right\}$

$$\implies \quad f \text{ is an } \left\{ \begin{array}{l} \text{even function if } n \text{ even} \\ \text{odd function if } n \text{ odd} \end{array} \right\}.$$

(N4) Range of $f(x) = x^n$: $\left\{ \begin{array}{ll} [0, \infty) & \text{if } n \text{ even} \\ (-\infty, \infty) & \text{if } n \text{ odd} \end{array} \right\}$ with $\quad f(0) = 0 \text{ and } f(1) = 1$

(N5) $0 < x < 1 \implies 0 < x^n < 1$ such that $0 < x^{n+1} < x^n < 1$.

Hence, if m and n are positive integers such that $m > n$, then $0 < x^m < x^n < 1$ whenever $0 < x < 1$.

(N6) Graph of $y = x^n$ for n odd, $n = 1, 3, 5$:

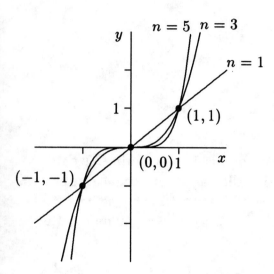

$f(x) = x^n$: odd function such that

$0 < x^5 < x^3 < x < 1$ whenever $0 < x < 1$

(N7) Graph of $y = x^n$ for n even, $n = 2, 4, 6$

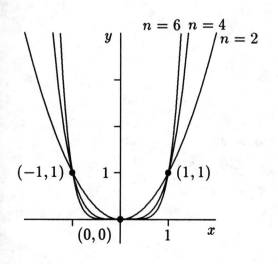

$f(x) = x^n$: even function such that
$0 < x^6 < x^4 < x^2 < 1$ whenever $0 < x < 1$;
$y = x^2$ is a parabola opening
upward from its vertex $(0, 0)$

(C2) $a = -1$: $f(x) = x^{-1} = \dfrac{1}{x}$

 Note!

(N1) $f(x) = \dfrac{1}{x}$: called the **reciprocal function** (since the rule maps the element in the domain to its reciprocal)

(N2) Domain of $f(x) = \dfrac{1}{x}$: $(-\infty, 0) \cup (0, \infty)$

(N3) $f(-x) = \dfrac{1}{-x} = -\dfrac{1}{x} = -f(x) \implies f$ is an odd function

(N4) Range of $f(x) = \dfrac{1}{x}$: $(-\infty, 0) \cup (0, \infty)$

(N5) Graph of $y = \dfrac{1}{x}$:

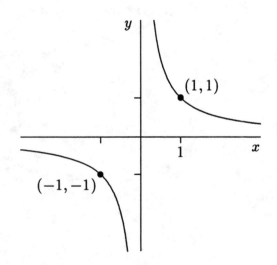

The graph of $y = 1/x$, sometimes written as $xy = 1$, is called an **equilateral hyperbola** and the coordinate axes serve as its asymptotes ($xy = 1 \implies x \neq 0$ and $y \neq 0$)

(C3) $a = \dfrac{1}{n}$, where n is a positive integer such that $n \geq 2$: $f(x) = x^{1/n}$

 Note!

(N1) $f(x) = x^{1/n} = \sqrt[n]{x}$: called the **$n$th root function**; $f(0) = 0$ and $f(1) = 1$

(N2) n even: requires $x \geq 0 \implies \left\{ \begin{array}{ll} \text{domain of } f(x) = x^{1/n} &: [0, \infty) \\ \text{range of } f(x) = x^{1/n} &: [0, \infty) \end{array} \right\}$

(N3) Graph of $y = \sqrt[n]{x}$ for n even, $n = 2, 4$:

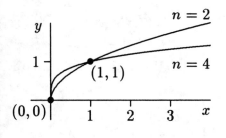

The graph of $y = \sqrt[2]{x} = \sqrt{x}$, the **square root function**, is the upper half of the parabola $x = y^2$ (which opens to the right from $(0,0)$);

$0 < \sqrt{x} < \sqrt[4]{x} < 1$ whenever $0 < x < 1$;

$1 < \sqrt[4]{x} < \sqrt{x}$ whenever $x > 1$

(N4) n odd $\implies$ domain of $f(x) = x^{1/n}$ is $(-\infty, \infty)$ with

$$f(-x) = (-x)^{1/n} = (-1)^{1/n}x^{1/n} = -x^{1/n} = -f(x)$$

Hence, $f(x) = x^{1/n}$ is an odd function with range $(-\infty, \infty)$

(N5) Graph of $y = \sqrt[n]{x}$ for n odd, $n = 3, 5$:

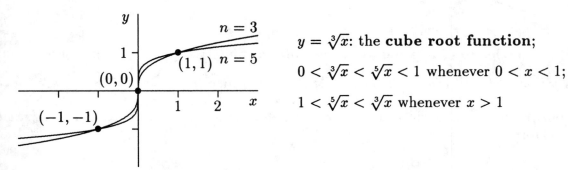

$y = \sqrt[3]{x}$: the **cube root function**;

$0 < \sqrt[3]{x} < \sqrt[5]{x} < 1$ whenever $0 < x < 1$;

$1 < \sqrt[5]{x} < \sqrt[3]{x}$ whenever $x > 1$

Definition: A function P is a **polynomial function** if

$$P(x) = a_n x^n + a_{n-1} x^{n-1} + \cdots + a_1 x + a_0,$$

where n is a nonnegative integer and the numbers $a_0, a_1, a_2, \ldots, a_n$, called **coefficients** of the polynomial, are constants.

Remarks:

(R1) a_0: the **constant term** of the polynomial

   Note! $a_0 = 0 \implies P(0) = 0$: the polynomial passes through the origin.

(R2) If $a_n \neq 0$, then a_n is the **leading coefficient** and n is the **degree** of the polynomial.

(R3) The domain of P: $(-\infty, \infty)$

(R4) If P has degree 0, then $P(x) = a_0$, where $a_0 \neq 0$, and P is a constant function.

Note! The constant function $P(x) = 0$ is called the **zero polynomial** and, by convention, is not assigned a degree.

(R5) A polynomial of the form $P(x) = a_1 x + a_0$, where $a_1 \neq 0$, is called a **linear function**.

Note!

(N1) The graph of $y = P(x)$ is a line with slope $a_1\ (\neq 0)$ and y intercept a_0

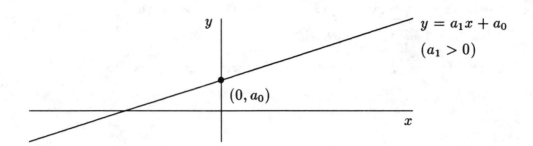

$$y = a_1 x + a_0$$
$$(a_1 > 0)$$
$$(0, a_0)$$

(N2) P is referred to as a **first-degree polynomial** since the degree of P is 1.

(N3) If $a_1 = 1\ (\neq 0)$, $a_0 = 0$, then $P(x) = x$ is the identity function.

(R6) A polynomial of the form $P(x) = a_2 x^2 + a_1 x + a_0$, where $a_2 \neq 0$, is called a **quadratic function**.

Note!

(N1) The graph of $y = P(x)$ is a parabola opening

$$\left\{ \begin{array}{l} \text{upward if } a_2 > 0 \\ \text{downward if } a_2 < 0 \end{array} \right\} \text{ from the vertex } \left(-\frac{a_1}{2}, -\frac{a_1^2 - 4a_2 a_0}{4a_2} \right)$$

(N2) P is referred to as a **second-degree polynomial** since the degree of P is 2.

(R7) A polynomial of the form $P(x) = a_3 x^3 + a_2 x^2 + a_1 x + a_0$, where $a_3 \neq 0$, is called a **cubic function**.

Definition: A **rational function** is the ratio (quotient) of two polynomial functions.

Remarks: If P and Q are polynomial functions, then a rational function f has the form

$$f(x) = \frac{P(x)}{Q(x)}.$$

(R1) Domain of f: $\{x \mid Q(x) \neq 0\}$ (i.e., all real numbers except the zeros of Q)

(R2) If $Q(x)$ is a nonzero constant function, then $f(x)$ is a polynomial function with domain $(-\infty, \infty)$.

(R3) The reciprocal function $f(x) = 1/x$ ($P(x) = 1$ and $Q(x) = x$) is a rational function.

Definition: A function is called **algebraic** if it can be constructed in a finite number of steps using only the algebraic operations (addition, subtraction, multiplication, division, and root extraction) starting with polynomial functions. If a function is not algebraic, then it is called **transcendental.**

Remarks: The trigonometric functions (see Appendix B of text) are transcendental.

(R1) Graph of $y = \sin x$ (x in radian measure):

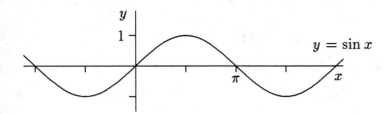

$y = \sin x$ *Note!* $f(x) = \sin x$ is an odd function

(R2) Graph of $y = \cos x$ (x in radian measure):

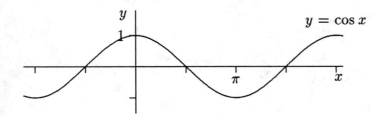

$y = \cos x$ *Note!* $f(x) = \cos x$ is an even function

Vertical and Horizontal Shifts (24): Suppose $c > 0$. To obtain the graph of

  $y = f(x) + c$, shift the graph of $y = f(x)$ upward c units.

  $y = f(x) - c$, shift the graph of $y = f(x)$ downward c units.

  $y = f(x - c)$, shift the graph of $y = f(x)$ to the right c units.

  $y = f(x + c)$, shift the graph of $y = f(x)$ to the left c units.

Remarks:

(R1) The transformation of either shifting a graph vertically or horizontally is called a **translation**.

(R2) Two graphs related by a translation have the same shape.

(R3) A way to determine the direction of a horizontal shift: say $x = a$ is in the domain of f.

$y = f(x)$: let $x = a$; plot the point $(a, f(a))$

$y = f(x - c)$: let $x - c = a \implies x = a + c$; plot $(a + c, f(a))$: shift $\rightarrow c$ units

$y = f(x + c)$: let $x + c = a \implies x = a - c$; plot $(a - c, f(a))$: shift $\leftarrow c$ units

Note! A choice: for $y = f(x)$, let $x = 0$ if 0 is in the domain of f

Illustration: Let $c > 0$.

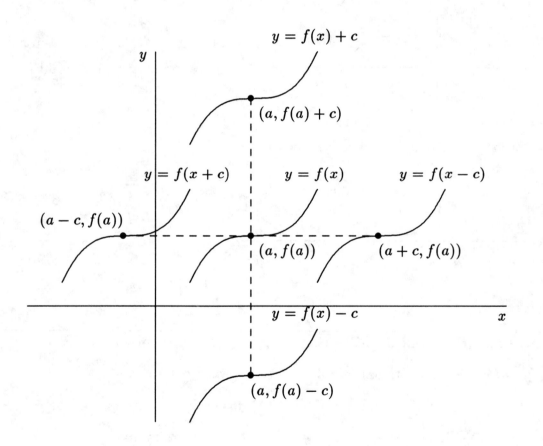

Example 1: Sketch the graph of $y = \sqrt{x+2} + 3$ via successive translations of the graph of $y = \sqrt{x}$.

$y = \sqrt{x}$:   sketch

$y = \sqrt{x+2}$:   shift 2 units $\leftarrow$

$y = \sqrt{x+2} + 3$:   shift 3 units $\uparrow$

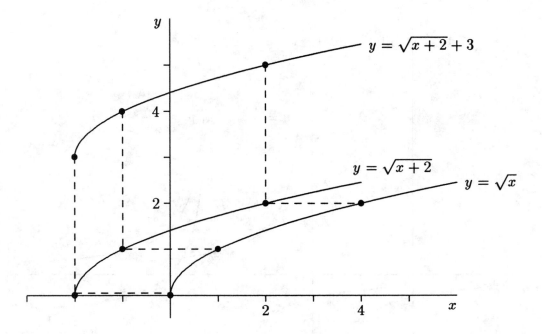

Example 2: Sketch the graph of $y = \dfrac{1}{x-2} - 1$ via successive translations of the graph of $y = \dfrac{1}{x}$.

$y = \dfrac{1}{x}$:   sketch

$y = \dfrac{1}{x-2}$:   shift 2 units $\rightarrow$

$y = \dfrac{1}{x-2} - 1$:   shift 1 unit $\downarrow$

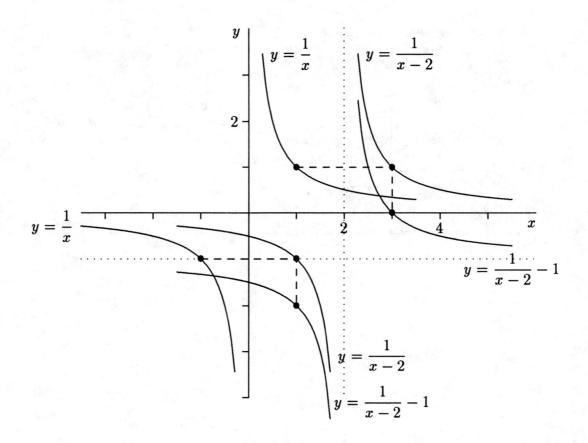

Note! $x = 0$ is not in the domain of $f(x) = 1/x$: chose $a = 1, -1$ due to the two branches of the hyperbola $y = 1/x$

Vertical and Horizontal Stretching and Reflecting (25): Suppose $c > 1$. To obtain the graph of

$y = cf(x)$, stretch the graph of $y = f(x)$ vertically by a factor of c.

$y = (1/c)f(x)$, compress the graph of $y = f(x)$ vertically by a factor of c.

$y = f(cx)$, compress the graph of $y = f(x)$ horizontally by a factor of c.

$y = f(x/c)$, stretch the graph of $y = f(x)$ horizontally by a factor of c.

$y = -f(x)$, reflect the graph of $y = f(x)$ about the x-axis.

$y = f(-x)$, reflect the graph of $y = f(x)$ about the y-axis.

Remarks:

(R1) Translation: transformation resulting from the addition or subtraction of a positive number

(R2) Vertical/horizontal stretching/compressing: transformation resulting from the multiplication by a positive number

(R3) Reflection: transformation resulting from the multiplication by the number -1

(R4) A way to determine whether a horizontal compression or a horizontal stretch: say $x = a$ is in the domain of f.

$y = f(x)$: let $x = a$; plot the point $(a, f(a))$

$y = f(cx)$: let $cx = a \implies x = \dfrac{a}{c}$; plot the point $\left(\dfrac{a}{c}, f(a)\right)$:

a compression since $\underbrace{\left|\dfrac{a}{c}\right| < |a|}_{(c > 1)}$

$y = f\left(\dfrac{x}{c}\right)$: let $\dfrac{x}{c} = a \implies x = ac$; plot the point $(ac, f(a))$:

a stretch since $\underbrace{|ac| > |a|}_{(c > 1)}$

Example 3: Sketch the graph of $y = 2\sqrt{-x} + 1$ via successive transformations of $y = \sqrt{x}$.

$y = \sqrt{x}$:   sketch

$y = \sqrt{-x}$:   reflection about y axis

$y = 2\sqrt{-x}$:   vertical stretch

$y = 2\sqrt{-x} + 1$:   shift 1 unit ↑

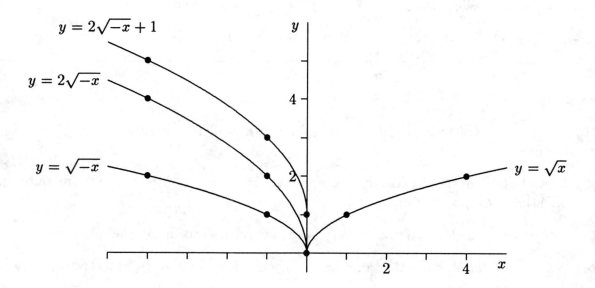

Example 4: Sketch the graph of $y = 1 - |x + 2|$ via successive transformations of $y = |x|$.

$y = |x|$:   sketch

$y = |x + 2|$:   shift 2 units ←

$y = -|x + 2|$:   reflect about x axis

$y = 1 - |x + 2|$:   shift 1 unit ↑

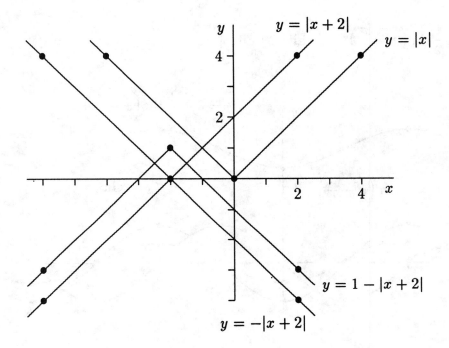

Example 5: Sketch the graph of $y = 3 + \dfrac{1}{2}\sin\left(x - \dfrac{\pi}{2}\right)$ via successive transformations of $y = \sin x$.

$y = \sin x$:   sketch

$y = \sin\left(x - \dfrac{\pi}{2}\right)$:   shift $\dfrac{\pi}{2}$ units $\rightarrow$

$y = \dfrac{1}{2}\sin\left(x - \dfrac{\pi}{2}\right)$:   vertical compression

$y = 3 + \dfrac{1}{2}\sin\left(x - \dfrac{\pi}{2}\right)$:   shift 3 units $\uparrow$

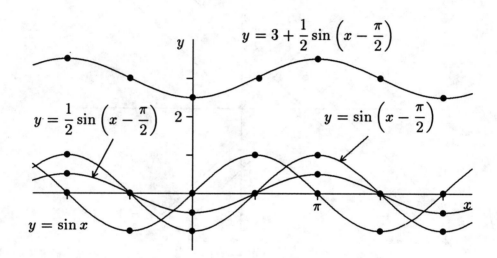

Example 6: Sketch the graph of $y = 2\cos\left(\dfrac{x}{2}\right)$ via successive transformations of $y = \cos x$.

$y = \cos x$:   sketch

$y = \cos\left(\dfrac{x}{2}\right)$:   horizontal stretch

$y = 2\cos\left(\dfrac{x}{2}\right)$:   vertical stretch

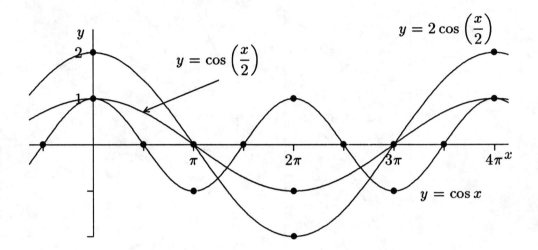

Example 7: Sketch the graph of $y = 3 - x^2$ via successive transformations of $y = x^2$.

$y = x^2$:   sketch

$y = -x^2$:   reflection about x axis

$y = -x^2 + 3$:   shift 3 units ↑

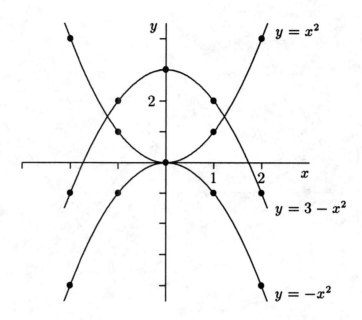

Example 8: Sketch the graph of $y = 3 - x^2 + 2x$ via successive transformations of $y = x^2$.

$$y = 3 - x^2 + 2x = -(x^2 - 2x) + 3 = -[x^2 - 2x + (-1)^2 - (-1)^2] + 3$$

↑

(completing the square, added zero)

$$= -[x + (-1)]^2 + 1 + 3 = -(x - 1)^2 + 4$$

$y = x^2$:   sketch

$y = (x - 1)^2$:   shift 1 unit →

$y = -(x - 1)^2$:   reflection about x axis

$y = -(x - 1)^2 + 4$:   shift 4 units ↑

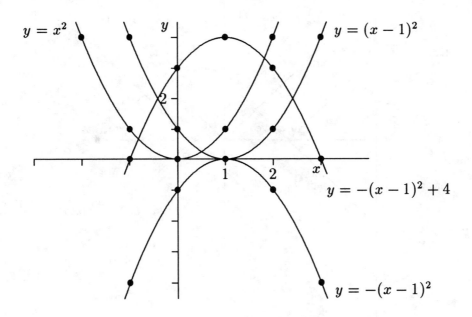

Recall (from Review and Preview Section 4): To sketch the graph of $y = |g(x)|$, sketch the graph of $y = g(x)$. Then

$$y = |g(x)| = \begin{cases} g(x) & \text{if } g(x) \geq 0 \\ \underline{-g(x)} & \text{if } g(x) < 0 \\ \text{(reflection about } x \text{ axis)} \end{cases}$$

Example 9: Sketch the graph of $y = \left|3 - |x - 1|\right|$ via successive transformations of $y = |x|$.

$y = |x|$:   sketch

$y = |x - 1|$:   shift 1 unit $\rightarrow$

$y = -|x - 1|$:   reflection about x axis

$y = 3 - |x - 1|$:   shift 3 units $\uparrow$

$y = \left|3 - |x - 1|\right|$:   reflect about x axis whenever $3 - |x - 1| < 0$

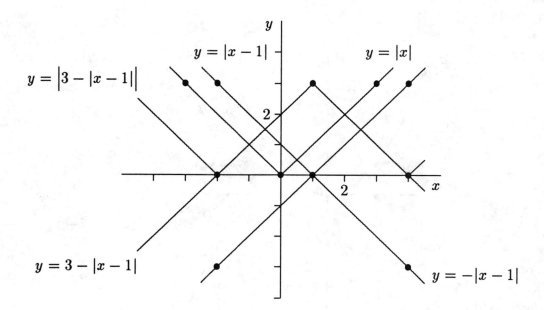

Example 10: Sketch the graph of $y = |3\cos 2x - 1|$ via successive transformations of $y = \cos x$.

$y = \cos x$:   sketch

$y = \cos 2x$:   horizontal compression

$y = 3\cos 2x$:   vertical stretch

$y = 3\cos 2x - 1$:   shift 1 unit $\downarrow$

$y = |3\cos 2x - 1|$:   reflect about x axis whenever $3\cos 2x - 1 < 0$

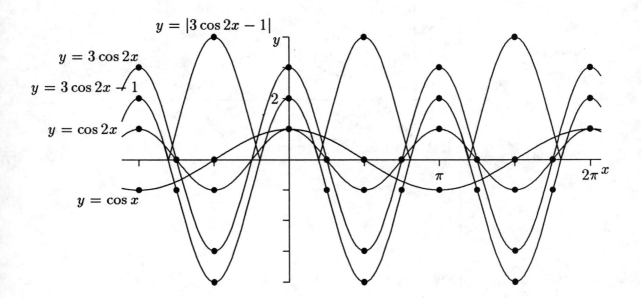

The graph of $y = (f - g)(x)$ can be obtained via graphical addition:

$$y = (f - g)(x) = f(x) - g(x) = f(x) + \underbrace{(-g(x))}$$

$(y = -g(x)$: reflection of $y = g(x)$ about x axis)

Illustration:

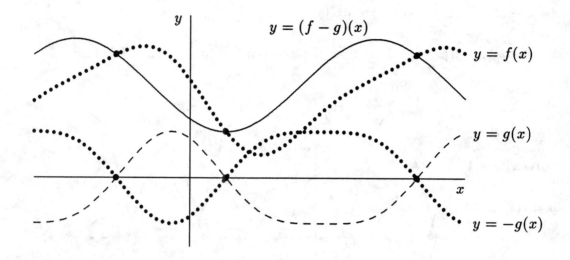

Example 11: Sketch the graph of $y = \cos \pi x - |x|$.

Note! Let $f(x) = \cos \pi x - |x|$

$$f(-x) = \cos \pi(-x) - |-x| = \cos \pi x - |x| = f(x) \quad \Longrightarrow \quad f \text{ is an even function}$$

$$\uparrow$$

(cosine function: even function; $|-x| = |x|$)

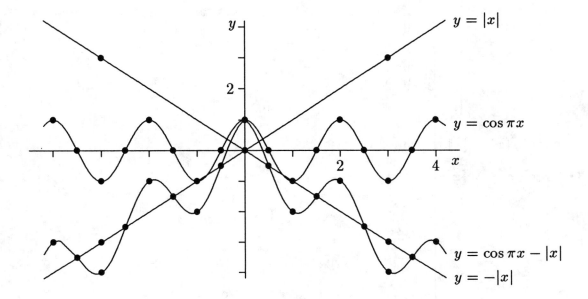

Definition: The function f defined by the rule

$$f(x) = \text{ largest integer less than or equal to } x,$$

denoted by $f(x) = [x]$, is called the **greatest integer function**.

Example 12:

(A1) $[4.3] = 4$

(A2) $[-4.3] = -5$

(A3) $[-4] = -4$

Remarks:

(R1) Domain of $f(x) = [x]$: $x \in (-\infty, \infty)$

(R2) Range of $f(x) = [x]$: $\underbrace{\{\ldots, -2, -1, 0, 1, 2, \ldots\}}_{\text{(integers)}}$

since $[x] = \begin{cases} \quad \vdots \\ -2 & \text{if } -2 \le x < -1 \\ -1 & \text{if } -1 \le x < 0 \\ \quad 0 & \text{if } 0 \le x < 1 \\ \quad 1 & \text{if } 1 \le x < 2 \\ \quad 2 & \text{if } 2 \le x < 3 \\ \quad \vdots \end{cases}$

Note! $f(x) = [x]$ is defined by cases.

(R3) Graph of $y = [x]$:

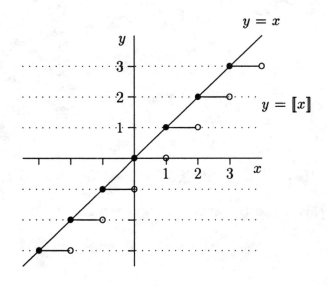

Procedure: A method to sketch the graph of $y = [\![g(x)]\!]$

(S1) Sketch the graph of $y = g(x)$.

(S2) Draw horizontal lines through each integer value located on the y axis.

(S3) Draw vertical lines through each point of intersection of the graph of $y = g(x)$ with the lines drawn in (S2).

(S4) Let n be an integer. For the adjacent pair of lines $y = n$ and $y = n + 1$ of (S2),

$$\underbrace{[\![g(x)]\!] = n}\ \text{ for all } x \text{ such that } n \le g(x) < n + 1$$

$\left(\begin{array}{l}\text{graph of } y = [\![g(x)]\!] \text{ coincides with}\\ \text{the line } y = n\text{: the lower line}\end{array}\right)$

Illustration:

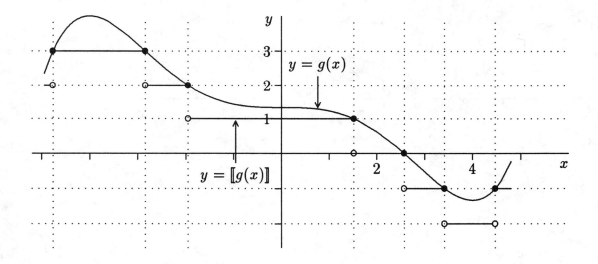

Example 13: Sketch the graph of $y = [\![1 + |x - 2|]\!]$.

$y = |x|$:   sketch

$y = |x - 2|$:   shift 2 units $\rightarrow$

$y = 1 + |x - 2|$:   shift 1 unit $\uparrow$

$y = [\![1 + |x - 2|]\!]$:   greatest integer method

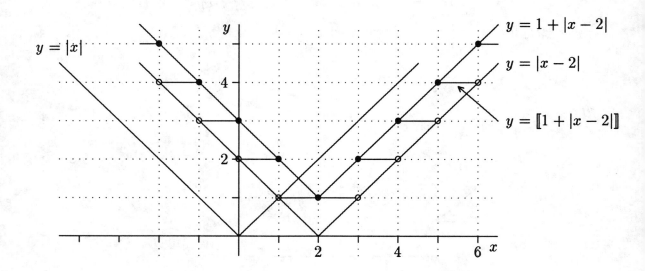

Example 14: Sketch the graph of $y = \left[2\sin\left(x + \dfrac{\pi}{2}\right)\right]$.

$y = \sin x$: sketch

$y = \sin\left(x + \dfrac{\pi}{2}\right)$: shift $\dfrac{\pi}{2}$ units $\leftarrow$

$y = 2\sin\left(x + \dfrac{\pi}{2}\right)$: vertical stretch

$y = \left[2\sin\left(x + \dfrac{\pi}{2}\right)\right]$: greatest integer method

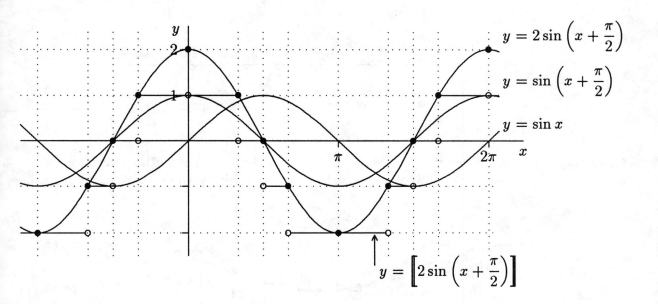

0.7 A Preview of Calculus

Area Problem: the central problem of *integral calculus*.

Let f be a function such that $f(x) \geq 0$ for $a \leq x \leq b$. Find the area A of the region bounded by the graph of $y = f(x)$, the x axis, and the lines $x = a$ and $x = b$.

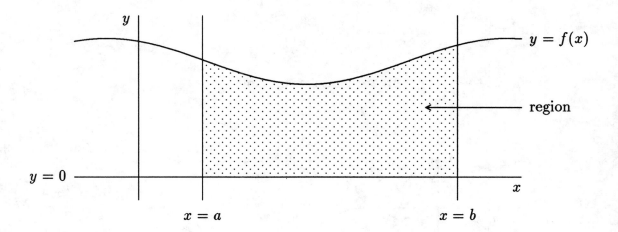

A Solution: Let n be a positive integer. Approximate the region via n nonoverlapping rectangles with equal width, but whose height is determined by the graph of $y = f(x)$.

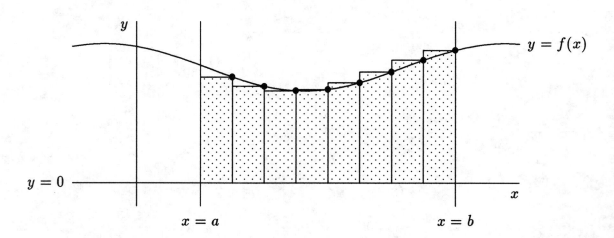

Note! $n = 8$ in illustration

Let A_n denote the sum of the area of the n rectangles. As the width of each rectangle decreases, the number of rectangles increase and the approximation

$$A \approx A_n \quad \text{should get better with} \quad A = \lim_{n \to \infty} A_n$$

(let n get larger and larger)

Tangent Problem: the central problem of *differential calculus*.

Let f be a function. Consider the point $P(a, f(a))$ on the graph of $y = f(x)$. Find an equation of the line tangent to the curve $y = f(x)$ at P.

Point P is on the tangent line, so the slope of the tangent line must be determined if the point-slope form for an equation of a line is to be used.

Let $Q(x, f(x))$ be a point on the graph of $y = f(x)$ near P.

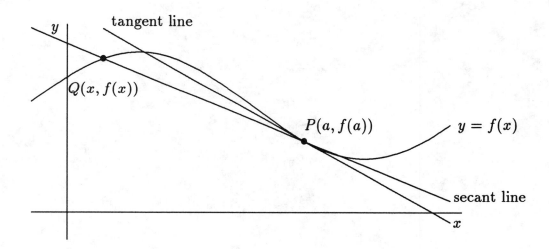

The slope of the secant line through P and Q is

$$m_{PQ} = \frac{f(x) - f(a)}{x - a} \quad \Longrightarrow \quad m_{PQ} \approx m,$$

where m represents the slope of the tangent line at P.

As Q moves closer to P the secant line through P and Q more closely approximates the tangent line at P and the approximation

$$m \approx m_{PQ} \quad \text{should get better with} \quad m = \lim_{\underbrace{Q \to P}} m_{PQ}$$

(let Q get closer and closer to P)

which, since x approaches a as Q approaches P, can be written as

$$m = \lim_{\underbrace{x \to a}} \frac{f(x) - f(a)}{x - a}$$

(x approaches a)

$\implies$ using the point slope form with $\left\{ \begin{array}{ll} \text{point} & (a, f(a)) \\ \text{slope} & m \end{array} \right\}$:

$y - f(a) = m(x - a)$ is an equation of the tangent line.

1

Limits and Rates of Change

1.1   The Tangent and Velocity Problems

Problem: Find the slope of the tangent line to a curve with equation $y = f(x)$ at the point $P(a, f(a))$.

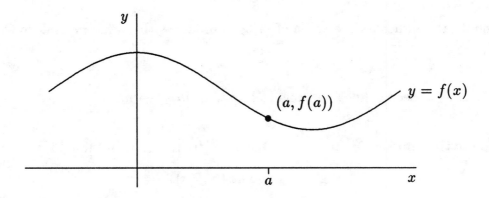

Question: What is a tangent line to a curve $y = f(x)$ at the point $P(a, f(a))$?

Definition: The **tangent line** to a curve $y = f(x)$ at $P(a, f(a))$ is the limiting position of a secant line joining P to nearby points $Q(x, f(x))$, $x \neq a$.

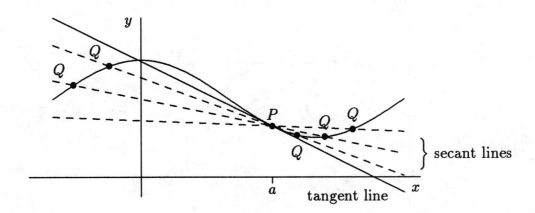

The slope of the secant line through the distinct points P and Q is given by

$$m_{PQ} = \frac{f(x) - f(a)}{x - a}$$

Remark: P and Q are defined such that $x \neq a$.

The slope of the tangent line through the point P is given by

$$m = \lim_{Q \to P} m_{PQ}$$

which symbolically indicates the value of m_{PQ} approaches the value m as Q moves closer to P.

Remark: As $x \to a$: secant lines $\to$ tangent line with $m_{PQ} \to m$.

An equation of the tangent line to $y = f(x)$ at $P(a, f(a))$ can be obtained via the point-slope form:

$$\text{given:} \quad \text{point } (a, f(a))$$

$$\text{determine:} \quad \text{slope } m = \lim_{Q \to P} m_{PQ}$$

$$\text{equation:} \quad y - f(a) = m(x - a)$$

Example 1: The point $P(2, 10)$ lies on the curve $y = 3x^2 - 4x + 6$.

(A1) If Q is the point $(x, 3x^2 - 4x + 6)$, find m_{PQ} for the following values of x near $a = 2$.

(C1) $x = 3$

$$Q(3, f(3)), P(2, f(2)): \quad m_{PQ} = \frac{f(3) - f(2)}{3 - 2} = \frac{21 - 10}{1} = 11$$

$$\left(f(x) = 3x^2 - 4x + 6 \implies f(3) = 2 \text{ and } f(2) = 10 \right)$$

(C2) $x = 2.5$

$$Q(2.5, f(2.5)): \quad m_{PQ} = \frac{f(2.5) - f(2)}{2.5 - 2} = \frac{14.75 - 10}{.5} = 9.5$$

$$\left(f(x) = 3x^2 - 4x + 6 \implies f(2.5) = 14.75 \right)$$

(C3) $x = 2.1$

$$Q(2.1, f(2.1)): \quad m_{PQ} = \frac{f(2.1) - f(2)}{2.1 - 2} = \frac{10.83 - 10}{.1} = 8.3$$

(A2) If Q is the point $(x, 3x^2 - 4x + 6)$, find the values of m_{PQ} and complete the following table:

x	m_{PQ}
3	11
2.5	9.5
2.1	8.3
2.01	8.03
2.001	8.003

x	m_{PQ}
1	5
1.5	6.5
1.9	7.7
1.99	7.97
1.999	7.997

(A3) Using the results of (A2), guess the value of m at $P(2, 10)$.

$m = 8$

(A4) Write an equation of the tangent line to the curve at $P(2, 10)$.

point-slope form with point $(2, 10)$ and slope $m = 8$:

$$y - 10 = 8(x - 2) \qquad \text{or, rewritten,} \qquad y = 8x - 6$$

(A5) On the same x, y coordinate system, sketch the curve, the tangent line at $P(2, 10)$, and the two secant lines determined by P and $Q(x, f(x))$ for the values $x = 1.5$ and $x = 2.5$.

$x = 1.5:$ $f(1.5) = 6.75$ $\implies$ $Q(1.5, 6.75)$

$x = 2.5:$ $f(2.5) = 14.75$ $\implies$ $Q(2.5, 14.75)$

$y = 3x^2 - 4x + 6$: parabola, opening upward from vertex: $\left(\frac{2}{3}, \frac{14}{3}\right)$; no real zeros

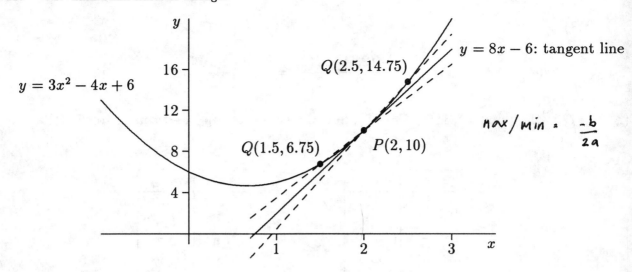

Example 2: The point $P(1, -3)$ lies on the curve $y = 2x^3 - 5x$.

(A1) If Q is the point $(x, 2x^3 - 5x)$, find an expression for m_{PQ} in terms of x.

$$Q(x, f(x)), P(1, f(1)): \quad m_{PQ} = \frac{f(x) - f(1)}{x - 1} = \frac{2x^3 - 5x - (-3)}{x - 1} = \frac{2x^3 - 5x + 3}{x - 1}$$

$$\uparrow$$

$$\left(f(x) = 2x^3 - 5x\right)$$

$$= 2x^2 + 2x - 3 \qquad\qquad (x \neq 1)$$

Note! Divided the expression $x - a$ from denominator of $m_{PQ} = \dfrac{f(x) - f(a)}{x - a}$.

(A2) Using the result of (A1), complete the following table for the given values of x:

x	m_{PQ}
2	9
1.5	4.5
1.1	1.62
1.01	1.0602
1.001	1.006002

x	m_{PQ}
0	-3
0.5	-1.5
0.9	0.42
0.99	0.9402
0.999	0.994002

$x = 2:$ $m_{PQ} = 2(2)^2 + 2(2) - 3 = 9$

$x = 1.5:$ $m_{PQ} = 2(1.5)^2 + 2(1.5) - 3 = 4.5$

Note! The procedure given in this example carried out the operation of division to obtain an expression for m_{PQ} in terms of x prior to the substitution of the particular value of x at Q.

Suggestion: Complete Example 1, (A2), using this procedure.

Problem: Suppose an object moves along a coordinate line (rectilinear motion) with its position at time t given by $s(t)$. Find the instantaneous velocity of the object when time $t = a$.

Definition: For $h \neq 0$, the **average velocity** of the object between the values of time $t = a$ and $t = a + h$ is

$$\text{average velocity} = \frac{s(a + h) - s(a)}{h} \quad \left(= \frac{\text{difference in position}}{\text{difference in time}} \right)$$

Remark: $h \neq 0$, so average velocity is determined at two distinct values of t.

Definition: The **instantaneous velocity** of the object when time $t = a$ is the limiting value of the average velocities over shorter and shorter time intervals (that is, as the time $t = a + h$ moves closer to $t = a$).

Remark: As $h \to 0$: average velocities $\to$ instantaneous velocity.

Example 3: A stone is thrown upward so its height, in feet, above the ground is given by $s = 80 + 64t - 16t^2$ after t seconds.

(A1) Find the average velocity over each time interval from $t = 0$ to:

   (C1) $t = 1$

   A choice: let $a = 0$

$$\text{average velocity} = \frac{s(0 + h) - s(0)}{h} = \frac{s(h) - s(0)}{h} = \frac{80 + 64h - 16h^2 - 80}{h}$$
$$\uparrow$$
$$\left(s(\text{\textbf{\textit{h}}}) = 80 + 64t - 16t^2 \right)$$

$$= \frac{64h - 16h^2}{h} = \frac{h(64 - 16h)}{h} = 64 - 16h \quad (\text{ft/sec})$$
$$\uparrow$$
$$(h \neq 0)$$

$\underbrace{a = 0,}_{(t=0)} \quad \underbrace{a + h = 1:}_{(t=1)} \quad h = 1, \quad \text{and}$$

average velocity $= 64 - 16(1) = 48 \quad \text{ft/sec}$

(C2) $t = 0.5$

$a = 0, \quad a + h = 0.5: \quad h = 0.5, \quad \text{and}$

average velocity $= 64 - 16(0.5) = 56 \quad \text{ft/sec}$
$\uparrow$

(from (C1): expression for average velocity with $a = 0$)

(C3) $t = 0.1$

$a = 0, \quad a + h = 0.1: \quad h = 0.1, \quad \text{and}$

average velocity $= 64 - 16(0.1) = 62.4 \quad \text{ft/sec}$

(C4) $t = 0.01$

$a = 0, \quad a + h = 0.01: \quad h = 0.01, \quad \text{and}$

average velocity $= 64 - 16(0.01) = 63.84 \quad \text{ft/sec}$

(A2) Find the instantaneous velocity at which the stone is thrown.

Want instantaneous velocity when $t = 0$ (time of release).

average velocity $= \underline{64 - 16h}$
      (from (A1) (C1) with $a = 0$)

so instantaneous velocity $= 64 \quad \text{ft/sec}$
$\uparrow$

(as $h \to 0$ in average velocity expression; see results of (A1))

(A3) Find the instantaneous velocity when the stone hits the ground.

hits ground $\implies s = 0:$ $80 + 64t - 16t^2 = 0$

$$\implies \quad -16(t^2 - 4t - 5) = -16(t - 5)(t + 1) = 0 \quad \implies \quad t = 5, \; -1 \quad \implies \quad t = 5 \; (\geq 0)$$

Want instantaneous velocity when $t = 5 \; (\geq 0)$: find the average velocity between the values of time $t = 5 + h$ where $h < 0$ (before stone hits ground), and $t = 5$.

$$\text{average velocity} = \frac{s(5 + h) - s(5)}{h}$$

$$= \frac{80 + 64(5 + h) - 16(5 + h)^2 - [80 + 64(5) - 16(5)^2]}{h}$$

$$= \frac{-96h - 16h^2}{h} = -96 - 16h \qquad \text{(ft/sec)}$$

As $h \to 0$: appears the instantaneous velocity $= -96$ ft/sec

(A4) Find the average velocity over the time interval $[1, 3]$.

A choice: $a = 1, \quad a + h = 3 \implies h = 2 \quad$ and

$$\text{average velocity} = \frac{s(3) - s(1)}{2}$$

$$= \frac{80 + 64(3) - 16(3)^2 - [80 + 64(1) - 16(1)^2]}{2} = 0 \;\; \text{ft/sec.}$$

$\implies \quad$ the stone is at the same position when $t = 3$ as when $t = 1$.

Another choice: $a = 3, \quad a + h = 1 \implies h = -2 \quad$ and

$$\text{average velocity} = \frac{s(1) - s(3)}{-2}$$

$$= \frac{80 + 64(1) - 16(1)^2 - [80 + 64(3) - 16(3)^2]}{-2} = 0 \;\; \text{ft/sec (as above)}$$

(A5) Find the instantaneous velocity when $t = 1$.

$$\text{average velocity} = \frac{s(1 + h) - s(1)}{h}$$
$$\uparrow$$

($a = 1$: when instantaneous velocity is desired) ✳

$$= \frac{80 + 64(1 + h) - 16(1 + h)^2 - [80 + 64(1) - 16(1)^2]}{h}$$

$$= 32 - 16h \;\; \text{ft/sec}$$

$$\implies \quad \text{instantaneous velocity} = 32 \;\; \text{ft/sec}$$
$$\uparrow$$
$$(\text{as } h \to 0)$$

Remark: In order to illustrate the connection between the problem of the slope of a tangent line and the problem of instantaneous velocity, graph the height function $s(t)$ in a t, s coordinate system.

$$\text{average velocity} = \frac{s(a+h) - s(a)}{h} = \frac{s(a+h) - s(a)}{(a+h) - a} = m_{PQ},$$

$$\uparrow$$

$$(h = (a+h) - a)$$

the slope of the secant line through the points $P(a, s(a))$ and $Q(a+h, s(a+h))$.

The instantaneous velocity when $t = a$ is the slope of the tangent line at $t = a$.

From Example 3, parts (A4) and (A5):

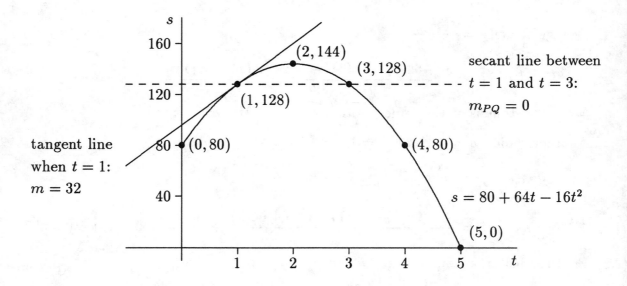

1.2   The Limit of a Function

Definition (1.3): We write

$$\lim_{x \to a} f(x) = L$$

and say

"the limit of $f(x)$, as x approaches a, equals L"

if we can make the values of $f(x)$ arbitrarily close to L (as close to L as we like) by taking x to be sufficiently close to a but not equal to a.

* $\underline{a}$ is a fixed number *

Remarks:

(R1) Alternative notation: $f(x) \rightarrow L$ as $x \rightarrow a$ for both $\left\{ \begin{array}{c} x < a \\ x > a \end{array} \right\}$ $x \neq a$.

$\qquad\qquad\qquad\qquad\quad \uparrow \qquad\qquad\quad \uparrow$

$\qquad\qquad\qquad\qquad\quad$ (approaches)

(R2) a, L: constants

(R3) The domain of f contains an open interval containing a, but f need not be defined at a. f is defined near a and $\lim_{x \to a} f(x)$ examines the behavior of $f(x)$ near a.

Definition (1.4): We write

$$\lim_{x \to a^-} f(x) = L$$

and say the **left-hand limit of $f(x)$ as x approaches a** (or the **limit of $f(x)$ as x approaches a from the left**) is equal to L if we can make the values of $f(x)$ arbitrarily close to L by taking x sufficiently close to a and $x < a$.

Similarly, we write

$$\lim_{x \to a^+} f(x) = L$$

and say the **right-hand limit of $f(x)$ as x approaches a** (or the **limit of $f(x)$ as x approaches a from the right**) is equal to L if we can make the values of $f(x)$ arbitrarily close to L by taking x sufficiently close to a and $x > a$.

Remarks:

(R1) $\left\{ \begin{array}{c} \text{Left-hand} \\ \text{Right-hand} \end{array} \right\}$ limits: one-sided limits.

(R2) $\lim_{x \to a^-} f(x)$: examines behavior of $f(x)$ near a but only for $x < a$;

$\qquad$ Alternative notation: $f(x) \rightarrow L$ as $x \rightarrow a$ for $x < a$, $x \neq a$.

(R3) $\lim_{x \to a^+} f(x)$: examines behavior of $f(x)$ near a but only for $x > a$;

$\qquad$ Alternative notation: $f(x) \rightarrow L$ as $x \rightarrow a$ for $x > a$, $x \neq a$.

Theorem (1.5):

$$\lim_{x \to a} f(x) = L \qquad \text{if and only if} \qquad \lim_{x \to a^-} f(x) = L \quad \text{and} \quad \lim_{x \to a^+} f(x) = L$$

Remark: For $\lim_{x \to a} f(x)$ to exist, both one-sided limits must exist and $\lim_{x \to a^-} f(x) = \lim_{x \to a^+} f(x)$; otherwise, $\lim_{x \to a} f(x)$ does not exist.

Example 1: Consider $f(x) = \begin{cases} -1 & \text{if } x \le 0 \\ 3 & \text{if } 0 < x < 2 \\ x + 1 & \text{if } x \ge 2 \end{cases}$

(A1) Sketch the graph of $y = f(x)$.

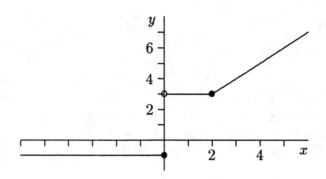

(A2) $\lim_{x \to 2} f(x) = ?$

Formula for $f(x)$ changes at $x = 2$:

$$\left. \begin{array}{l} \lim_{x \to 2^-} f(x) = \lim_{x \to 2^-} 3 = 3 \\[2mm] \lim_{x \to 2^+} f(x) = \lim_{x \to 2^+} (x + 1) = 3 \end{array} \right\} \implies \lim_{x \to 2} f(x) = 3$$

(A3) $\lim_{x \to 0} f(x) = ?$

Formula for $f(x)$ changes at $x = 0$:

$$\left. \begin{array}{l} \lim_{x \to 0^-} f(x) = \lim_{x \to 0^-} -1 = -1 \\[2mm] \lim_{x \to 0^+} f(x) = \lim_{x \to 0^+} 3 = 3 \end{array} \right\} \implies \begin{array}{l} \lim_{x \to 0} f(x) \quad \underline{\text{does not exist}} \\[2mm] \left(\lim_{x \to 0^-} f(x) \ne \lim_{x \to 0^+} f(x) \right) \end{array}$$

✱   a case function must have at least
     2 equations involved   ✱

(A4) $\lim\limits_{x \to 4} f(x) = ?$

$$\lim_{x \to 4} f(x) = \lim_{x \to 4}(x + 1) = 5$$

Note! Did not separately consider the one-sided limits since the formula for $f(x)$ is the same on both sides of and near $x = 4$.

Example 2: $\lim\limits_{x \to -2} \dfrac{x^2 - 4}{x + 2} = ?$

$x = -2$ is not in the domain of $f(x) = \dfrac{x^2 - 4}{x + 2}$; the domain of f is $(-\infty, -2) \cup (-2, \infty)$.

Complete the following table by determining values of $f(x)$ (correct to six decimal places) for the values of x which approach -2:

$x < -2$	$f(x)$
-3	-5
-2.5	-4.5
-2.1	-4.1
-2.01	-4.01
-2.001	-4.001

$x > -2$	$f(x)$
-1	-3
-1.5	-3.5
-1.9	-3.9
-1.99	-3.99
-1.999	-3.999

Appears, from the table, that $\lim\limits_{x \to -2} \dfrac{x^2 - 4}{x + 2} = -4$.

Note!

(N1) Sketch the graph of $y = \dfrac{x^2 - 4}{x + 2}$.

$x \neq -2:$ $y = \dfrac{x^2 - 4}{x + 2} = \dfrac{(x - 2)(x + 2)}{x + 2} = x - 2$: a line

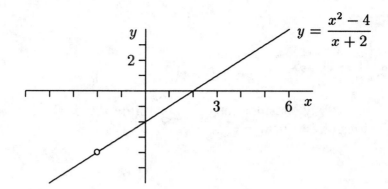

Appears, from the sketch, that $\displaystyle\lim_{x\to-2}\frac{x^2-4}{x+2}=-4$.

(N2) Suppose that b is a constant. Consider the function $g(x)=\begin{cases} f(x) & \text{if } x\neq-2 \\ b & \text{if } x=-2 \end{cases}$

The domain of $g(x)$ is $(-\infty,\infty)$; $g(x)$ differs from $f(x)$ only at $x=-2$, but

$$\lim_{x\to-2}g(x)=-4=\lim_{x\to-2}f(x).$$

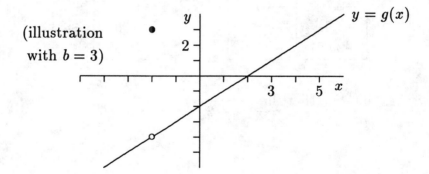

(illustration
with $b=3$)

Example 3: $\displaystyle\lim_{x\to1}\frac{x^2+|x-1|-1}{|x-1|}= ?$

$x=1$ is not in the domain of $f(x)=\dfrac{x^2+|x-1|-1}{|x-1|}$;

the domain of f is $(-\infty,1)\cup(1,\infty)$.

Complete the following table by determining values of $f(x)$ (correct to six decimal places) for the values of x which approach 1:

$x < 1$	$f(x)$
0	0
0.5	−0.5
0.9	−0.9
0.99	−0.99
0.999	−0.999

$x > 1$	$f(x)$
2	4
1.5	3.5
1.1	3.1
1.01	3.01
1.001	3.001

Appears, from the table,
$$\left\{ \begin{array}{l} \lim_{x \to 1^-} f(x) = -1 \\ \lim_{x \to 1^+} f(x) = 3 \end{array} \right\} \implies \lim_{x \to 1} f(x) \underbrace{\text{does not exist}}$$
$$\left(\lim_{x \to 1^-} f(x) \neq \lim_{x \to 1^+} f(x) \right)$$

Note! Sketch the graph of $y = \dfrac{x^2 + |x - 1| - 1}{|x - 1|}$.

Recall: $|x - 1| = \begin{cases} x - 1 & \text{if } x \geq 1 \\ -(x - 1) & \text{if } x < 1 \end{cases}$ so express $f(x)$ in cases.

For $x \neq 1$:

$$x < 1 : f(x) = \frac{x^2 + [-(x-1)] - 1}{-(x-1)} = \frac{x(x-1)}{-(x-1)} = -x$$
$$\uparrow$$
$$(x \neq 1)$$

$$x > 1 : f(x) = \frac{x^2 + x - 1 - 1}{x - 1} = \frac{x^2 + x - 2}{x - 1} = \frac{(x-1)(x+2)}{x-1} = x + 2$$
$$\uparrow$$
$$(x \neq 1)$$

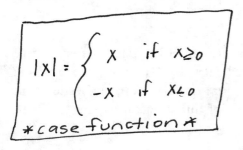

$$|x| = \begin{cases} x & \text{if } x \geq 0 \\ -x & \text{if } x \leq 0 \end{cases}$$
* case function *

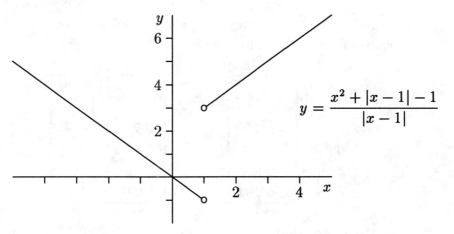

$$y = \frac{x^2 + |x - 1| - 1}{|x - 1|}$$

Appears, from the sketch, that $\lim_{x \to 1^-} f(x) = -1 \neq 3 = \lim_{x \to 1^+} f(x)$; hence,

$\lim_{x \to 1} f(x)$ does not exist.

Example 4: $\lim_{x \to 1} \dfrac{x - 1}{\sqrt{x + 24} - 5} = ?$

$x = 1$ is not in the domain of $f(x) = \dfrac{x - 1}{\sqrt{x + 24} - 5}$; the domain of f is $[-24, 1) \cup (1, \infty)$.

Complete the following table by determining values of $f(x)$ (correct to five decimal places) for the values of x which approach 1:

$x < 1$	$f(x)$
0	9.89903
0.5	9.95025
0.9	9.99001
0.99	10
0.999	10

$x > 1$	$f(x)$
2	10.09897
1.5	10.05025
1.1	10.01001
1.01	10
1.001	10

Appears, from the table, that $\lim_{x \to 1} \dfrac{x - 1}{\sqrt{x + 24} - 5} = 10$.

Note! If values of x even closer to 1 are taken, say 1.0000001, then the results from my calculator indicate division by zero. Does this indicate that 10 is not the answer? The difficulty is related to the subtraction of nearly equal quantities in the denominator as x approaches 1. See the section named The Perils of Subtraction in Appendix D.

Example 5: $\lim\limits_{x\to 0}\dfrac{\tan x}{x} = ?$

$x = 0$ is not in the domain of $f(x) = \dfrac{\tan x}{x}$; the domain of f contains $\left(-\dfrac{\pi}{2}, 0\right) \cup \left(0, \dfrac{\pi}{2}\right)$.

Complete the following table by determining values of $f(x)$ (correct to five decimal places) for the values of x which approach 0.

$x < 0$	$f(x)$
-1	1.55741
-0.5	1.0926
-0.1	1.0033
-0.01	1
-0.001	1

$x > 0$	$f(x)$
1	1.55741
0.5	1.0926
0.1	1.0033
0.01	1
0.001	1

Note! Be sure to use radian measure when calculating $\tan x$.

Appears, from the table, that $\lim\limits_{x\to 0}\dfrac{\tan x}{x} = 1$

Question: Rather than using a calculator and/or a graph to guess the value of a limit, since such methods may not always lead to the correct conclusion, is there another way to determine the value of a limit?

θ radians	$\theta\,\dot{}$	$\sin\theta$
0	0	$\frac{\sqrt{0}}{2} = 0$
$\pi/6$	30$\,\dot{}$	$\frac{\sqrt{1}}{2} = \frac{1}{2}$
$\pi/4$	45$\,\dot{}$	$\frac{\sqrt{2}}{2}$
$\pi/3$	60$\,\dot{}$	$\frac{\sqrt{3}}{2}$
$\pi/2$	90$\,\dot{}$	$\frac{\sqrt{4}}{2} = 1$

1.3   Calculating Limits using the Limit Laws

The following eleven properties of limits can be applied when calculating limits.

Limit Laws (1.6): Suppose that c is a constant and the limits

$$\lim_{x \to a} f(x) \quad \text{and} \quad \lim_{x \to a} g(x)$$

exist. Then

1. $\lim_{x \to a}[f(x) + g(x)] = \lim_{x \to a} f(x) + \lim_{x \to a} g(x)$

2. $\lim_{x \to a}[f(x) - g(x)] = \lim_{x \to a} f(x) - \lim_{x \to a} g(x)$

3. $\lim_{x \to a} cf(x) = c \lim_{x \to a} f(x)$

4. $\lim_{x \to a}[f(x)g(x)] = \lim_{x \to a} f(x) \cdot \lim_{x \to a} g(x)$

5. $\lim_{x \to a} \dfrac{f(x)}{g(x)} = \dfrac{\lim_{x \to a} f(x)}{\lim_{x \to a} g(x)} \quad \text{if} \quad \lim_{x \to a} g(x) \neq 0$

6. $\lim_{x \to a}[f(x)]^n = \left[\lim_{x \to a} f(x)\right]^n \quad$ where n is a positive integer.

7. $\lim_{x \to a} c = c \qquad$ 8. $\lim_{x \to a} x = a$

9. $\lim_{x \to a} x^n = a^n \quad$ where n is a positive integer.

10. $\lim_{x \to a} \sqrt[n]{x} = \sqrt[n]{a} \quad$ where n is a positive integer.

    (If n is even, we assume that $a > 0$.)

11. $\lim_{x \to a} \sqrt[n]{f(x)} = \sqrt[n]{\lim_{x \to a} f(x)} \quad$ where n is a positive integer.

    (If n is even, we assume that $\lim_{x \to a} f(x) > 0$.)

Remarks:

(R1) Limit laws 1-5 presented above may be stated verbally as follows: If the limit of each component function exists, then:

   1. The limit of a sum is the sum of the limits. (Sum Law)

2. The limit of a difference is the corresponding difference of the limits. (Difference Law)

3. The limit of a constant times a function is the constant times the limit of the function. (Constant Multiple Law)

4. The limit of a product is the product of the limits. (Product Law)

5. The limit of a quotient is the quotient of the limits, provided the limit of the denominator is not zero. (Quotient Law)

(R2) The proofs to Limit Laws 1-5, 7, 8, and 10 follow from the precise definition of a limit given in Section 1.4.

(R3) Limit Law 6 (Power Law) is obtained by repeatedly applying the Product Law with $g(x) = f(x)$.

(R4) Limit Law 11 (Root Law) is proved in Section 1.5 as a special case of Theorem 1.23.

Example 1: Evaluate $\lim\limits_{x \to a} \dfrac{x^2 - x - 2}{x^2 + x - 6}$, if it exists.

Justify each step by indicating the appropriate Limit Laws, given:

(A1) $a = 1$

$$\lim_{x \to 1} \frac{x^2 - x - 2}{x^2 + x - 6} = \underset{\uparrow}{\frac{\lim\limits_{x \to 1}(x^2 - x - 2)}{\lim\limits_{x \to 1}(x^2 + x - 6)}} = \underset{\uparrow}{\frac{\lim\limits_{x \to 1} x^2 - \lim\limits_{x \to 1} x - \lim\limits_{x \to 1} 2}{\lim\limits_{x \to 1} x^2 + \lim\limits_{x \to 1} x - \lim\limits_{x \to 1} 6}} = \underset{\uparrow}{\frac{(1)^2 - 1 - 2}{(1)^2 + 1 - 6}} = \frac{1}{2}$$

$\qquad$ (Law 5) $\qquad\qquad$ (Laws 1 and 2) $\qquad\qquad$ (Laws 9, 8, and 7)

Note! Each law is applied provided the limit of each component function exists (which can be observed when the limits are actually found).

(A2) $a = -1$

$$\lim_{x \to -1} \frac{x^2 - x - 2}{x^2 + x - 6} = \underset{\uparrow}{\frac{\lim\limits_{x \to -1} x^2 - \lim\limits_{x \to -1} x - \lim\limits_{x \to -1} 2}{\lim\limits_{x \to -1} x^2 + \lim\limits_{x \to -1} x - \lim\limits_{x \to -1} 6}} = \underset{\uparrow}{\frac{(-1)^2 - (-1) - 2}{(-1)^2 + (-1) - 6}} = \frac{0}{6} = 0$$

$\quad$ (as in (A1): Laws 5, 1, and 2) $\qquad$ (Laws 9, 8, and 7)

(A3) $a = 2$

$$\lim_{x \to 2} \frac{x^2 - x - 2}{x^2 + x - 6} = \underset{\uparrow}{\frac{2^2 - 2 - 2}{2^2 + 2 - 6}} = \frac{0}{0} \quad : \quad$$ the zero in the denominator indicates Law 5 is not applicable.

(as in (A1): Laws 5, 1, 2, 9, 8, and 7)

Note! The form $\frac{0}{0}$ signifies you must *do something* further in order to determine

$$\left\{ \begin{array}{l} \text{the value of the limit, if it exists,} \\ \text{or} \\ \text{the limit does not exist.} \end{array} \right\}$$

$f(x) = (x^2 - x - 2)/(x^2 + x - 6)$ is a rational function such that both the numerator and denominator are zero at $x = 2$; hence, $x - 2$ is a common factor.

$$f(x) = \frac{x^2 - x - 2}{x^2 + x - 6} = \frac{(x - 2)(x + 1)}{(x - 2)(x + 3)} = \frac{x + 1}{x + 3}, \quad x \neq 2 \quad \text{and}$$

$$\lim_{x \to 2} \frac{x^2 - x - 2}{x^2 + x - 6} = \underset{\uparrow}{\lim_{x \to 2}} \frac{(x - 2)(x + 1)}{(x - 2)(x + 3)} = \underset{\uparrow}{\lim_{x \to 2}} \frac{x + 1}{x + 3}$$

$$\left(\frac{0}{0} : \text{do} \right) \qquad\qquad (\text{in limit: } x \neq a = 2 \Longrightarrow x - 2 \neq 0)$$

$$\underset{\uparrow}{=} \frac{2 + 1}{2 + 3} = \frac{3}{5}$$

(Laws 5, 1, 8, and 7)

(A4) $a = -3$

$$\lim_{x \to -3} \frac{x^2 - x - 2}{x^2 + x - 6} = \underset{\uparrow}{\frac{(-3)^2 - (-3) - 2}{(-3)^2 + (-3) - 6}} = \frac{10}{0} \quad : \quad$$ zero in the denominator indicates Law 5 is not applicable.

(Laws 5, 1, 2, 9, 8, and 7)

Note! The form $\frac{10}{0}$, that is $\frac{\neq 0}{0}$, signifies that the limit does not exist (in the sense, the limit is not a finite number).

Therefore, $\lim_{x \to -3} \dfrac{x^2 - x - 2}{x^2 + x - 6}$ does not exist.

Note!

(N1) Consider $f(x) = \dfrac{x^2 - x - 2}{x^2 + x - 6}$. Determine the domain of f:

$$x^2 + x - 6 = 0 \quad \Longrightarrow \quad (x-2)(x+3) = 0 \quad \Longrightarrow \quad x = 2, -3$$

$\Longrightarrow$ domain of f: $(-\infty, -3) \cup (-3, 2) \cup (2, \infty)$

The graph of $y = f(x)$, where f is a rational function, is discussed in Chapter 3 but, for illustrative purposes, the graph is given below:

$$y = \frac{x^2 - x - 2}{x^2 + x - 6} = \frac{(x-2)(x+1)}{(x-2)(x+3)} = \frac{x+1}{x+3}, \quad x \neq 2.$$

$$\underset{\uparrow}{(x - 2 \neq 0)}$$

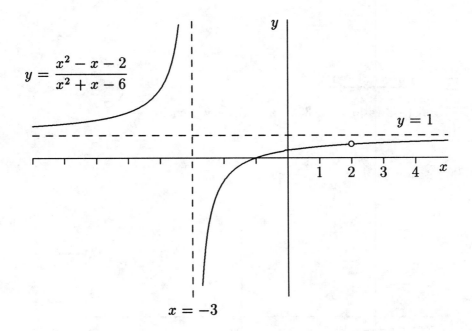

(N2) The function $g(x) = \dfrac{x+1}{x+3}$, whose domain is $(-\infty, -3) \cup (-3, \infty)$,

agrees with $f(x)$ on their common domain and $\lim\limits_{x \to a} f(x) = \lim\limits_{x \to a} g(x)$ for all $a \neq -3$.

(N3) The graph of $y = f(x)$ is "complicated," so the Limit Laws become important when evaluating limits.

Example 2: Evaluate $\lim\limits_{x \to a} \dfrac{2x^2 + x}{x^2 - x}$, if it exists. Justify each step by indicating the appropriate Limit Laws.

(A1) $a = -2$

$$\lim_{x \to -2} \frac{2x^2 + x}{x^2 - x} \underset{\uparrow}{=} \frac{2(-2)^2 + (-2)}{(-2)^2 - (-2)} = 1$$

(Laws 5, 1, 2, 3, 9, and 8)

(A2) $a = 0$

$$\lim_{x \to 0} \frac{2x^2 + x}{x^2 - x} \underset{\uparrow}{=} \lim_{x \to 0} \frac{x(2x + 1)}{x(x - 1)} \underset{\uparrow}{=} \lim_{x \to 0} \frac{2x + 1}{x - 1} \underset{\uparrow}{=} -1$$

$\left(\text{Laws 5, 1, 2, 3, 9, and 8 } \frac{0}{0} : \text{ do}\right)(x \neq 0)$ (Laws 5, 1, 2, 3, 8, and 7)

(A3) $a = 1$

$$\lim_{x \to 1} \frac{2x^2 + x}{x^2 - x} \quad \underline{\text{does not exist}}$$

$$\left(\text{Laws 5, 1, 2, 3, 9, and 8; } \frac{3}{0}, \text{which is } \frac{\neq 0}{0}\right)$$

Note! The graph of $y = \dfrac{2x^2 + x}{x^2 - x}$ is:

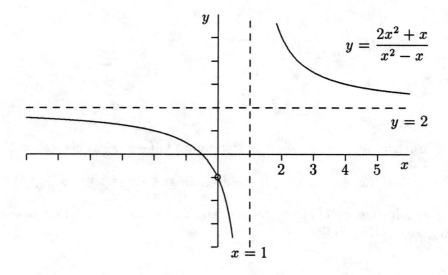

Remark: Limit Laws 1, 2, 3, 5, 7, 8, and 9 can be used to show that the limits in the two preceding examples may be examined via the direct substitution of the value of a into the given function.

(1.7): If f is a polynomial or a rational function and a is in the domain of f, then

$$\lim_{x \to a} f(x) = f(a)$$

Remarks: If a is not in the domain of a rational function (i.e., the denominator is zero) then direct substitution would yield the two cases:

(R1) $\dfrac{0}{0}$: do something further;

      factor $x - a$ from numerator and denominator and divide out this factor;

      then substitute $x = a$.

(R2) $\dfrac{\neq 0}{0}$: limit does not exist.

Example 3: Evaluate $\lim\limits_{x \to 5} \dfrac{\sqrt{x-1}-2}{x-5}$. Justify each step by indicating the appropriate Limit Laws.

$$\lim_{x \to 5} \frac{\sqrt{x-1}-2}{x-5} \underset{\uparrow}{=} \frac{\sqrt{\lim\limits_{x \to 5}(x-1)} - \lim\limits_{x \to 5} 2}{\lim\limits_{x \to 5} x - \lim\limits_{x \to 5} 5} \underset{\uparrow}{=} ?$$

$$\text{(Laws 5, 2, and 11)} \left(\text{Laws 2, 7, and 8}; \frac{0}{0} : \text{ do} \right)$$

Note! Rationalize the numerator $\sqrt{x-1} - 2$:

multiply by its "conjugate" $\sqrt{x-1} + 2 \;\; (\neq 0)$

$$\lim_{x \to 5} \frac{\sqrt{x-1}-2}{x-5} = \lim_{x \to 5} \frac{\sqrt{x-1}-2}{x-5} \cdot \frac{\sqrt{x-1}+2}{\sqrt{x-1}+2} \underset{\uparrow}{=} \lim_{x \to 5} \frac{(x-1)-4}{(x-5)(\sqrt{x-1}+2)}$$

$$\text{(multiply out terms involving the conjugates)}$$

$$= \lim_{x \to 5} \frac{x-5}{(x-5)(\sqrt{x-1}+2)} = \lim_{x \to 5} \frac{1}{\sqrt{x-1}+2} = \frac{1}{2+2} = \frac{1}{4}$$

$$(x \neq 5) \quad (\text{Laws } 5, 1, 7, 11, 2, 8)$$

Example 4: Evaluate $\lim_{h \to 0} \dfrac{\sqrt{h+9}-3}{h}$.

$$\lim_{h \to 0} \frac{\sqrt{h+9}-3}{h} = \lim_{h \to 0} \frac{\sqrt{h+9}-3}{h} \cdot \frac{\sqrt{h+9}+3}{\sqrt{h+9}+3} = \lim_{h \to 0} \frac{(h+9)-9}{h(\sqrt{h+9}+3)}$$

$$\left(\frac{0}{0} : \text{ do} \right)$$

$$= \lim_{h \to 0} \frac{h}{h(\sqrt{h+9}+3)} = \lim_{h \to 0} \frac{1}{\sqrt{h+9}+3} = \frac{1}{6}$$

Example 5: Evaluate $\lim_{h \to 0} \dfrac{\dfrac{1}{\sqrt{1+h}} - 1}{h}$.

$$\lim_{h \to 0} \frac{\dfrac{1}{\sqrt{1+h}} - 1}{h} = \lim_{h \to 0} \frac{1-\sqrt{1+h}}{h\sqrt{1+h}} = \lim_{h \to 0} \frac{1-\sqrt{1+h}}{h\sqrt{1+h}} \cdot \frac{1+\sqrt{1+h}}{1+\sqrt{1+h}}$$

$$\left(\frac{0}{0} : \text{ do} \right)$$

$$= \lim_{h \to 0} \frac{1-(1+h)}{h\sqrt{1+h}(1+\sqrt{1+h}} = \lim_{h \to 0} \frac{-1}{\sqrt{1+h}(1+\sqrt{1+h})} = \frac{-1}{2}$$

The following two theorems, which involve the comparison of functions via inequalities, give two additional properties of limits.

Theorem (1.9): If $f(x) \leq g(x)$ for all x is an open interval that contains a (except possibly at a) and the limits of f and g both exist as x approaches a, then

$$\lim_{x \to a} f(x) \leq \lim_{x \to a} g(x)$$

The Squeeze Theorem (1.10): If $f(x) \leq g(x) \leq h(x)$ for all x in an open interval that contains a (except possibly at a) and

$$\lim_{x \to a} f(x) = \lim_{x \to a} h(x) = L$$

then

$$\lim_{x \to a} g(x) = L$$

Illustration: the Squeeze Theorem

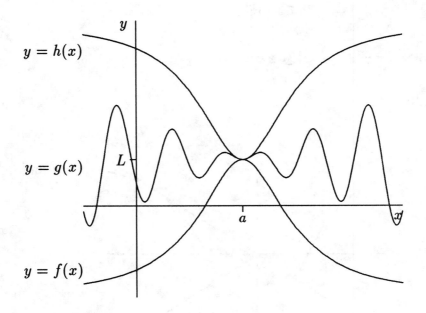

Example 6: $\displaystyle\lim_{x \to 0} x \sin \frac{1}{x} = ?$

Cannot use Limit Law 4: $\displaystyle\lim_{x \to 0} x \sin \frac{1}{x} = \underbrace{\left(\lim_{x \to 0} x\right)}_{(\,=\,0)} \underbrace{\left(\lim_{x \to 0} \sin \frac{1}{x}\right)}_{\text{(does not exist: Ex 4, Sec 1.2, p. 64)}}$

$|\sin \theta| \leq 1$ for all real θ $\implies$ $-1 \leq \sin \theta \leq 1$ for all real θ

$\underset{\uparrow}{\implies}$ $-1 \leq \sin \dfrac{1}{x} \leq 1$ for all $x \neq 0$

$\left(\theta = \dfrac{1}{x}\right)$

$$x > 0: \quad -x \le x \sin \frac{1}{x} \le x \implies -|x| \le x \sin \frac{1}{x} \le |x|$$
$$\uparrow$$
$$(|x| = x)$$

$$x < 0: \quad -x \ge x \sin \frac{1}{x} \ge x \implies -(-|x|) \ge x \sin \frac{1}{x} \ge -|x|$$
$$\uparrow$$
$$(|x| = -x)$$

or, rewritten $-|x| \le x \sin \dfrac{1}{x} \le |x|$

Hence, for all $x \ne 0$, $-|x| \le x \sin \dfrac{1}{x} \le |x|$.

Graphically:

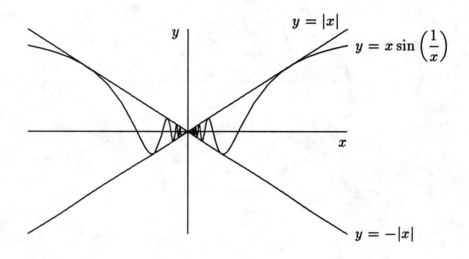

From graph of $y = |x|$ above:

$$\lim_{x \to 0} |x| = 0 \implies \lim_{x \to 0} -|x| = 0$$
$$\uparrow$$
$$\text{(Limit Law 4)}$$

$$\implies \lim_{x \to 0} x \sin \frac{1}{x} = 0$$
$$\uparrow$$

$$\left(\text{Squeeze Thm.: } f(x) = -|x|,\ g(x) = x \sin \frac{1}{x},\ h(x) = |x|,\ a = 0,\ L = 0 \right)$$

Comment: The eleven Limit Laws and the Squeeze Theorem also hold for one-sided limits.

Example 7: $\lim\limits_{x\to 1}\dfrac{x^2+|x-1|-1}{|x-1|}=\,?$

$x=1$ is not in the domain of $f(x)=\dfrac{x^2+|x-1|-1}{|x-1|}$

$|x-1|=\begin{cases}x-1 & \text{if } x\ge 1\\ -(x-1) & \text{if } x<1\end{cases}$ so consider one-sided limits at $x=1$.

$\lim\limits_{x\to 1^+}\dfrac{x^2+|x-1|-1}{|x-1|}=\lim\limits_{x\to 1^+}\dfrac{x^2+x-1-1}{x-1}$ $(x>1\Longrightarrow x-1>0)$

$=\lim\limits_{x\to 1^+}\dfrac{x^2+x-2}{x-1}=\lim\limits_{x\to 1^+}\dfrac{(x-1)(x+2)}{x-1}=\lim\limits_{x\to 1^+}(x+2)=3$

$\left(\text{rational function; }\dfrac{0}{0}:\text{ do}\right)$ (polynomial function)

$\lim\limits_{x\to 1^-}\dfrac{x^2+|x-1|-1}{|x-1|}=\lim\limits_{x\to 1^-}\dfrac{x^2+[-(x-1)]-1}{-(x-1)}$ $(x<1\Longrightarrow x-1<0)$

$=\lim\limits_{x\to 1^-}\dfrac{x^2-x}{-(x-1)}=\lim\limits_{x\to 1^-}\dfrac{x(x-1)}{-(x-1)}=\lim\limits_{x\to 1^-}(-x)=-1$

$\left(\text{rational function; }\dfrac{0}{0}:\text{ do}\right)$ (polynomial function)

$\lim\limits_{x\to 1^+}f(x)\ne\lim\limits_{x\to 1^-}f(x)\Longrightarrow\lim\limits_{x\to 1}\dfrac{x^2+|x-1|-1}{|x-1|}$ does not exist.

Example 8: Consider $f(x)=\begin{cases}3x^2+7 & \text{if } x\le 1\\ \dfrac{x-1}{\sqrt{x+24}-5} & \text{if } x>1\end{cases}$.   Does $\lim\limits_{x\to 1}f(x)$ exist?

$\lim\limits_{x\to 1^-}f(x)=\lim\limits_{x\to 1^-}(3x^2+7)=3(1)^2+7=10$

(polynomial function)

$$\lim_{x \to 1+} f(x) = \lim_{x \to 1+} \frac{x-1}{\sqrt{x+24}-5} = \lim_{x \to 1+} \frac{x-1}{\sqrt{x+24}-5} \cdot \frac{\sqrt{x+24}+5}{\sqrt{x+24}+5}$$

$$\left(\text{Limit Laws, } \frac{0}{0}: \text{ do; rationalize denominator}\right)$$

$$= \lim_{x \to 1+} \frac{(x-1)(\sqrt{x+24}+5)}{(x+24)-25} = \lim_{x \to 1+} (\sqrt{x+24}+5) = 5+5 = 10$$

$$(x \neq 1) \qquad\qquad (\text{Limit Laws})$$

$$\lim_{x \to 1-} f(x) = 10 = \lim_{x \to 1+} f(x) \implies \lim_{x \to 1} f(x) = 10$$

Recall: The *greatest integer function*: $[x]$ = the largest integer less than or equal to x.

(R1) $[6] = 6$, $[6.1] = 6$, $[5.9] = 5$, $[-3] = -3$, $[-3.1] = -4$, $[-2.9] = -3$

(R2) Graph of $y = [x]$:   *step function*

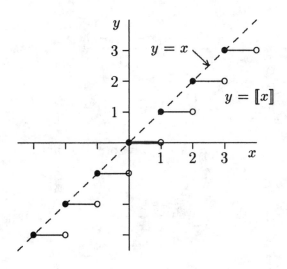

Note! The range of $[x]$ is the integers.

choose largest or = to   less than the #

$$[5] = 5 \qquad\qquad [4.9] = 4$$
$$[5.1] = 5 \qquad\qquad [-4.9] = -5$$
$$[-5.] = -6$$

Example 9: Evaluate $\lim_{x \to a} [\![x]\!]$, provided the limit exists, if:

(A1) $a = 2.7$

$$\lim_{x \to 2.7^-} [\![x]\!] = \lim_{x \to 2.7^-} 2 = 2; \qquad\qquad \lim_{x \to 2.7^+} [\![x]\!] = \lim_{x \to 2.7^+} 2 = 2$$
$$\qquad\quad \uparrow \qquad\qquad\qquad\qquad\qquad\qquad\qquad \uparrow$$
$$([\![x]\!] = 2 \text{ for } 2 \le x \le 2.7) \qquad\qquad ([\![x]\!] = 2 \text{ for } 2.7 \le x < 3)$$

$$\lim_{x \to 2.7^-} [\![x]\!] = 2 = \lim_{x \to 2.7^+} [\![x]\!] \implies \lim_{x \to 2.7} [\![x]\!] = 2$$

Note! Need not consider the one-sided limits since the formula for $[\![x]\!]$ is the same on both sides of and near $x = 2.7$.

(A2) $a = 2$

$$\lim_{x \to 2^-} [\![x]\!] = \lim_{x \to 2^-} 1 = 1; \qquad\qquad \lim_{x \to 2^+} [\![x]\!] = \lim_{x \to 2^+} 2 = 2$$
$$\qquad\quad \uparrow \qquad\qquad\qquad\qquad\qquad\qquad\qquad \uparrow$$
$$([\![x]\!] = 1 \text{ for } 1 \le x < 2) \qquad\qquad ([\![x]\!] = 2 \text{ for } 2 \le x < 3)$$

$$\lim_{x \to 2^-} [\![x]\!] \ne \lim_{x \to 2^+} [\![x]\!] \implies \lim_{x \to 2} [\![x]\!] \text{ does not exist.}$$

(A3) $a = -2$

$$\lim_{x \to -2^-} [\![x]\!] = \lim_{x \to -2^-} (-3) = -3; \qquad\qquad \lim_{x \to -2^+} [\![x]\!] = \lim_{x \to -2^+} (-2) = -2$$
$$\qquad\quad \uparrow \qquad\qquad\qquad\qquad\qquad\qquad\qquad\qquad \uparrow$$
$$([\![x]\!] = -3 \text{ for } -3 \le x < 2\ ) \qquad\qquad ([\![x]\!] = -2 \text{ for } -2 \le x < -1)$$

$$\lim_{x \to -2^-} [\![x]\!] \ne \lim_{x \to -2^+} [\![x]\!] \implies \lim_{x \to -2} [\![x]\!] \text{ does not exist.}$$

Example 10: Consider the function $f(x) = \left[\!\left[\dfrac{x}{3} + 1\right]\!\right]$.

(A1) Sketch the graph of $y = f(x)$.

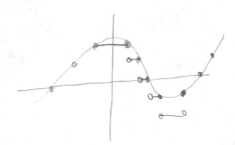

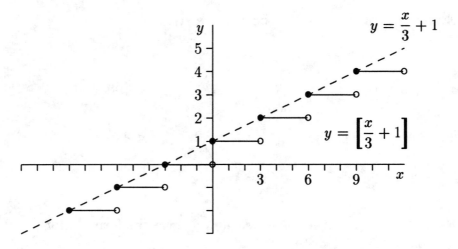

(A2) $\lim\limits_{x \to 2} \left[\!\left[\dfrac{x}{3} + 1 \right]\!\right] = ?$

$\lim\limits_{x \to 2} \left[\!\left[\dfrac{x}{3} + 1 \right]\!\right] \underset{\uparrow}{=} \lim\limits_{x \to 2} 1 = 1$

$\left(\left[\!\left[\dfrac{x}{3} + 1 \right]\!\right] = 1 \text{ for } 0 \le x < 3 \right)$

(A3) $\lim\limits_{x \to 3} \left[\!\left[\dfrac{x}{3} + 1 \right]\!\right] = ?$

$\lim\limits_{x \to 3^-} \left[\!\left[\dfrac{x}{3} + 1 \right]\!\right] \underset{\uparrow}{=} \lim\limits_{x \to 3^-} 1 = 1$

$\left(\left[\!\left[\dfrac{x}{3} + 1 \right]\!\right] = 1 \text{ for } 0 \le x < 3 \right)$

$\lim\limits_{x \to 3^+} \left[\!\left[\dfrac{x}{3} + 1 \right]\!\right] \underset{\uparrow}{=} \lim\limits_{x \to 3^-} 2 = 2$

$\left(\left[\!\left[\dfrac{x}{3} + 1 \right]\!\right] = 2 \text{ for } 3 \le x < 6 \right)$

$\lim\limits_{x \to 3^-} \left[\!\left[\dfrac{x}{3} + 1 \right]\!\right] \ne \lim\limits_{x \to 3^+} \left[\!\left[\dfrac{x}{3} + 1 \right]\!\right] \implies \lim\limits_{x \to 3} \left[\!\left[\dfrac{x}{3} + 1 \right]\!\right]$ does not exist.

Example 11: $\lim\limits_{x \to 3} \left(\left[\!\left[\dfrac{x}{3} + 1 \right]\!\right] + \left[\!\left[9 - x^2 \right]\!\right] \right) = ?$

From Example 10 above: $\lim\limits_{x \to 3} \left[\!\left[\dfrac{x}{3} + 1 \right]\!\right]$ does not exist

$\Longrightarrow$ cannot use Limit Law 1: $\lim\limits_{x\to 3}\left(\left[\!\left[\frac{x}{3}+1\right]\!\right]+[\![9-x^2]\!]\right)=\lim\limits_{x\to 3}\left[\!\left[\frac{x}{3}+1\right]\!\right]+\lim\limits_{x\to 3}[\![9-x^2]\!]$

Try one-sided limits:

Form Example 10 above: $\lim\limits_{x\to 3^-}\left[\!\left[\frac{x}{3}+1\right]\!\right]=1$

$\lim\limits_{x\to 3^-}[\![9-x^2]\!]=?$

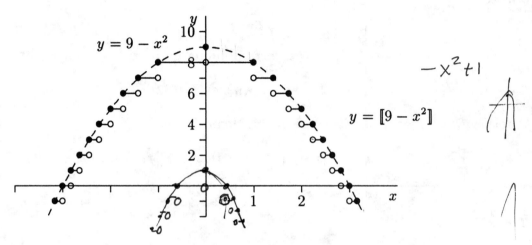

$y = 9 - x^2$

$-x^2+1$

$y = [\![9-x^2]\!]$

$[\![9-x^2]\!]=0$ for $\sqrt{8}<x\leq 3 \implies \lim\limits_{x\to 3^-}[\![9-x^2]\!]=0$

$\lim\limits_{x\to 3^-}\left(\left[\!\left[\frac{x}{3}+1\right]\!\right]+[\![9-x^2]\!]\right)=\lim\limits_{x\to 3^-}\left[\!\left[\frac{x}{3}+1\right]\!\right]+\lim\limits_{x\to 3^-}[\![9-x^2]\!]=1+0=1$

$\uparrow$
(Limit Law 1)

From Example 10 above: $\lim\limits_{x\to 3^+}\left[\!\left[\frac{x}{3}+1\right]\!\right]=2$

$[\![9-x^2]\!]=-1$ for $3<x\leq\sqrt{10} \implies \lim\limits_{x\to 3^+}[\![9-x^2]\!]=-1$

$\lim\limits_{x\to 3^+}\left(\left[\!\left[\frac{x}{3}+1\right]\!\right]+[\![9-x^2]\!]\right)=\lim\limits_{x\to 3^+}\left[\!\left[\frac{x}{3}+1\right]\!\right]+\lim\limits_{x\to 3^+}[\![9-x^2]\!]=2+(-1)=1$

$\uparrow$
(Limit Law 1)

$\lim\limits_{x\to 3^-}\left(\left[\!\left[\frac{x}{3}+1\right]\!\right]+[\![9-x^2]\!]\right)=1=\lim\limits_{x\to 3^+}\left(\left[\!\left[\frac{x}{3}+1\right]\!\right]+[\![9-x^2]\!]\right)$

$$\implies \quad \lim_{x \to 3}\left(\left[\frac{x}{3}+1\right]+[9-x^2]\right)=1$$

1.4   The Precise Definition of a Limit

A precise definition of a limit, which is used to prove the Limit Laws given in Section 1.3, is:

Definition (1.12): Let f be a function defined on some open interval that contains the number a, except possibly at a itself. Then we say that the **limit of $f(x)$ as x approaches a is L**, and we write

$$\lim_{x \to a} f(x) = L$$

if for every number $\epsilon > 0$ there is a corresponding number $\delta > 0$ such that

$$|f(x) - L| < \epsilon \qquad \text{whenever} \qquad 0 < |x - a| < \delta$$

Another way to write the last line of this definition is

$$\text{if } \; 0 < |x - a| < \delta \qquad \text{then} \qquad |f(x) - L| < \epsilon$$

Remarks:

(R1) Consider $0 < |x - a| < \delta$.

$$|x - a| < \delta \quad \implies \quad -\delta < x - a < \delta$$

$$\implies \quad a - \delta < x < a + \delta: \; x \text{ is in the open interval } (a - \delta, a + \delta).$$

$$0 < |x - a| \quad \implies \quad x \neq a$$

Note! $x \neq a$ indicates $f(a)$ need not be defined.

(R2) $|f(x) - L| < \epsilon \quad \implies \quad -\epsilon < f(x) - L < \epsilon$

$$\implies \quad L - \epsilon < f(x) < L + \epsilon: \; f(x) \text{ is in the open interval } (L - \epsilon, L + \epsilon).$$

(R3) The definition, in words: $f(x)$ can be made as close to L as desired by taking x close enough to a, but not equal to a.

(R4) Geometrically: choose arbitrary $\epsilon > 0$; can a $\delta > 0$ be found such that

$$L - \epsilon < f(x) < L + \epsilon \ \text{ whenever } \ a - \delta < x < a + \delta, \ \ x \neq a \ ?$$

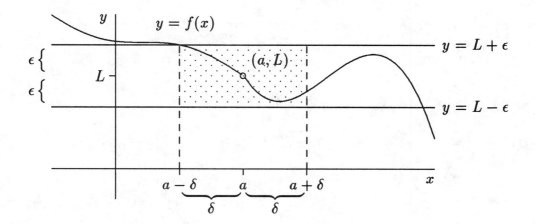

Note!

(N1) The graph of $y = f(x)$ lies between the lines $y = L + \epsilon$ and $y = L - \epsilon$ for all x satisfying $a - \delta < x < a + \delta$, $x \neq a$.

(N2) Any smaller, non-zero δ will suffice for the chosen ϵ.

(N3) Usually $\underbrace{\delta = g(\epsilon)}$ in a way that δ must get smaller as ϵ gets smaller.
     (δ depends on ϵ)

(R5) The definition indicates if L is a limit, but not how to find L.

A Procedure: Show $\lim\limits_{x \to a} f(x) = L$.

(S1) Let $\epsilon > 0$ be arbitrary

(S2) Begin with the expression $|f(x) - L|$ for $x \neq a$

and obtain (through manipulation) $|f(x) - L| \leq k|x - a| < \epsilon$

where k is a positive constant.

(S3) From (S2): $|x - a| < \dfrac{\epsilon}{k}$

(S4) Choose $\delta > 0$ such that $\delta \leq \dfrac{\epsilon}{k}$

Note! Must find a δ that works, so usually let $\delta = \dfrac{\epsilon}{k}$.

(S5) Show the δ found in step (S4) works.

Example 1: Using the ϵ, δ definition of a limit, show $\lim\limits_{x \to 1}(2x + 3) = 5$.

Note! $a = 1 \implies x - a = x - 1; \quad f(x) = 2x + 3; \quad L = 5$

Let $\epsilon > 0$ be arbitrary.

$$|f(x) - L| = |(2x + 3) - 5| = |2x - 2| = |2(x - 1)| = 2\,|x - 1| < \epsilon$$
$$\uparrow$$
$$(k)$$

$$\implies \quad |x - 1| < \frac{\epsilon}{2}; \quad \text{a choice: } \delta = \frac{\epsilon}{2}.$$

Show $\delta = \dfrac{\epsilon}{2}$ works:

$$|f(x) - L| = |(2x + 3) - 5| = 2|x - 1| < 2\left(\frac{\epsilon}{2}\right) = \epsilon$$

Note!

(N1) Graphically: let $\epsilon = 2$; then $\delta = \dfrac{\epsilon}{2} = 1$

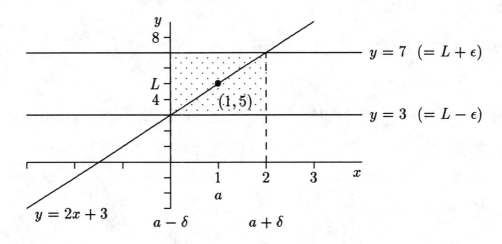

$$0 < |x - a| < \delta \implies 0 < |x - 1| < 1 \implies 0 < x < 2, \quad x \neq 1$$

(N2) $f(1) = 2(1) + 3 = 5,$ so $|f(a) - L| = |f(1) - 5| = 0 < \epsilon$

(N3) As ϵ gets smaller, $\delta = \dfrac{\epsilon}{2}$ gets smaller.

Example 2: Consider $f(x) = \begin{cases} 2x + 3 & \text{if } x \neq 1 \\ 8 & \text{if } x = 1 \end{cases}$

Using the ϵ, δ definition of a limit, show $\lim\limits_{x \to 1} f(x) = 5$.

   Note! $a = 1 \implies x - a = x - 1;$ $L = 5$

   Let $\epsilon > 0$ be arbitrary.

   For $0 < |x - 1|$ $\hspace{8cm}$ $(x \neq 1 = a)$

   $|f(x) - L| = |(2x + 3) - 5| = \underbrace{2}_{(k)} |x - 1| < \epsilon \implies |x - 1| < \dfrac{\epsilon}{2};$ a choice: $\delta = \dfrac{\epsilon}{2}$

   Show $\delta = \dfrac{\epsilon}{2}$ works:

   $|f(x) - L| = |(2x + 3) - 5| = 2|x - 1| < 2\left(\dfrac{\epsilon}{2}\right) = \epsilon$

   Note!

   (N1) Graphically: as in Example 1 above except, now, $f(1) = 8$.

   (N2) $|f(a) - L| = |f(1) - 5| = |8 - 5| = 3 \not< \epsilon$ if $\epsilon \leq 3$,

   $\hspace{1cm}$ but interested in x near $a = 1$, $\quad x \neq 1$.

Example 3: Using the ϵ, δ definition of a limit, show $\lim\limits_{x \to -1} (x^2 - x) = 2$.

   Note! $a = -1 \implies x - a = x + 1;$ $f(x) = x^2 - x;$ $L = 2$

   Let $\epsilon > 0$ be arbitrary.

   $|f(x) - L| = |(x^2 - x) - 2| = |x^2 - x - 2| = |(x - 2)(x + 1)| = \underbrace{|x - 2|}_{\text{(depends on } x\text{: not a constant)}} \underbrace{|x + 1|}_{(|x - a|)}$

Agree to restrict δ, say $\delta \leq 1$ (the value 1 is an arbitrary choice)

$\Longrightarrow \quad |x - a| = |x + 1| < \delta \leq 1 \quad \Longrightarrow \quad |x + 1| < 1 \quad \Longrightarrow \quad -1 < x + 1 < 1$

$\Longrightarrow \quad -2 < x < 0 \quad \Longrightarrow \quad -4 < x - 2 < -2 \quad \Longrightarrow \quad \underbrace{-4 < x - 2 < 4}$

                                      ($-2 < 4$; obtained $\; -b < x - 2 < b$, where $b > 0$)

$\Longrightarrow \; |x - 2| < 4$

$\uparrow$

$(-b < x - 2 < b \iff |x - 2| < b)$

$\Longrightarrow \quad |f(x) - L| = |x - 2||x + 1| < 4\,|x + 1| < \epsilon$

                                       $\uparrow$

                                       (k)

$\Longrightarrow \quad |x + 1| < \dfrac{\epsilon}{4}; \quad$ a choice: $\delta = \min\left\{\dfrac{\epsilon}{4}, 1\right\} \quad \left(\begin{array}{l}\text{the smaller value,}\\ \text{depending upon the value of } \epsilon\end{array}\right)$

Note!

(N1) $\delta = \min\left\{\dfrac{\epsilon}{4}, 1\right\} = \begin{cases} \epsilon/4 & \text{if } \epsilon < 4 \\ 1 & \text{if } \epsilon \geq 4 \end{cases}$

(N2) Graphically: $\epsilon = 2 \quad \Longrightarrow \quad \epsilon < 4, \quad$ hence $\quad \delta = \dfrac{\epsilon}{4} = \dfrac{1}{2} \;\; (< 1)$

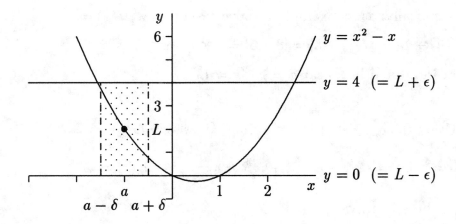

The value of δ could be slightly larger.

Example 4: Using the ϵ, δ definition of a limit, show $\lim\limits_{x \to 3}(x-1)^2 = 4$.

    Note! $a = 3 \implies x - a = x - 3$; $f(x) = (x-1)^2$; $L = 4$

Let $\epsilon > 0$ be arbitrary.

$$|f(x) - L| = |(x-1)^2 - 4| = |x^2 - 2x - 3| = |(x+1)(x-3)| = \underbrace{|x+1|}\,|x-3|$$

$$\text{(depends on } x\text{: not a constant)}$$

Restrict δ, say $\delta \le 1$

$\implies$ $|x - a| = |x - 3| < \delta \le 1$ $\implies$ $|x - 3| < 1$ $\implies$ $-1 < x - 3 < 1$

$\implies$ $2 < x < 4$ $\implies$ $3 < x + 1 < 5$ $\implies$ $\underbrace{-5 < x + 1 < 5}$ $\implies$ $|x + 1| < 5$

$$(-5 < 3; \text{ obtained } -b < x + 1 < b, \text{ where } b > 0)$$

$\implies$ $|f(x) - L| = |x+1||x-3| < 5|x-3| < \epsilon$

$\implies$ $|x - 3| < \dfrac{\epsilon}{5}$;   a choice: $\delta = \min\left\{\dfrac{\epsilon}{5}, 1\right\}$

Example 5: Using the ϵ, δ definition of a limit, show $\lim\limits_{x \to 1}(x-1)^2 = 0$

    Note! $a = 1 \implies x - a = x - 1$; $f(x) = (x-1)^2$; $L = 0$

Let $\epsilon > 0$ be arbitrary.

$$|f(x) - L| = |(x-1)^2 - 0| = \underbrace{|x-1|^2} < \epsilon$$

$$\left(|x - a|^2\right)$$

$\implies$ $|x - 1| < \sqrt{\epsilon}$;   a choice: $\delta = \sqrt{\epsilon}$.

Note! Step(S2) of the procedure was modified since, for a positive integer n,

$$|f(x) - L| = |x - a|^n < \epsilon \quad \text{yields} \quad |x - a| < \sqrt[n]{\epsilon} \quad (= g(\epsilon))$$

Example 6: Using the ϵ, δ definition of a limit, show $\lim\limits_{x \to 1}\sqrt{x + 3} = 2$.

Note! $a = 1 \implies x - a = x - 1; \quad f(x) = \sqrt{x+3}; \quad L = 2$

$$|f(x) - L| = |\sqrt{x+3} - 2| = |\sqrt{x+3} - 2| \cdot \underset{\uparrow}{\frac{|\sqrt{x+3} + 2|}{|\sqrt{x+3} + 2|}} = \underset{\uparrow}{\frac{|(x+3) - 4|}{\sqrt{x+3} + 2|}}$$

$$\qquad\qquad\qquad\qquad \text{(rationalize numerator)} \qquad\qquad \left(\sqrt{x+3} + 2 > 0\right)$$

$$= \frac{|x-1|}{\sqrt{x+3} + 2} \underset{\uparrow}{\leq} \frac{|x-1|}{2} = \frac{1}{2}|x-1| < \epsilon$$

$$\left(\sqrt{x+3} \geq 0 \implies \sqrt{x+3} + 2 \geq 2\right)$$

$$\implies \quad |x-1| < 2\epsilon$$

However, $\sqrt{x+3}$ requires $x + 3 \geq 0$

$$\implies \quad x \geq -3 \implies \underbrace{x - 1}_{(x - a)} \geq -4 \implies \underbrace{|x - 1|}_{(|x - a|)} \leq 4$$

A choice: $\delta = \min\{2\epsilon, 4\}$

A Procedure: Show $\lim\limits_{x \to a} f(x)$ does not exist. Use an indirect proof.

(S1) Suppose $\lim\limits_{x \to a} f(x) = L$.

(S2) Choose a particular value for ϵ such that $\epsilon > 0$.

(S3) From $|f(x) - L| < \epsilon$ whenever $0 < |x - a| < \delta$, obtain a contradiction.

Example 7: Show $\lim\limits_{x \to 1} \dfrac{1}{x - 1}$ does not exist.

Note! $a = 1 \implies x - a = x - 1; \quad f(x) = \dfrac{1}{x - 1}.$

Suppose $\lim\limits_{x \to 1} \dfrac{1}{x - 1} = L$.

A choice: let $\epsilon = 1$. Then there is a $\delta > 0$ such that

$$|f(x) - L| = \left| \frac{1}{x-1} - L \right| < 1 \text{ whenever } 0 < \underbrace{|x-1|}_{(|x-a|)} < \delta$$

$$(\epsilon)$$

$$\implies L - 1 < \frac{1}{x-1} < L + 1 \text{ whenever } 0 < |x-1| < \delta$$

For $x - 1 > 0$, the value of $\dfrac{1}{x-1}$ gets arbitrarily large as $x - 1 \to 0$ (i.e., as $x \to 1$)

$$\implies \frac{1}{x-1} > L + 1 \text{ for any finite number } L: \text{ contradiction.}$$

Note! Graphically: $\epsilon = 1$

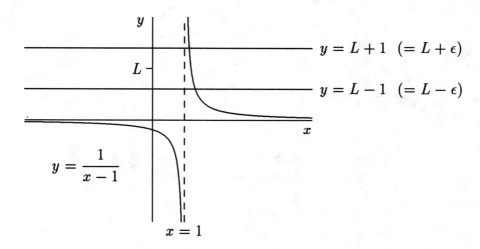

$y = L + 1 \;\; (= L + \epsilon)$

$y = L - 1 \;\; (= L - \epsilon)$

$y = \dfrac{1}{x-1}$

$x = 1$

For any choice of L with $\epsilon = 1$, there will be points $(x, f(x)) = \left(x, \dfrac{1}{x-1} \right)$ for values of x near $a = 1$ which do not lie between the lines $y = L + 1$ and $y = L - 1$.

Example 8: Consider $f(x) = \left\{ \begin{array}{ll} -1 & \text{if } x \le 0 \\ 1 & \text{if } x > 0 \end{array} \right\}$.   Show $\lim\limits_{x \to 0} f(x)$ does not exist.

Note! $a = 0 \implies x - a = x$

Suppose $\lim\limits_{x \to 0} f(x) = L$.

A choice: let $\epsilon = 1$. Then there is a $\delta > 0$ such that

$$|f(x) - L| < 1 \text{ whenever } 0 < |x - 0| < \delta \qquad\qquad (|x - a| = |x - 0|)$$

$$\Longrightarrow \quad L - 1 < f(x) < L + 1 \quad \text{whenever} \quad 0 < |x| < \delta.$$

$$x = \frac{\delta}{2} \quad (> 0) \text{ satisfies } 0 < |x| < \delta$$

$$\Longrightarrow \quad L - 1 < f\left(\frac{\delta}{2}\right) < L + 1 \quad \Longrightarrow \quad L - 1 < \underbrace{1 < L + 1}_{\text{(requires } L > 0)}$$

$$x = -\frac{\delta}{2} \quad (< 0) \text{ satisfies } 0 < |x| < \delta$$

$$\Longrightarrow \quad L - 1 < f\left(-\frac{\delta}{2}\right) < L + 1 \quad \Longrightarrow \quad \underbrace{L - 1 < -1}_{\text{(requires } L < 0)} < L + 1$$

$L > 0$ and $L < 0$ (simultaneously): not possible.

Note! Graphically: $L > 0$, $\epsilon = 1$

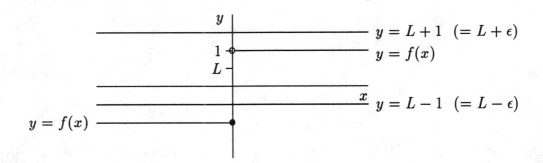

For any $x < 0$, the point $(x, f(x))$ does not lie between the lines $y = L + 1$ and $y = L - 1$.

Use the ϵ, δ definition of a limit to prove the Sum Limit Law.

Sum Law: If $\lim\limits_{x \to a} f(x) = L$ and $\lim\limits_{x \to a} g(x) = M$, then $\underbrace{\lim\limits_{x \to a}[f(x) + g(x)]}_{\text{(limit of sum)}} = \underbrace{L + M}_{\text{(sum of limits)}}$.

Proof: Let $\epsilon > 0$ be arbitrary $\implies \dfrac{\epsilon}{2} > 0$ is arbitrary.

$\lim\limits_{x \to a} f(x) = L$

$\implies$ there is a $\delta_1 > 0$ such that $|f(x) - L| < \dfrac{\epsilon}{2}$ whenever $0 < |x - a| < \delta_1$.

$\lim\limits_{x \to a} g(x) = M$

$\implies$ there is a $\delta_2 > 0$ such that $|g(x) - M| < \dfrac{\epsilon}{2}$ whenever $0 < |x - a| < \delta_2$.

Let $\delta = \min\{\delta_1, \delta_2\}$

$\implies$ if $0 < |x - a| < \delta$, then $\underbrace{0 < |x - a| < \delta \le \delta_1, \quad 0 < |x - a| < \delta \le \delta_2,}_{\text{(both conditions on } |x - a| \text{ hold)}}$

and $|[f(x) + g(x)] - (L + M)| = |[f(x) - L] + [g(x) - M]|$

$$\le \underset{\uparrow}{|f(x) - L|} + |g(x) - M| \underset{\uparrow}{<} \frac{\epsilon}{2} + \frac{\epsilon}{2} = \epsilon$$

(triangle inequality)   (both conditions hold)

$\underset{\uparrow}{\implies} \lim\limits_{x \to a}[f(x) + g(x)] = L + M$

(ϵ, δ definition of limit)

Remark: The proofs of other Limit Laws and the Squeeze Theorem are given in Appendix C of the text, pp. A23-A27.

A precise definition of each one-sided limit is given as follows:

Definition of Left-Hand Limit (1.13):

$$\lim_{x \to a^-} f(x) = L$$

if for every number $\epsilon > 0$ there is a corresponding number $\delta > 0$ such that

$$|f(x) - L| < \epsilon \qquad \text{whenever} \qquad a - \delta < x < a$$

Definition of Right-Hand Limit (1.14):

$$\lim_{x \to a^+} f(x) = L$$

if for every number $\epsilon > 0$ there is a corresponding number $\delta > 0$ such that

$$|f(x) - L| < \epsilon \qquad \text{whenever} \qquad a < x < a + \delta$$

Note! These ϵ, δ definitions are the same as the definition for $\lim_{x \to a} f(x)$ except x is now restricted to lie to one side of a:

(N1) left-hand limit: $a - \delta < x < a \implies x$ is to the left of a, $x \neq a$.

(N2) right-hand limit: $a < x < a + \delta \implies x$ is to the right of a, $x \neq a$.

Remark: Imitate the procedures given earlier to show that a one-sided limit equals a value or that it does not exist.

Example 9: Using the ϵ, δ definition of a left-hand limit, show $\lim_{x \to 3^-} \sqrt{3 - x} = 0$.

   Note!

   (N1) $a = 3 \implies x - a = x - 3; \quad f(x) = \sqrt{3 - x}; \quad L = 0.$

   (N2) $f(x) = \sqrt{3 - x}$ requires $3 - x \geq 0 \implies x \leq 3$,

       so only a left-hand limit can be considered at $x = 3$.

   Let $\epsilon > 0$ be arbitrary.

$$|f(x) - L| = |\sqrt{3-x} - 0| = |\sqrt{3-x}| = \sqrt{3-x} < \epsilon$$

$$\underset{\left(\sqrt{3-x} \geq 0\right)}{\uparrow}$$

$$\implies 3 - x < \epsilon^2 \implies 0 < 3 - x < \epsilon^2 \implies -3 < -x < -3 + \epsilon^2$$

$$\uparrow$$

(left-hand limit: $x < 3 = a$)

$$\implies 3 > x > 3 - \epsilon^2 \quad \text{or, rewritten,} \quad 3 - \epsilon^2 < x < 3;$$

a choice: $\delta = \epsilon^2$ (left-hand limit: $a - \delta < x < a$, $a = 3$)

Example 10: Using the ϵ, δ definition of a right-hand limit, show $\lim\limits_{x \to 3^+} (\sqrt{x-3} + 1) = 1$

Note!

(N1) $a = 3 \implies x - a = x - 3;$ $f(x) = \sqrt{x-3} + 1;$ $L = 1$

(N2) $f(x) = \sqrt{x-3} + 1$ requires $x - 3 \geq 0 \implies x \geq 3$, so only a right-hand limit can be considered at $x = 3$

Let $\epsilon > 0$ be arbitrary.

$$|f(x) - L| = |(\sqrt{x-3} + 1) - 1| = |\sqrt{x-3}| = \sqrt{x-3} < \epsilon$$

$$\underset{\left(\sqrt{x-3} \geq 0\right)}{\uparrow}$$

Proceed as in Example 9 above to show:

a choice: $\delta = \epsilon^2$, since $3 < x < 3 + \epsilon^2$ (right-hand limit: $a < x < a + \delta$, $a = 3$)

1.5 Continuity

Definition (1.16): A function f is **continuous at a number a** if

$$\lim_{x \to a} f(x) = f(a)$$

Remarks:

* graphs do not show continuit

(R1) Three requirements for continuity at $x = a$:

   (1) $f(a)$ is defined (i.e., a is in the domain of f)

   (2) $\lim_{x \to a} f(x)$ exists

   Note! (1) and (2) require f to be defined on an open interval containing a.

   (3) $\lim_{x \to a} f(x) = f(a)$

   Note! Need a specific value for $\lim_{x \to a} f(x)$.

(R2) Intuitively: the graph of $y = f(x)$ has no break or jump at $x = a$.

(R3) If a break or jump in the graph of $y = f(x)$ occurs at $x = a$, at least one of the requirements (1)-(3) given in (R1) above is violated

   $\implies$ f is **discontinuous at a** or f has a **discontinuity at a**.

Illustration: Geometrically, consider the following graph:

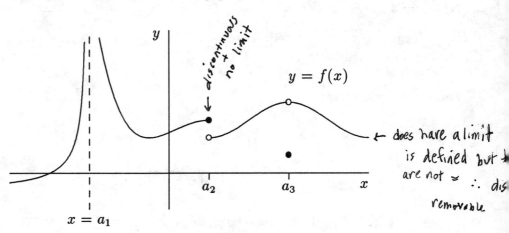

Discontinuities of f: ïnfinite

$x = a_1$: $f(a_1)$ is not defined   (violates (1) of (R1))

$x = a_2$: $\lim_{x \to a_2} f(x)$ does not exist. jump   (violates (2) of (R1), since $\lim_{x \to a_2^-} f(x) \neq \lim_{x \to a_2^+} f(x)$)

$x = a_3$: $\lim_{x \to a_3} f(x) \neq f(a_3)$ removable   (violates (3) of (R1))

Note! f is continuous at all other a in the domain of f since no breaks or jumps occur:

$$\lim_{x \to a} f(x) = f(a) \quad \text{for all} \quad a \neq a_1, a_2, a_3.$$

The three kinds of discontinuities illustrated above are identified as follows:

(K1) f has a **removable** discontinuity at a if $\lim\limits_{x \to a} f(x)$ exists but

$$\left\{ \begin{array}{c} f(a) \text{ is not defined} \\ \text{or} \\ \text{if } f(a) \text{ is defined, but } \lim\limits_{x \to a} f(x) \neq f(a) \end{array} \right\}$$

Note!

(N1) Since $\lim\limits_{x \to a} f(x)$ exists, define f at $x = a$ to have the value $f(a) = \lim\limits_{x \to a} f(x)$

$\implies$ f is now continuous at a (i.e., removed discontinuity at a).

(N2) In the illustration above, $x = a_3$ is a removable discontinuity.

(K2) f has a **jump** discontinuity at a if both one-sided limits exist, but

$$\lim_{x \to a^-} f(x) \neq \lim_{x \to a^+} f(x) \quad (\text{hence, } \lim_{x \to a} f(x) \text{ does not exist})$$

Note!

(N1) Unable to define f at the single value $x = a$ in order for $\lim\limits_{x \to a} f(x)$ to exist.

(N2) In the illustration above, $x = a_2$ is a jump discontinuity.

(K3) f has an **infinite** discontinuity at a if at least one of the one-sided limits is not a finite number in the sense $f(x)$ becomes arbitrarily large, positively or negatively, as x approaches a from at least one side (hence $\lim\limits_{x \to a} f(x)$ does not exist).

Note!

(N1) Unable to define f at the single value $x = a$ in order for $\lim\limits_{x \to a} f(x)$ to exist.

(N2) In the illustration above, $x = a_1$ is an infinite discontinuity.

Example 1: Show $f(x) = \dfrac{x^2 + x - 6}{x^2 - 4}$ is discontinuous:

(A1) at $x = 2$ and identify the kind of discontinuity.

$f(2) = \dfrac{0}{0}$: not defined $\implies$ f discontinuous at $x = 2$.

$$\lim_{x \to 2} \frac{x^2 + x - 6}{x^2 - 4} = \lim_{x \to 2} \frac{(x + 3)(x - 2)}{(x - 2)(x + 2)} = \lim_{x \to 2} \frac{x + 3}{x + 2} = \frac{5}{4}$$
$$\uparrow \qquad\qquad\qquad\qquad \uparrow$$
$$\left(\frac{0}{0} : \text{do}\right) \qquad\qquad (x \ne 2)$$

$\implies$ $x = 2$: removable discontinuity.

Note! Define $f(2) = \dfrac{5}{4}$ $\quad\left(= \lim_{x \to 2} f(x)\right)$ to remove the discontinuity at $x = 2$.

(A2) at $x = -2$ and identify the kind of discontinuity.

$f(-2) = \dfrac{-4}{0}$: not defined $\implies$ f discontinuous at $x = -2$.

$$\lim_{x \to -2} \frac{x^2 + x - 6}{x^2 - 4} \quad \text{does not exist} \qquad\qquad \left(\text{of the form } \frac{\ne 0}{0}\right)$$

with $f(x)$ becoming arbitrarily large, positively and negatively, as x approaches -2.

$\implies$ $x = -2$: infinite discontinuity.

Note! A sketch of the graph of $y = \dfrac{x^2 + x - 6}{x^2 - 4}$, a rational function (A method for graphing is discussed in Chapter 3.) is:

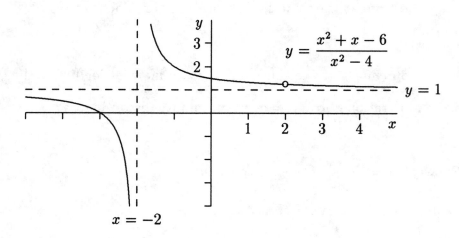

Example 2: Consider $f(x) = \begin{cases} x + 2 & \text{if } x < 0 \\ |x - 1| & \text{if } x \geq 0 \end{cases}$ *case function*

(A1) Show $f(x)$ is discontinuous at $x = 0$ and identify the kind of discontinuity.

$f(0) = |0 - 1| = |-1| = 1$

$\lim_{x \to 0} f(x) = ?$: $f(x)$ changes cases at $x = 0$, so use one-sided limits.

$\lim_{x \to 0^-} f(x) = \lim_{\substack{\uparrow \\ (x < 0)}} (x + 2) = 0 + 2 = 2$

$\lim_{x \to 0^+} f(x) = \lim_{\substack{\uparrow \\ (x > 0)}} |x - 1| = \lim_{\substack{x \to 0^+ \\ \uparrow}} [-(x - 1)] = -(0 - 1) = 1$

$$\left(|x - 1| = \begin{cases} x - 1 & \text{if } x \geq 1 \;\; (x - 1 \geq 0) \\ -(x - 1) & \text{if } x < 1 \;\; (x - 1 < 0) \end{cases} \right)$$

$\lim_{x \to 0^-} f(x) \neq \lim_{x \to 0^+} f(x) \implies \lim_{x \to 0} f(x)$ does not exist

$\implies \;\; x = 0$ is a jump discontinuity.

(A2) Show f is continuous at $x = 1$.

$f(1) = |1 - 1| = 0$

$\lim_{x \to 1} f(x) = \lim_{x \to 1} |x - 1|$: $f(x)$ ~~changes~~ doesn't cases at $x = 1$, so use one-sided limits.

* *when absolute val dividing look at both sides* *

$\lim_{x \to 1^-} |x - 1| = \lim_{\substack{x \to 1^- \\ \uparrow}} [-(x - 1)] = -(1 - 1) = 0$

$$\left(x < 1; \;\; |x - 1| = \begin{cases} x - 1 & \text{if } x \geq 1 \;\; (x - 1 \geq 0) \\ -(x - 1) & \text{if } x < 1 \;\; (x - 1 < 0) \end{cases} \right)$$

$\lim_{x \to 1^+} |x - 1| = \lim_{\substack{x \to 1^+ \\ \uparrow \\ (x > 1)}} (x - 1) = 1 - 1 = 0$

$\lim_{x \to 1^-} |x - 1| = 0 = \lim_{x \to 1^+} |x - 1| \implies \lim_{x \to 1} |x - 1| = 0$

$$\lim_{x \to 1} |x - 1| = 0 = f(1), \text{ so } f \text{ is continuous at } x = 1.$$

(A3) Sketch the graph of $y = f(x)$.

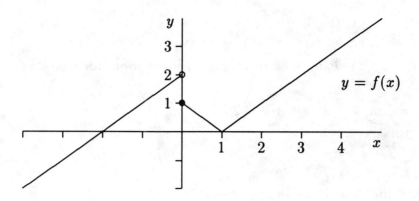

Note!

(N1) $x < 0$: $y = x + 2$ is a line.

(N2) $0 \le x \le 1$: $y = -(x - 1)$ is a line.

(N3) $x > 1$: $y = x - 1$ is a line.

Definition (1.17): A function f is **continuous from the right at a number a** if

$$\lim_{x \to a^+} f(x) = f(a)$$

and f is **continuous from the left at a number a** if

$$\lim_{x \to a^-} f(x) = f(a)$$

Example 3: Consider $f(x) = \begin{cases} x + 2 & \text{if } x < 0 \\ |x - 1| & \text{if } x \ge 0 \end{cases}$, the function of Example 2 above.

(A1) Is f continuous from the right at $x = 0$?

$$f(0) = 1$$

$$\lim_{x\to 0^+} f(x) = \lim_{x\to 0^+} |x-1| = \lim_{x\to 0^+} [-(x-1)] = 1$$

$\uparrow$

$(x > 0)$ (see (A1), Example 2 above)
$\uparrow$

$\lim_{x\to 0^+} f(x) = 1 = f(0)$, so f is continuous from the right at $x = 2$.

(A2) Is f continuous from the left at $x = 0$?

$f(0) = 1$

$$\lim_{x\to 0^-} f(x) = \lim_{x\to 0^-} (x+2) = 2$$

$\uparrow$

$(x < 0)$

$\lim_{x\to 0^-} f(x) \neq f(0)$, so f is not continuous from the left at $x = 0$.

The following theorem, which follows from Limit Laws 1-5 given in Section 1.3, combines the notion of continuity and arithmetic operations on continuous functions.

Theorem (1.19): If f and g are continuous at a and c is a constant, then the following functions are also continuous at a:

1. $f+g$ 2. $f-g$ 3. cf 4. fg 5. $\dfrac{f}{g}$ if $g(a) \neq 0$

Remark: In 5, if $g(a) = 0$, then $\dfrac{f}{g}$ is not defined at $x = a \implies \dfrac{f}{g}$ is discontinuous at $x = a$.

Proof of 4: Given f and g continuous at a

$\implies \lim_{x\to a} f(x) = f(a)$ and $\lim_{x\to a} g(x) = g(a)$

$\implies \lim_{x\to a}(fg)(x) = \lim_{x\to a} f(x)g(x) = \left[\lim_{x\to a} f(x)\right]\left[\lim_{x\to a} g(x)\right] = f(a)\cdot g(a) = (fg)(a)$

$\uparrow$ $\uparrow$

(Limit Law 4) $\begin{pmatrix} f \text{ and } g \text{ each} \\ \text{continuous at } a \end{pmatrix}$

$\implies$ fg is continuous at a.

Remark: The remaining proofs follow in a similar fashion.

Example 4: Consider $f(x) = \left\{ \begin{array}{ll} x + 2 & \text{if } x < 0 \\ |x - 1| & \text{if } x \geq 0 \end{array} \right\}$, the function of Example 2 above, and

$g(x) = \left\{ \begin{array}{ll} -1 & \text{if } x < 0 \\ 0 & \text{if } x \geq 0 \end{array} \right.$.

(A1) Have shown in (A2) of Example 2 above that f is continuous at $x = 1$. Show g is continuous at $x = 1$.

$g(1) = 0$

$\displaystyle\lim_{x \to 1} g(x) = \lim_{x \to 1} 0 = 0$
              $\uparrow$
(x near 1, so $x > 0$)

$\displaystyle\lim_{x \to 1} g(x) = 0 = g(0)$, so g is continuous at $x = 1$.

(A2) Use the results of (A1) to conclude $f + g$ is continuous at $x = 1$.

f, g continuous at $x = 1$ $\implies$ $f + g$ continuous at $x = 1$.
                          $\uparrow$
              (Theorem (1.19) 1)

(A3) Sketch the graph of $y = (f + g)(x)$,

$(f + g)(x) = f(x) + g(x) = \left\{ \begin{array}{ll} x + 2 + (-1) = x + 1 & \text{if } x < 0 \\ |x - 1| + 0 = |x - 1| & \text{if } x \geq 0 \end{array} \right.$
              $\uparrow$
(domain of both f and g is $(-\infty, \infty)$)

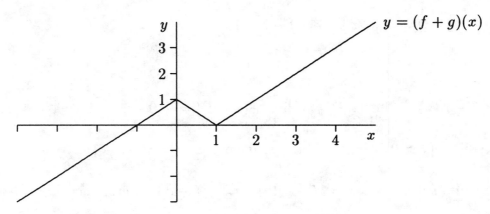

$y = (f + g)(x)$

(A4) Have shown in (A2) of Example 2 above that f is discontinuous at $x = 0$. Show g is discontinuous at $x = 0$.

$g(0) = 0$

$\lim\limits_{x \to 0} g(x) = ?$: $g(x)$ changes cases at $x = 0$, so consider one-sided limits.

$$\left.\begin{array}{c} \lim\limits_{x \to 0^-} g(x) = \lim\limits_{x \to 0^-} (-1) = -1 \\ \uparrow \\ (x < 0) \\ \lim\limits_{x \to 0^+} g(x) = \lim\limits_{x \to 0^+} (0) = 0 \\ \uparrow \\ (x > 0) \end{array}\right\} \implies \lim\limits_{x \to 0} g(x) \text{ does not exist.}$$

$\implies$ g is discontinuous at $x = 0$.

Note! $\lim\limits_{x \to 0^+} g(x) = g(0)$, so g is continuous from the right at $x = 0$.

(A5) Does $f + g$ have a discontinuity at $x = 0$?

From (A3) above $(f + g)(x) = f(x) + g(x) = \begin{cases} x + 1 & \text{if } x < 0 \\ |x - 1| & \text{if } x \geq 0 \end{cases}$

$(f + g)(0) = |0 - 1| = 1$

$\lim\limits_{x \to 0} (f + g)(x) = ?$: $(f + g)(x)$ changes cases at $x = 0$, so consider one-sided limits.

$$\lim_{x \to 0^-} (f + g)(x) = \lim_{x \to 0^-} (x + 1) = 0 + 1 = 1$$

$$\uparrow$$

$$(x < 0)$$

$$\lim_{x \to 0^+} (f + g)(x) = \lim_{x \to 0^+} |x - 1| = \lim_{x \to 0^+} [-(x - 1)] = -(0 - 1) = 1$$

$$\uparrow \qquad\qquad\qquad\qquad\qquad \uparrow$$

$$(x > 0) \qquad\qquad (x \text{ near } 0, \text{ so } x < 1 \implies x - 1 < 0)$$

Hence, $\lim_{x \to 0}(f + g)(x) = 1 = (f + g)(0) \implies f + g$ is continuous at $x = 0$.

Note! If f and g are continuous at a, then Theorem (1.19) indicates what to expect. If f and/or g are discontinuous at a, then Theorem (1.19) does not indicate what to expect.

Definition (1.18): A function f is **continuous on the open interval** (a, b) [or (a, ∞) or $(-\infty, b)$ or $(-\infty, \infty)$] if it is continuous at every number in the interval. It is **continuous on the closed interval** $[a, b]$ if it is continuous on (a, b) and is also continuous from the right at a and continuous from the left at b.

Similarly, the definition of continuity on intervals including only one endpoint requires continuity on the open interval and the appropriate one-sided continuity at the included endpoint. For instance, continuity on $[a, b)$ requires both

continuity on (a, b), and continuity from the right at a (included endpoint)

Recall: From Section 1.3, statement (1.7):

(1.7): If f is a polynomial or a rational function and a is in the domain of f, then

$$\lim_{x \to a} f(x) = f(a)$$

A restatement of this result yields:

Theorem (1.20):

(a) Any polynomial is continuous everywhere; that is, it is continuous on $\mathbf{R} = (-\infty, \infty)$.

(b) Any rational function is continuous wherever it is defined; that is, it is continuous on its domain.

Proof:

(a) Consider the polynomial

$$f(x) = c_n x^n + c_{n-1} x^{n-1} + \cdots + c_1 x + c_0, \qquad \text{where } c_i, \ i = 0, 1, 2, \cdots, n, \text{ is a constant.}$$

The domain of f is $(-\infty, \infty)$. Fix a in the domain of f.

$$\lim_{x \to a} f(x) = \lim_{x \to a} (c_n x^n + c_{n-1} x^{n-1} + \cdots + c_1 x + c_0)$$

$$= c_n \lim_{x \to a} x^n + c_{n-1} \lim_{x \to a} x^{n-1} + \cdots + c_1 \lim_{x \to a} x + \lim_{x \to a} c_0$$
↑
(Limit Laws 1 and 3)

$$= c_n a^n + c_{n-1} a^{n-1} + \cdots + c_1 a + c_0 = f(a)$$
↑
(Limit Laws 9 and 7)

$$\implies \ f \text{ is continuous at } a \implies f \text{ is continuous on } (-\infty, \infty)$$
↑
$$(a: \text{any real number in } (-\infty, \infty))$$

(b) Consider the rational function $f(x) = \dfrac{P(x)}{Q(x)}$, where P and Q are polynomials.

The domain of f is $D = \{x \mid Q(x) \neq 0\}$.

Fix $a \in D$. Then $Q(a) \neq 0 \implies f = \dfrac{P}{Q}$ is continuous at a
↑
$$\left(P, Q: \text{polynomials} \implies \underline{P, Q: \text{continuous at } a}; \text{ apply Theorem (1.19) 5} \right)$$
$$(\text{part (a) of theorem})$$

$\implies$ f is continuous at each $a \in D$

$\uparrow$

(a: any real number in D)

Example 5: Find the value(s) of x at which $f(x) = \dfrac{x+1}{x^2-1}$ is discontinuous.

f is a rational function of the form $f(x) = \dfrac{P(x)}{Q(x)}$, where $P(x) = x+1$ and $Q(x) = x^2-1$.

$Q(x) = 0 \implies x^2 - 1 = 0 \implies (x-1)(x+1) = 0 \implies x = 1,\ -1$

$\implies$ f is discontinuous at $x = 1,\ -1$.

$\uparrow$

(Theorem $(1.20)(b)$)

Note!

(N1) Consider $x = -1$.

$$\lim_{x \to -1} f(x) = \lim_{x \to -1} \frac{x+1}{x^2-1} \underset{\uparrow}{=} \lim_{x \to -1} \frac{x+1}{(x-1)(x+1)} \underset{\uparrow}{=} \lim_{x \to -1} \frac{1}{x-1} = \frac{1}{-1-1} = -\frac{1}{2}$$

$\left(\dfrac{0}{0} : \text{do}\right)$ $(x \neq -1,\ \text{but near } x = -1)$

$\implies$ $x = -1$: removable discontinuity.

(N2) Consider $x = 1$.

$\lim_{x \to 1} f(x) = \lim_{x \to 1} \dfrac{x+1}{x^2-1}$ does not exist $\left(\text{of the form } \dfrac{\neq 0}{0}\right)$

$\implies$ $x = 1$: infinite discontinuity.

(N3) Sketch the graph of $y = \dfrac{x+1}{x^2-1}$.

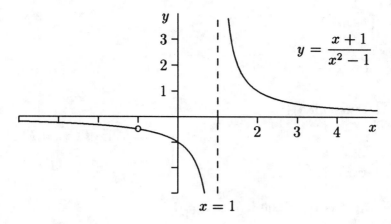

$$y = \frac{x+1}{x^2-1}$$

$$x = 1$$

Break/jump occurs at $x = -1, 1$.

Example 6: Consider the function $f(x) = \sqrt{1-x}$.

(A1) State the domain of f. *values of independent variable*

$$1 - x \geq 0 \implies x \leq 1 \quad \text{or, restated,} \quad x \in (-\infty, 1]. = \quad -\infty < x \leq 1$$

(A2) Show f is continuous on $(-\infty, 1)$.

Fix $a \in (-\infty, 1)$.

$$f(a) = \underbrace{\sqrt{1-a} \quad \text{is defined}}_{(1-a > 0)}$$

$$\lim_{x \to a} f(x) = \lim_{x \to a} \sqrt{1-x} = \sqrt{\lim_{x \to a}(1-x)} = \sqrt{1-a} = f(a)$$
$$\qquad\qquad\qquad\qquad \uparrow \qquad\qquad\quad \uparrow$$
$$\text{(Limit Law 11)} \qquad \underbrace{\left(P(x) = 1-x: \text{polynomial} \implies \lim_{x \to a} P(x) = P(a) \right)}_{\text{(Thm (1.20)(a))}}$$

$$\implies f \text{ is continuous at } a \implies f \text{ is continuous on } (-\infty, 1).$$
$$\uparrow$$
$$(a: \text{any } a \in (-\infty, 1))$$

(A3) Show f is continuous from the left at $x = 1$.

1 is rt. endpoint

$$f(1) = \sqrt{1-1} = \sqrt{0} = 0$$

$$\lim_{x \to 1^-} f(x) = \lim_{x \to 1^-} \sqrt{1-x} = \sqrt{\lim_{x \to 1^-}(1-x)} = \sqrt{1-1} = 0 = f(1)$$

$\uparrow$

(Limit Law 11 (one-sided))

$\uparrow$

$$\left(\begin{array}{l} P(x) = 1-x\text{: polynomial} \\ \implies \lim_{x \to 1} P(x) = P(1) \\ \implies \lim_{x \to 1^-} P(x) = P(1) \end{array} \right)$$

$\implies$ f is continuous from the left at $x = 1$.

Note! Combining (A2) and (A3): f is continuous on $(-\infty, 1]$.

(A4) Sketch the graph of $y = \sqrt{1-x}$.

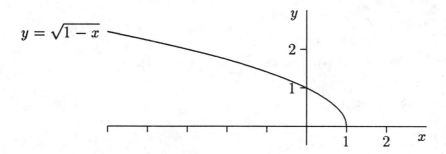

Note!

(N1) The graph of $y = \sqrt{1-x}$ is the upper half of the graph of the parabola $x = 1-y^2$.

(N2) No breaks or jumps occur in the graph.

Theorem (1.22):

If n is a positive even integer, then $f(x) = \sqrt[n]{x}$ is continuous on $[0, \infty)$.

If n is a positive odd integer, then f is continuous on $(-\infty, \infty)$.

Remarks: Consider $f(x) = \sqrt[n]{x}$, where n is a positive integer.

(R1) For n even, the domain of f is $[0, \infty)$ and the proof is similar to the procedure given in Example 6, (A2) and (A3), above.

(R2) For n odd, the domain of f is $(-\infty, \infty)$.

(R3) A sketch of $y = f(x)$ for $n = 2$:

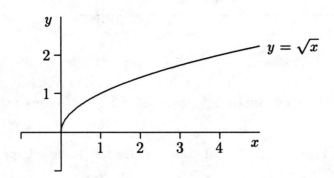

Note!

(N1) The graph of $y = \sqrt{x}$ is the upper half of the graph of $x = y^2$.

(N2) No breaks or jumps occur in $[0, \infty)$.

(R4) A sketch of $y = f(x)$ for $n = 3$.

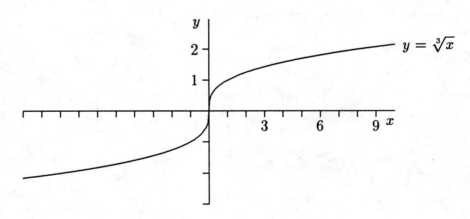

Note! No breaks or jumps occur in $(-\infty, \infty)$.

Recall: If f is continuous at a, then $\lim\limits_{x \to a} f(x) = f(a)$

$\Longrightarrow$ $\lim\limits_{x \to a} f(x) = f\left(\lim\limits_{x \to a} x\right)$: "moved the limit through."

$\uparrow$

(Limit Law 8)

Theorem (1.23): If f is continuous at b and $\lim\limits_{x \to a} g(x) = b$, then

$$\lim_{x \to a} f(g(x)) = f(b) = f(\lim_{x \to a} g(x))$$

Remarks:

(R1) When finding the limit of composite functions, the theorem indicates one can move the limit through the outer function f provided f is continuous at the limit of the inner function g.

(R2) For a proof of Theorem (1.23) using an ϵ, δ argument, see Appendix C.

Example 7: Consider $f(x) = x^2 - 1$ and $g(x) = \dfrac{x^2 - 4}{x + 1}$.

(A1) $\lim\limits_{x \to 1} f(g(x)) = ?$

$$\lim_{x \to 1} f(g(x)) = f\left(\lim_{x \to 1} g(x)\right)$$
$\uparrow$

(f: polynomial, so continuous everywhere)

$$= f(g(1))$$
$\uparrow$

$$\left(\begin{array}{l} g\text{: rational function with domain } (-\infty, -1) \cup (-1, \infty) \\ \implies \quad g \text{ continuous at } x = 1 \quad \implies \quad \lim_{x \to 1} g(x) = g(1) \end{array}\right)$$

$$= f\left(-\frac{3}{2}\right) = \left(-\frac{3}{2}\right)^2 - 1 = \frac{5}{4}$$
$\uparrow$

$$\left(g(1) = -\frac{3}{2}\right)$$

(A2) $\lim\limits_{x \to 1} g(f(x)) = ?$

$$\lim_{x \to 1} g(f(x)) = g\left(\lim_{x \to 1} f(x)\right)$$
$\uparrow$

$$\left(\text{provided } g \text{ continuous at } \lim_{x \to 1} f(x)\right)$$

$$= g(f(1))$$
↑
$$\left(\text{f: polynomial, so continuous everywhere} \implies \lim_{x \to 1} f(x) = f(1) \right)$$

$$= g(0) = \frac{0^2 - 4}{0 + 1} = -4$$
↑
$$\left(\begin{array}{l} \text{g: rational function continuous on its domain } (-\infty, -1) \cup (-1, \infty); \\ f(1) = 1^2 - 1 = 0 \quad \text{in domain of } g \end{array} \right)$$

Example 8: Consider $f(x) = |x|$.

(A1) State the domain of f.

$$f(x) = \left\{ \begin{array}{ll} x & \text{if } x \geq 0 \\ -x & \text{if } x < 0 \end{array} \right. \implies \text{domain of } f \text{ is } (-\infty, \infty).$$

(A2) Show f is continuous on its domain.

Fix $a > 0$:

$$\lim_{x \to a} f(x) = \lim_{x \to a} |x| = \lim_{x \to a} x = a = |a| = f(a) \implies f \text{ continuous on } (0, \infty).$$
$$\qquad\qquad\qquad\qquad\quad \uparrow \qquad\qquad \uparrow$$
$$(a > 0, x \text{ near } a: x > 0) \qquad (a > 0)$$

Fix $a < 0$:

$$\lim_{x \to a} f(x) = \lim_{x \to a} |x| = \lim_{x \to a} (-x) = -a = |a| = f(a) \implies f \text{ continuous on } (-\infty, 0).$$
$$\qquad\qquad\qquad\qquad\quad \uparrow \qquad\qquad\quad \uparrow$$
$$(a < 0, x \text{ near } a: x < 0) \qquad (a < 0)$$

Let $a = 0$. Use one-sided limits.

$$\lim_{x \to 0^+} f(x) = \lim_{x \to 0^+} |x| = \lim_{x \to 0^+} x = 0$$
$$\qquad\qquad\qquad\qquad \uparrow$$
$$(x > 0)$$

Absolute values are always continuous

$$\lim_{x \to 0^-} f(x) = \lim_{x \to 0^-} |x| = \lim_{x \to 0^-} (-x) = 0$$

$$\underset{\uparrow}{} $$
$$(x < 0)$$

Hence, $\lim_{x \to 0} f(x) = 0 = f(0) \implies f$ is continuous at $x = 0$

Thus f is continuous on $(-\infty, \infty)$.

Example 9: $\lim_{x \to -1} \left| \dfrac{x+1}{x^2-1} \right| = ?$

Let $f(g(x)) = \left| \dfrac{x+1}{x^2-1} \right|$, where $f(x) = |x|$ and $g(x) = \dfrac{x+1}{x^2-1}$.

$$\lim_{x \to -1} \left| \dfrac{x+1}{x^2-1} \right| \underset{\uparrow}{=} \left| \lim_{x \to -1} \dfrac{x+1}{x^2-1} \right|$$

$(f(x) = |x|$: continuous everywhere)

$$\underset{\uparrow}{=} \left| \lim_{x \to -1} \dfrac{x+1}{(x-1)(x+1)} \right| \underset{\uparrow}{=} \left| \lim_{x \to -1} \dfrac{1}{x-1} \right|$$

$\left(\dfrac{0}{0} : \text{do} \right)$ $(x \neq -1 \implies x+1 \neq 0)$

$$\underset{\uparrow}{=} \left| \dfrac{1}{-1-1} \right| = \left| -\dfrac{1}{2} \right| = \dfrac{1}{2}$$

$\left(\dfrac{1}{x-1} : \text{rational function continuous at } x = -1 \right)$

Note! A sketch of the graph of $y = \left| \dfrac{x+1}{x^2-1} \right|$.

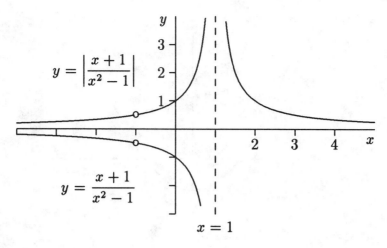

$$x > 1 : \quad \left| \frac{x+1}{x^2-1} \right| = \frac{x+1}{x^2-1}$$

Consider $f(x) = \sqrt[n]{x}$, where n is a positive integer.

Then $f(g(x)) = \sqrt[n]{g(x)}$ and, provided all the roots exist, $\lim\limits_{x \to a} \sqrt[n]{g(x)} = \sqrt[n]{\lim\limits_{x \to a} g(x)}$.

$\Big($Theorem (1.22): continuity of $\sqrt[n]{x}\Big)$

Example 10: $\lim\limits_{x \to a} \sqrt[n]{x^2 - 1} = ?$

(A1) $a = 3, \quad n = 2.$

$$\lim\limits_{x \to 3} \sqrt{x^2 - 1} = ?$$

Find the domain of $\sqrt{x^2 - 1} \implies$ determine values of x such that $x^2 - 1 \ge 0$:

(W1) Sketch graph of $y = x^2 - 1$: a parabola opening upward

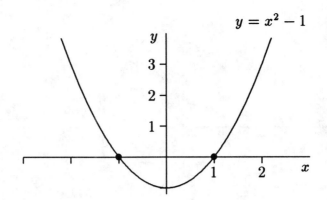

$$x^2 - 1 \geq 0 \implies x \leq -1 \text{ or } x \geq 1$$

(W2) $x^2 - 1 = 0 \implies (x-1)(x+1) = 0 \implies x = 1, -1$

$$\begin{array}{ccccc} + & 0 & - & 0 & + \\ \end{array} \qquad (x-1)(x+1)$$

$$\begin{array}{cc} -1 & 1 \end{array} \qquad x$$

$$x^2 - 1 = (x-1)(x+1) \geq 0 \implies x \leq -1 \text{ or } x \geq 1$$

Domain of $\sqrt{x^2 - 1}$: $(-\infty, -1] \cup [1, \infty)$.

$$\lim_{x \to 3} \sqrt{x^2 - 1} = \sqrt{\lim_{x \to 3} (x^2 - 1)} = \sqrt{3^2 - 1} = \sqrt{8}$$

$(x^2 - 1 \geq 0$ near $x = 3)$ $(x^2 - 1$: polynomial)

(A2) $a = 0, \quad n = 3.$

$$\lim_{x \to 0} \sqrt[3]{x^2 - 1} = ?$$

Find the domain of $\sqrt[3]{x^2 - 1}$:

$x^2 - 1$, a polynomial, exists everywhere $\implies \sqrt[3]{x^2 - 1}$ exists everywhere

$(n = 3$: odd)

$$\lim_{x \to 0} \sqrt[3]{x^2 - 1} = \sqrt[3]{\lim_{x \to 0}(x^2 - 1)} = \sqrt[3]{0^2 - 1} = -1$$

$$\uparrow \qquad\qquad\qquad\qquad \uparrow$$

(domain of $x^2 - 1$ is $(-\infty, \infty)$) ($x^2 - 1$: polynomial)

Theorem (1.24): If g is continuous at a and f is continuous at $g(a)$, then $(f \circ g)(x) = f(g(x))$ is continuous at a.

Remark: In words, the composite of continuous functions is continuous.

Proof:

g continuous at a $\implies$ $\lim_{x \to a} g(x) = g(a)$

f continuous at $g(a)$ $\implies$ $\lim_{x \to a} f(g(x)) = f(g(a))$ $\qquad$ ($b = g(a)$ in Thm. (1.23))

$\implies$ $\lim_{x \to a}(f \circ g)(x) = (f \circ g)(a)$ $\implies$ $f \circ g$ is continuous at $x = a$.

Example 11: Consider $h(x) = \left| \dfrac{x + 1}{x^2 - 1} \right|$.

(A1) State the domain of h.

$x^2 - 1 = 0$ $\implies$ $x = 1, \ -1$

$\implies$ domain of h: $(-\infty, -1) \cup (-1, 1) \cup (1, \infty)$.

(A2) State why h is continuous on its domain.

$h(x) = f(g(x))$, where $f(x) = |x|$ and $g(x) = \dfrac{x + 1}{x^2 - 1}$.

g: rational function is continuous on its domain $(-\infty, -1) \cup (-1, 1) \cup (1, \infty)$.

f: continuous everywhere.

Hence, using Theorem (1.24), $f(g(x))$ is continuous on its domain.

Example 12: Consider $h(x) = x + \dfrac{1}{\sqrt{x-1}}$.

(A1) State the domain of h.

$\sqrt{x-1}$ requires $x - 1 \geq 0 \implies x \geq 1$

$\dfrac{1}{\sqrt{x-1}}$ requires $x - 1 > 0 \implies x > 1$

$\implies$ domain of h: $(1, \infty)$

(A2) State why h is continuous on its domain.

$x - 1$: polynomial, so continuous everywhere

$\implies \sqrt{x-1}$ is continuous for $x - 1 \geq 0$ or, restated, $x \geq 1$.
$\uparrow$
(Theorem (1.22) and Theorem (1.24))

$\implies \dfrac{1}{\sqrt{x-1}}$ is continuous for $x > 1$
$\uparrow$
($f(x) = 1$: continuous everywhere; Theorem (1.19) 5)

$\implies h(x) = x + \dfrac{1}{\sqrt{x-1}}$ is continuous for $x > 1$
$\uparrow$
(x: polynomial, so continuous everywhere; Theorem (1.19) 1)

The following is a property of functions continuous on a closed interval.

The Intermediate Value Theorem (1.25): Suppose that f is continuous on the closed interval $[a, b]$ and let N be any number strictly between $f(a)$ and $f(b)$. Then there exists a number c in (a, b) such that $f(c) = N$.

Remarks:

(R1) Graphically:

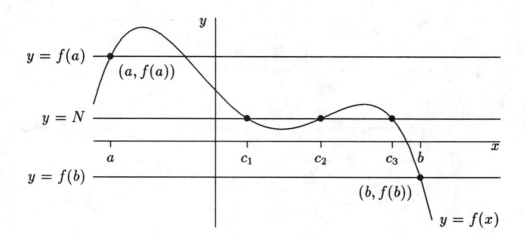

If $y = N$ is any line between the horizontal lines $y = f(a)$ and $y = f(b)$, then the graph of $y = f(x)$ must intersect $y = N$ for at least one value of x in (a, b), call it c (i.e., $f(c) = N$ for $c \in (a, b)$).

Note! For the chosen intermediate value N in the sketch above, there are three values of x in (a, b), denoted by c_1, c_2, c_3, such that $f(c_i) = N$, $i = 1, 2, 3$.

(R2) If $[a, b]$ is contained in the domain of f, but f is not continuous on $[a, b]$, then:

(C1) the conclusion of the Intermediate Value Theorem could hold.

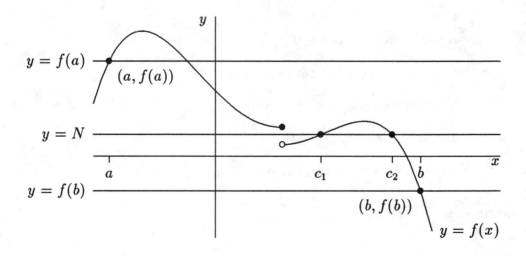

There exist c_1, $c_2 \in (a, b)$ with $f(c_1) = N = f(c_2)$.

(C2) the conclusion of the Intermediate Value Theorem need not hold.

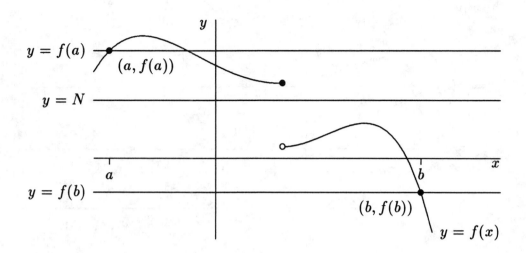

There is no x in (a, b) such that $f(x) = N$.

The following corollary is an application of the Intermediate Value Theorem to solving for roots of equations.

Corollary: Let f be continuous on $[a, b]$. If $f(a)f(b) < 0$, then there exists $c \in (a, b)$ such that $f(c) = 0$.

Remarks:

(R1) c is a root of the equation $f(x) = 0$.

(R2) $f(a)f(b) < 0 \implies f(a)$, $f(b)$ have opposite signs

  (i.e., one is positive while the other is negative)

  $\implies N = 0$ is an intermediate value

  $\implies$ there exists at least one $c \in (a, b)$ such that $f(c) = 0$.

  $\uparrow$

  (Intermediate Value Theorem)

(R3) Graphically:

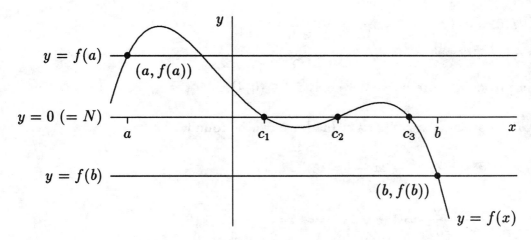

$f(x) = 0$ for three values c_1, c_2, $c_3 \in (a, b)$.

Example 13: Consider $f(x) = x^2 - 2x - 1$.

(A1) Show that there is a number c such that $f(c) = 2$.

Given: $N = 2$.

Choose a and b such that N is between $f(a)$ and $f(b)$: a choice is $\left\{ \begin{array}{l} f(-2) = 7 \\ f(0) = -1 \end{array} \right\}$ with

$7 = f(-2) > 2 > f(0) = -1$
$\quad\;\uparrow \qquad\;\; \uparrow \qquad\quad\; \uparrow$
$\quad (a) \quad\;\; (N) \qquad (b)$

$f(x)$: polynomial, so continuous everywhere $\implies$ f is continuous on $[-2, 0]$

By Intermediate Value Theorem: there exists $c \in (-2, 0)$ such that $f(c) = 2 \; (= N)$.

Note! For the given $f(x)$ a value for c can be found:

$f(x) = 2 \implies x^2 - 2x - 1 = 2 \implies x^2 - 2x - 3 = 0$

$\implies (x - 3)(x + 1) = 0 \implies x = 3, \; -1 \implies c = -1 \in (-2, 0)$

(A2) Does $f(x)$ have a zero in $[0, 4]$?

$f(x)$: polynomial, so continuous on $[0, 4]$

$$\left\{ \begin{array}{l} f(0) = -1 \\ f(4) = 7 \end{array} \right\} \implies f(0)f(4) < 0$$

By preceding corollary: there exists $c \in (0,4)$ such that $f(c) = 0 \; (= N)$.

Note! For the given $f(x)$ a value for c can be found:

$$f(x) = 0 \implies x^2 - 2x - 1 = 0$$

$$\implies \underset{\uparrow}{x} = \frac{2 \pm \sqrt{8}}{2} = 1 \pm \sqrt{2} \implies \underset{\uparrow}{c} = 1 + \sqrt{2} \in (0,4)$$

(using the quadratic formula) $\left(1 - \sqrt{2} < 0\right)$

(A3) Sketch $y = f(x)$ and display the solutions to (A1) and (A2).

$y = x^2 - 2x - 1$:   parabola opening upward

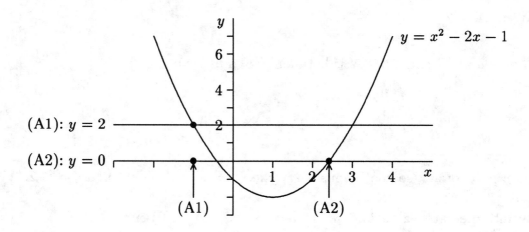

(A1): $(a, b) = (-2, 0)$ with $c = -1$

(A2): $(a, b) = (0, 4)$ with $c = 1 + \sqrt{2}$

1.6 Tangents, Velocities, and Other Rates of Change

Now that limits have been defined and related properties and techniques have been introduced, this section returns to the ideas of rate of change encountered in Section 1.1.

Definition (1.26): The **tangent line** to the curve $y = f(x)$ at the point $P(a, f(a))$ is the line through P with slope

$$m = \lim_{x \to a} \frac{f(x) - f(a)}{x - a}$$

provided that this limit exists.

Example 1: Find an equation of the tangent line to the parabola $y = 3x^2 - 4x + 6$ at the point $P(2, 10)$.

Given: $f(x) = 3x^2 - 4x + 6$ and $(a, f(a)) = (2, 10) \implies a = 2$.

$$m = \lim_{x \to 2} \frac{f(x) - f(2)}{x - 2} = \lim_{x \to 2} \frac{(3x^2 - 4x + 6) - 10}{x - 2}$$

$$= \lim_{x \to 2} \frac{3x^2 - 4x - 4}{x - 2} = \lim_{x \to 2} \frac{(x - 2)(3x + 2)}{x - 2} = \lim_{x \to 2}(3x + 2) = 3(2) + 2 = 8$$

$$\left(\frac{0}{0} : \text{do} \right) \qquad\qquad (x \neq 2)$$

$\implies$ an equation of the tangent line:

$$\left(\text{point-slope form: } \left\{ \begin{array}{ll} \text{point} & P(2, 10) \\ \text{slope} & m = 8 \end{array} \right. \right)$$

$$y - 10 = 8(x - 2) \qquad \text{or, rewritten,} \qquad y = 8x - 6$$

Note! When finding $\lim_{x \to 2} \dfrac{3x^2 - 4x - 4}{x - 2} = ?$

$$\left(\frac{0}{0} : \text{do} \right)$$

Numerator equals zero $\implies$ $x = 2$ is a zero of the polynomial $3x^2 - 4x - 4$

$\implies$ $x - 2$ is a factor of $3x^2 - 4x - 4$; use long division (if necessary) to find the remaining factor: $(3x^2 - 4x - 4) \div (x - 2) = 3x + 2$

Another expression for m, obtained via the substitution $h = x - a$

$$\left\{ \begin{array}{l} x = a + h \\ h \to 0 \quad \text{as} \quad x \to a \end{array} \right\}, \quad \text{is}$$

$$m = \lim_{h \to 0} \frac{f(a + h) - f(a)}{h}, \qquad \text{provided this limit exists.}$$

Remarks:

(R1) Geometrically: fix the point value a; if $h = x - a$, the point $Q(x, f(x))$ can also be designated by $Q(a + h, f(a + h))$

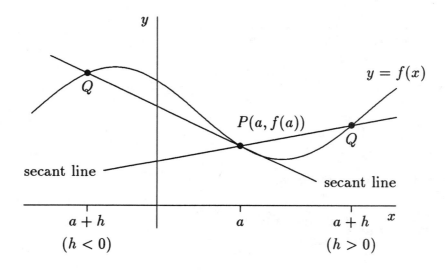

m_{PQ}: slope of secant line through P and Q

As $h \to 0$: $Q \to P$ and

$$m = \lim_{h \to 0} m_{PQ} = \lim_{h \to 0} \frac{f(a + h) - f(a)}{(a + h) - a} = \lim_{h \to 0} \frac{f(a + h) - f(a)}{h}$$

(R2) $m = \lim_{h \to 0} \dfrac{f(a + h) - f(a)}{h}$ may be easier to use.

Example 2: Redo Example 1 above via the h-expression for m.

Given: $f(x) = 3x^2 - 4x + 6$ and $(a, f(a)) = (2, 10)$ $\implies$ $a = 2$

$$m = \lim_{h \to 0} \frac{f(a+h) - f(a)}{h} = \lim_{h \to 0} \frac{f(2+h) - f(2)}{h}$$

$$= \lim_{\substack{h \to 0 \\ \uparrow}} \frac{[3(2+h)^2 - 4(2+h) + 6] - [3(2)^2 - 4(2) + 6]}{h}$$

(substitute into $f(x)$)

$$= \lim_{h \to 0} \frac{12 + 12h + 3h^2 - 8 - 4h + 6 - (12 - 8 + 6)}{h}$$

$$= \lim_{\substack{h \to 0 \\ \uparrow}} \frac{3h^2 + 8h}{h} = \lim_{h \to 0} \frac{h(3h + 8)}{h}$$

(all "non h-terms" in numerator add out)

$$= \lim_{\substack{h \to 0 \\ \uparrow}} (3h + 8) = 3(0) + 8 = 8 \qquad \text{(as in Example 1 above)}$$

($h \neq 0$: must divide out original h in denominator)

$$\implies y - 10 = 8(x - 2) \quad \text{or, rewritten,} \quad y = 8x - 6\text{: an equation of tangent line}$$
$$\uparrow$$

(point-slope form)

Example 3: Find the slope of the tangent line to the curve $y = \dfrac{1}{\sqrt{1 - x}}$ when the x coordinate is -3.

Given: $f(x) = \dfrac{1}{\sqrt{1 - x}}$ and $a = -3$

$$m = \lim_{\substack{h \to 0 \\ \uparrow}} \frac{f(-3 + h) - f(-3)}{h} = \lim_{h \to 0} \frac{\dfrac{1}{\sqrt{1 - (-3 + h)}} - \dfrac{1}{\sqrt{1 - (-3)}}}{h}$$

($a = -3$)

$$= \lim_{h \to 0} \frac{\dfrac{1}{\sqrt{4-h}} - \dfrac{1}{2}}{h} = \lim_{h \to 0} \frac{2 - \sqrt{4-h}}{2h\sqrt{4-h}}$$

$$\uparrow$$

$$\left(\frac{0}{0} : \text{do} \right)$$

$$= \lim_{h \to 0} \frac{2 - \sqrt{4-h}}{2h\sqrt{4-h}} \cdot \frac{2 + \sqrt{4-h}}{2 + \sqrt{4-h}} \qquad \text{(rationalize numerator)}$$

$$= \lim_{h \to 0} \frac{h}{2h\sqrt{4-h}(2 + \sqrt{4-h})} \qquad \text{(add out all non } h\text{-terms in numerator)}$$

$$= \lim_{h \to 0} \frac{1}{2\sqrt{4-h}(2 + \sqrt{4-h})} \qquad \text{(divide out original } h \text{ in denominator)}$$

$$= \frac{1}{2\sqrt{4}(2 + \sqrt{4})} = \frac{1}{2 \cdot 2 \cdot 4} = \frac{1}{16}$$

Example 4: Find the slopes of the tangent lines to the graph of $y = \dfrac{x+1}{x}$ at the x coordinates -1, 1, and 3.

Since more than one slope for the given function is requested, find the slope for a general x coordinate a (in the domain of the function).

Given: $f(x) = \dfrac{x+1}{x}$

$$m = \lim_{h \to 0} \frac{f(a+h) - f(a)}{h} = \lim_{h \to 0} \frac{\dfrac{a+h+1}{a+h} - \dfrac{a+1}{a}}{h}$$

$$= \lim_{h \to 0} \frac{a(a+h+1) - (a+h)(a+1)}{h(a+h)a} \qquad \left(\frac{0}{0} : \text{do} \right)$$

$$= \lim_{h \to 0} \frac{a^2 + ah + a - (a^2 + a + ah + h)}{ah(a+h)}$$

$$= \lim_{h \to 0} \frac{-h}{ah(a+h)} \qquad \text{(add out all non } h\text{-terms in numerator)}$$

$$= \lim_{h \to 0} \frac{-1}{a(a+h)} = -\frac{1}{a^2}$$
$\uparrow$

(divide out original h in denominator)

$$a = -1: \quad m = -\frac{1}{(-1)^2} = -1$$

$$a = 1: \quad m = -\frac{1}{1^2} = -1$$

$$a = 3: \quad m = -\frac{1}{3^2} = -\frac{1}{9}$$

Note! Could have evaluated $\lim_{h \to 0} \dfrac{f(a+h) - f(a)}{h}$, separately, for each $a = -1, 1, 3$;

saved calculations by using a general a to find m in terms of a and, then, substituting the particualr values of a into the expression for m (after the limit has been taken).

Definition: Suppose an object moves along a coordinate line with its position at time t given by $s(t)$. The **instantaneous velocity** of the object when $t = a$, denoted $v(a)$, is

$$v(a) = \lim_{h \to 0} \frac{s(a+h) - s(a)}{h}$$

provided that this limit exists.

Remarks:

(R1) From Section 1.1: $\dfrac{s(a+h) - s(a)}{h}$ represents the average velocity of the object between the times $t = a$ and $t = a + h$.

(R2) $v(a) = m$, where m is the slope of the tangent line to the graph of $s = s(t)$ (in a t, s coordinate system) at the point $P(a, s(a))$.

Example 5: A stone is thrown upward so its height, in feet, above the ground is given by $s = 80 + 64t - 16t^2$ after t $(t \geq 0)$ seconds. Find the (instantaneous) velocity:

(A1) when the stone is thrown. $a = 0$

(A2) when the stone hits the ground.   when position = 0

(A3) after 3 seconds. $a = 3$

Since more than one velocity for the stone is requested, find the velocity for a general time $t = a$ (in the domain of $s(t)$).

Given: $s(t) = 80 + 64t - 16t^2$.

$$v(a) = \lim_{h \to 0} \frac{s(a + h) - s(a)}{h} = \lim_{h \to 0} \frac{80 + 64(a + h) - 16(a + h)^2 - (80 + 64a - 16a^2)}{h}$$

$$= \lim_{h \to 0} \frac{64h - 32ah - 16h^2}{h} = \lim_{h \to 0}(64 - 32a - 16h) = 64 - 32a.$$

(A1) When released: $t = 0 \implies a = 0$

$v(0) = 64 - 32(0) = 64$ ft/s.

(A2) Stone hits the ground $\implies s = 0$

$s = 0 \implies 80 + 64t - 16t^2 = 0$

$\implies -16(t^2 - 4t - 5) = -16(t - 5)(t + 1) = 0 \implies t = 5, -1 \implies t = 5 \ (\geq 0)$

$\implies v(5) = 64 - 32(5) = -96$ ft/s

(A3) $a = 3$.

$v(3) = 64 - 32(3) = -16$ ft/s

Generalize the ideas of slope of the tangent line to a curve and the instantaneous velocity of an object (in rectilinear motion): $\underbrace{y = f(x)}$.

$(y$ depends upon x in some fashion$)$

Let x change from x_1 to x_2. The **increment** of x, denoted by Δx, is

$$\Delta x = \underbrace{x_2 - x_1};$$
$$((\text{to}) - (\text{from}))$$

the corresponding change in y, denoted by Δy, is

$$\Delta y = f(x_2) - f(x_1).$$

The **average rate of change of y with respect to x** between the values x_1 and x_2 is given by the **difference quotient**

$$\frac{\Delta y}{\Delta x} = \frac{f(x_2) - f(x_1)}{x_2 - x_1}.$$

Remark: $\dfrac{\Delta y}{\Delta x}$ corresponds to the slope of the secant line in graph of $y = f(x)$ or the average velocity in rectilinear motion with position by given by $y = f(x)$.

Definition: The **instantaneous rate of change of y with respect to x** at $x = x_1$ is given by

$$\lim_{\Delta x \to 0} \frac{\Delta y}{\Delta x} = \lim_{\Delta x \to 0} \frac{f(x_2) - f(x_1)}{x_2 - x_1} = \lim_{\Delta x \to 0} \frac{f(x_1 + \Delta x) - f(x_1)}{\Delta x},$$

where $\Delta x = x_2 - x_1$

Remark: $\displaystyle \lim_{\Delta x \to 0} \frac{\Delta y}{\Delta x}$ corresponds to the slope of the tangent line in graph of $y = f(x)$ at $x = x_1$ or the instantaneous velocity in rectilinear motion with position given by $y = f(x)$ when $x = x_1$, where Δx has replaced h.

Example 6: See Section 1.6, Example 5 in text on pp. 101, 102

2

Derivatives

2.1 Derivatives

> **Definition (2.2):** The **derivative of a function f at a number a**, denoted $f'(a)$, is
> $$f'(a) = \lim_{h \to 0} \frac{f(a+h) - f(a)}{h}$$
> if this limit exists.

Remarks:

(R1) For a fixed number a, let $x = a + h \implies h = x - a$. Then
$$h \to 0 \iff x \to a,$$
so an equivalent expression for $f'(a)$ is
$$f'(a) = \lim_{x \to a} \frac{f(x) - f(a)}{x - a}$$
provided that this limit exists.

(R2) Interpretations of $f'(a)$:

    (I1) $f'(a) = m$, the slope of the tangent line to the graph of $y = f(x)$ at the point $(a, f(a))$.

    (I2) $f'(a) = v(a)$, the (instantaneous) velocity when time $t = a$ for rectilinear motion of an object with position given by $s = f(t)$.

    (I3) $f'(a)$ is the instantaneous rate of change of y with respect to x at $x = a$ if $y = f(x)$ and $\Delta y = f(x) - f(a)$, $\Delta x = x - a$:
$$f'(a) = \lim_{\Delta x \to 0} \frac{\Delta y}{\Delta x} \qquad \text{(from (R1) above).}$$

(R3) To find an equation of the tangent line in (R2) (I1), use the point-slope form: determine $m = f'(a)$ at the point $(a, f(a))$; then $y - f(a) = f'(a)(x - a)$.

(R4) In (R2) (I2) above, the **speed** of the object when $t = a$ is
$$\text{speed} = |v(a)| = |f'(a)| \qquad (\geq 0).$$

Example 1: Find an equation of the tangent line to the curve $y = \sqrt{1-x}$ at the point $(-3, 2)$.

Given: $f(x) = \sqrt{1-x};$ point $(-3, 2)$ $\implies$ $a = -3$

$$f'(-3) = \lim_{h \to 0} \frac{f(-3+h) - f(-3)}{h}$$

$$= \lim_{h \to 0} \frac{\sqrt{1-(-3+h)} - \sqrt{1-(-3)}}{h} = \lim_{h \to 0} \frac{\sqrt{4-h} - 2}{h}$$

$$\left(\frac{0}{0} : \quad \text{do; rationalize numerator} \right)$$

$$= \lim_{h \to 0} \frac{\sqrt{4-h} - 2}{h} \cdot \frac{\sqrt{4-h} + 2}{\sqrt{4-h} + 2}$$

$$= \lim_{h \to 0} \frac{4 - h - 4}{h(\sqrt{4-h} + 2)} = \lim_{h \to 0} \frac{-1}{\sqrt{4-h} + 2} = \frac{-1}{4} = -\frac{1}{4}$$

(add out non h-terms in numerator; divide out original h in denominator)

$$\implies \quad \underbrace{y - 2 = -\frac{1}{4}[x - (-3)]}_{\text{(point-slope form)}} \quad \implies \quad x + 4y - 5 = 0$$

Example 2: If an object is projected directly upward from the ground with a velocity of 96 ft/s, its height, in feet, after t seconds is given by $s = 96t - 16t^2$.

(A1) Find the velocity when $t = a$.

Given: $s(t) = 96t - 16t^2$

$$v(a) = s'(a) = \lim_{h \to 0} \frac{s(a+h) - s(a)}{h}$$

$$= \lim_{h \to 0} \frac{96(a+h) - 16(a+h)^2 - (96a - 16a^2)}{h}$$

$$= \lim_{h \to 0} \frac{96h - 32ah - 16h^2}{h} = \lim_{h \to 0}(96 - 32a - 16h) = 96 - 32a$$

(add out non h-terms in numerator; divide out original h in denominator)

(A2) Find the velocity when the object strikes the ground.

Strikes ground: $s = 0 \implies 96t - 16t^2 = 0 \implies -16t(t - 6) = 0$

$\implies t = 0,\ 6 \implies t = 6\ (> 0): \quad v(6) = s'(6) = 96 - 32(6) = -96\ $ ft/s
$$\uparrow$$
$$\text{(A1)}$$

Note! $t = 0$: initial time of projection.

$$s'(0) = 96 - 32(0) = 96 \quad \text{ft/s} \qquad\qquad \text{(as stated in given information)}$$
$$\uparrow$$
$$\text{(A1)}$$

(A3) Find the velocity after 3 s.

$$v(3) = s'(3) = 96 - 32(3) = 0\ \text{ft/s}$$
$$\uparrow$$
$$\text{(A1)}$$

Note! $t = 3$: object is at rest (i.e., at the instant $t = 3$, the object is not moving).

(A4) Sketch the graph of $s = s(t)$ in a t, s coordinate system.

$s = 96t - 16t^2$: a parabola opening downward

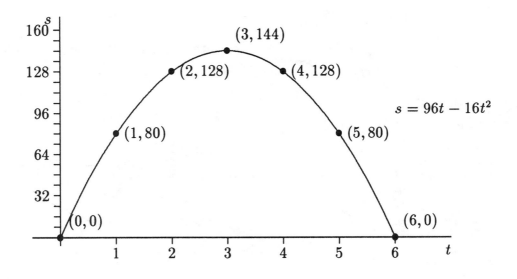

Note!

(N1) Object reaches maximum height when $t = 3$:

$$s(3) = 144 \text{ ft} \quad \text{with} \quad s'(3) = 0 \text{ ft/s} \quad \text{(object at rest)}$$

(N2) Graphically, the slope of the tangent line is positive for each $a \in (0,3)$.

$\implies$ $s'(a) > 0$ for each $a \in (0,3)$ $\implies$ $v(a) > 0$ for each $a \in (0,3)$:

object going upwards (i.e., height of the object is increasing) for $t \in (0,3)$.

Check: $v(a) = 96 - 32a > 0$ $\implies$ $96 > 32a$ $\implies$ $a < 3$ $\implies$ $a \in (0,3)$
            $\uparrow$
            (A1)

(N3) Graphically, the slope of the tangent line is negative for each $a \in (3,6)$.

$\implies$ $s'(a) < 0$ for each $a \in (3,6)$ $\implies$ $v(a) < 0$ for each $a \in (3,6)$:

object going downwards (i.e., height of the object is decreasing) for $t \in (3,6)$.

Check: $v(a) = 96 - 32a < 0$ $\implies$ $96 < 32a$ $\implies$ $a > 3$ $\implies$ $a \in (3,6)$
            $\uparrow$
            (A1)

Find the speed of the object after 4 s.

$$v(4) = 96 - 32(4) = -32 \text{ ft/s} \quad \implies \quad \text{speed when } t = 4: \quad |v'(4)| = |-32| = 32 \text{ ft/s}$$
$\uparrow$
(A1)

Given a function f, let $D = \{a \mid \underbrace{f'(a) \text{ exits}}\}$. Associate the value $f'(x)$ with each $x \in D$.

$$\left(\lim_{h \to 0} \frac{f(a+h) - f(a)}{h} \quad \text{exists} \right)$$

Then f' is a function with domain D.

Definition: The **derivative of f** is the function f' whose value at x is

$$f'(x) = \lim_{h \to 0} \frac{f(x+h) - f(x)}{h}$$

Remarks:

(R1) D, the domain of f' is a subset of the domain of f since the expression $f'(x)$ requires f to be defined at x.

(R2) x replaced a in the definition of $f'(a)$, so x is a fixed, yet arbitrary, element of D.

(R3) Sometimes refer to $f'(x)$ as the **derivative of $f(x)$ with respect to x**: specifies x as the independent variable.

(R4) The process of calculating $f'(x)$ from a given $f(x)$ is called **differentiation**.

(R5) Other notations for the derivative of $y = f(x)$:

$$f'(x) = y' = \frac{dy}{dx} = \frac{df}{dx} = \frac{d}{dx}\, f(x) = Df(x) = D_x f(x).$$

Example 3: Consider $f(x) = \sqrt{2-x} + 3$.

(A1) State the domain of f.

$$\sqrt{2-x} \text{ requires } 2 - x \geq 0 \implies x \leq 2: \qquad \text{domain of } f$$

(A2) Find $f'(x)$.

$$f'(x) = \lim_{h \to 0} \frac{f(x+h) - f(x)}{h}$$

$$= \lim_{h \to 0} \frac{\sqrt{2-(x+h)} + 3 - (\sqrt{2-x} + 3)}{h} = \lim_{h \to 0} \frac{\sqrt{2-x-h} - \sqrt{2-x}}{h}$$

$$= \lim_{h \to 0} \frac{\sqrt{2-x-h} - \sqrt{2-x}}{h} \cdot \frac{\sqrt{2-x-h} + \sqrt{2-x}}{\sqrt{2-x-h} + \sqrt{2-x}} \qquad \left(\frac{0}{0} : \text{ do; rat. num.}\right)$$

$$= \lim_{h \to 0} \frac{2-x-h-(2-x)}{h(\sqrt{2-x-h} + \sqrt{2-x})}$$

$$= \lim_{h \to 0} \frac{-h}{h(\sqrt{2-x-h} + \sqrt{2-x})} \qquad \text{(add out all non } h\text{-terms in numerator)}$$

$$= \lim_{h \to 0} \frac{-1}{\sqrt{2-x-h} + \sqrt{2-x}} \qquad \text{(divide out original } h \text{ in denominator)}$$

$$= \frac{-1}{\sqrt{2-x}+\sqrt{2-x}} = -\frac{1}{2\sqrt{2-x}}$$

(A3) State the domain of f'.

$$\frac{1}{\sqrt{2-x}} : \begin{cases} \sqrt{2-x} & \text{requires } 2-x \geq 0 \implies x \leq 2 \\ 2-x & \text{in denominator requires } 2-x \neq 0 \implies x \neq 2 \end{cases}$$

$$\implies \quad x < 2: \quad \text{domain of } f'$$

Note! $\underbrace{\text{domain of } f'}_{(-\infty, 2)}$ is a proper subset of the $\underbrace{\text{domain of } f}_{(-\infty, 2]}$

(A4) Find an equation of the tangent line to the graph of $y = f(x)$ at $x = -2$.

$$\text{At } a = -2 : \begin{cases} f(-2) = \sqrt{2-(-2)} + 3 = 5 \\ f'(-2) = -\frac{1}{\underset{\uparrow}{2\sqrt{2-(-2)}}} = -\frac{1}{4} \\ \quad\quad\text{(A2)} \end{cases}$$

Using point-slope form: $\underbrace{y - 5 = -\frac{1}{4}[x-(-2)]}_{y - f(a) = f'(a)(x-a)} \implies x + 4y - 18 = 0$

Definition (2.7): A function f is **differentiable at a** if $f'(a)$ exists. It is **differentiable on an open interval** (a, b) [or (a, ∞) or $(-\infty, a)$ or $(-\infty, \infty)$] if it is differentiable at every number in the interval.

Remarks: Let $y = f(x)$.

(R1) The Leibniz notation $\dfrac{dy}{dx}$ for the derivative reflects the increment notation for the difference quotient:

$$\frac{dy}{dx} = \lim_{\Delta x \to 0} \frac{\Delta y}{\Delta x}.$$

(R2) If f is differentiable at a, Leibniz notation indicating the value $f'(a)$ is

$$\left.\frac{dy}{dx}\right|_{x=a} \qquad \text{or} \qquad \left.\frac{dy}{dx}\right]_{x=a}.$$

Example 4: Consider $f(x) = \dfrac{1}{x-2}$.

(A1) State the domain of f.

$\qquad$ $x - 2$ in denominator requires $x - 2 \neq 0$ $\implies$ domain of f: $(-\infty, 2) \cup (2, \infty)$

(A2) Show f is differentiable on the domain of f.

$$\frac{df}{dx} = \lim_{h \to 0} \frac{f(x+h) - f(x)}{h}$$

$$= \lim_{h \to 0} \frac{\dfrac{1}{x+h-2} - \dfrac{1}{x-2}}{h} = \lim_{h \to 0} \frac{x - 2 - (x + h - 2)}{h(x + h - 2)(x - 2)}$$

$$\underset{\uparrow}{=} \lim_{h \to 0} \frac{-1}{(x + h - 2)(x - 2)} = -\frac{1}{(x-2)^2} \qquad\qquad \text{which exists if } x - 2 \neq 0$$

$\qquad$ (add out non h-terms in numerator; divide out original h in denominator)

$\implies \dfrac{df}{dx}$ exists for x in $(-\infty, 2) \cup (2, \infty)$, the domain of f.

Example 5: Consider $f(x) = |x|$.

(A1) State the domain of f.

$$|x| = \begin{cases} x & \text{if } x \geq 0 \\ -x & \text{if } x < 0 \end{cases} \implies \text{domain of } f\text{: } (-\infty, \infty)$$

(A2) Sketch the graph of $y = f(x)$.

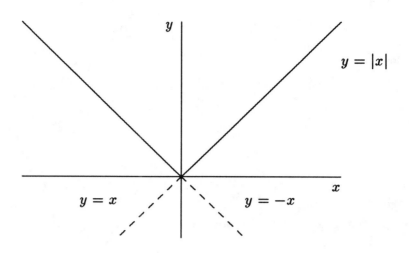

(A3) Where is f differentiable?

$$f'(x) = \lim_{h \to 0} \frac{f(x+h) - f(x)}{h} = \lim_{h \to 0} \frac{|x+h| - |x|}{h} = ?$$

$$\uparrow$$

$$\left(\frac{0}{0} : \text{ do}\right)$$

Note! Need to remove the absolute value symbol before dividing out original h in denominator.

(C1) Fix $x > 0$. $|x| = x$.

$$h \to 0 \implies h \text{ near } 0 \implies x + h > 0 \implies |x+h| = x + h$$

$$\uparrow$$

$$(x > 0)$$

$$\implies f'(x) = \lim_{h \to 0} \frac{x + h - x}{h} = \lim_{h \to 0} \frac{h}{h} = \lim_{h \to 0} 1 = 1$$

$$\implies f \text{ differentiable on } (0, \infty)$$

(C2) Fix $x < 0$. $|x| = -x$.

$$h \to 0 \implies h \text{ near } 0 \implies x + h < 0 \implies |x + h| = -(x + h)$$
$$\uparrow$$
$$(x < 0)$$

$$\implies f'(x) = \lim_{h \to 0} \frac{-(x + h) - (-x)}{h} = \lim_{h \to 0} \frac{-h}{h} = \lim_{h \to 0}(-1) = -1$$

$$\implies f \text{ differentiable on } (-\infty, 0)$$

(C3) Let $x = 0$.

Note! $f(x)$ changes cases at $x = 0$.

$$f'(0) = \lim_{h \to 0} \frac{|0 + h| - |0|}{h} = \lim_{h \to 0} \frac{|h|}{h}$$

Use one-sided limits:

$$\lim_{h \to 0^-} \frac{|h|}{h} = \lim_{h \to 0^-} \frac{-h}{h} = \lim_{h \to 0^-}(-1) = -1$$
$$\uparrow$$
$$(h < 0)$$

$$\lim_{h \to 0^+} \frac{|h|}{h} = \lim_{h \to 0^+} \frac{h}{h} = \lim_{h \to 0^+} 1 = 1$$
$$\uparrow$$
$$(h > 0)$$

$$\implies \lim_{h \to 0^-} \frac{|h|}{h} \neq \lim_{h \to 0^+} \frac{|h|}{h}$$

$$\implies \lim_{h \to 0} \frac{|h|}{h} \text{ does not exist } \implies f \text{ is not differentiable at } 0.$$

Combining cases (C1)-(C3): f is differentiable at all x except 0.

Note!

(N1) An equation of the line tangent to the graph of $y = |x|$ at $x = a$:

$$a > 0 \implies f'(a) = 1 \text{ (see (A3)(C1)): } \underbrace{y - |a| = (1)(x - a)}_{\text{(point-slope form)}} \implies y - a = x - a$$

$$\implies y = x, \text{ which coincides with } y = |x| \ (= f(x)) \text{ for } x > 0.$$

$$a < 0 \implies f'(a) = -1 \text{ (see (A3)(C2)):}$$

$$y - |a| = (-1)(x - a) \implies y - (-a) = -x + a$$

$$\implies y = -x, \text{ which coincides with } y = |x| \ (= f(x)) \text{ for } x < 0.$$

$a = 0$: $f'(0)$ does not exist (see (A3)(C3)) and the curve $y = |x|$ does not have a tangent line at $(0,0)$ (see graph in (A2)).

(N2) The domain of f' is $(-\infty, 0) \cup (0, \infty)$ and $f'(x) = \begin{cases} -1 & \text{if } x < 0 \\ 1 & \text{if } x > 0 \end{cases}$

or, rewritten, $f'(x) = \dfrac{|x|}{x}$

(N3) The interpretation of $f'(x)$ as the slope of the tangent line to the graph of $y = f(x)$ at $(x, f(x))$ can be viewed via the graphs sketched (on identical x, y coordinate systems with $y = f'(x)$ sketch directly beneath $y = f(x)$) below.

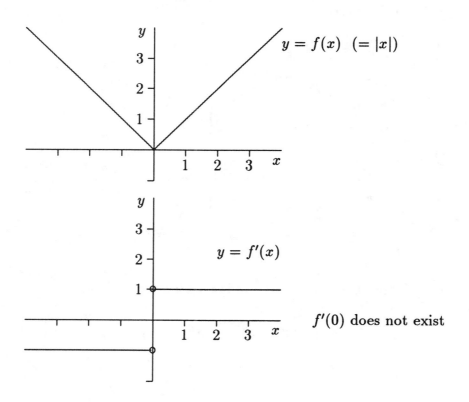

Example 6: Given the graph of $y = f(x)$, use the interpretation of $f'(x)$ as the slope of the tangent line to the graph of $y = f(x)$ at $(x, f(x))$ to sketch the graph of $y = f'(x)$ in the given x, y coordinate system.

(A1)

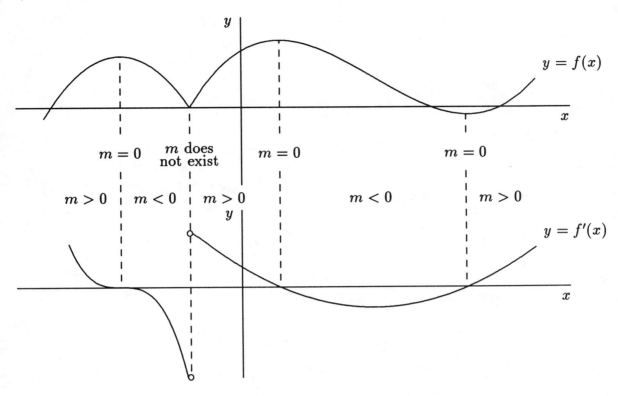

(A2)

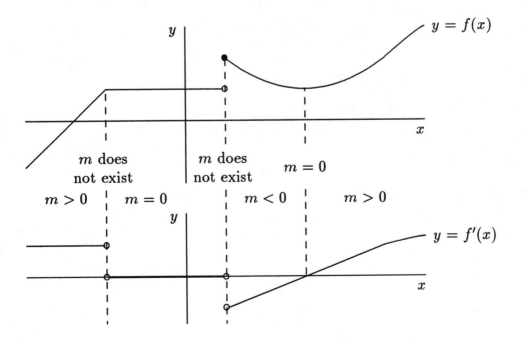

Theorem (2.8): If f is differentiable at a, then f is continuous at a.

Proof: Show $\displaystyle\lim_{x \to a} f(x) = f(a)$

Given f is differentiable at a $\implies$ $f'(a) = \displaystyle\lim_{x \to a} \frac{f(x) - f(a)}{x - a}$ exists

For $x \neq a$: $f(x) = f(x) + \underbrace{f(a) - f(a)}_{\text{(added 0)}} = f(a) + \dfrac{f(x) - f(a)}{x - a}\underset{\uparrow}{(x - a)}$

(rearranged; $x \neq a$: multiplied by 1)

$\implies \displaystyle\lim_{x \to a} f(x) = \lim_{x \to a} \left[f(a) + \frac{f(a) - f(a)}{x - a}(x - a) \right]$

$= \displaystyle\lim_{x \to a} f(a) + \lim_{x \to a} \frac{f(x) - f(a)}{x - a} \cdot \underset{\uparrow}{\lim_{x \to a}(x - a)} = f(a) + f'(a) \cdot 0 = f(a)$

($f(a)$ is a constant; $f'(a)$ exists)

Remarks:

(R1) Theorem (2.8) indicates the existence of the derivative yields continuity.

(R2) The converse of the theorem (i.e., If $f(x)$ is continuous at $x = a$, then $f'(a)$ exists.) is not true.

Example 7: Consider $f(x) = |x| + x$.

$$f(x) = \begin{cases} x + x = 2x & \text{if } x \geq 0 \\ -x + x = 0 & \text{if } x < 0 \end{cases}$$

A sketch of the graph of $y = f(x)$ is:

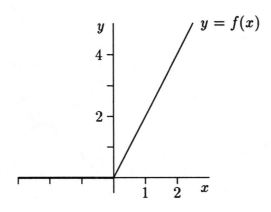

Consider $a = 0$.

Continuity: $f(0) = 2(0) = 0$ $\lim\limits_{x \to 0} f(x) = ?$

$$\left.\begin{array}{l} \lim\limits_{x \to 0^+} f(x) = \lim\limits_{x \to 0^+} 2x = 2(0) = 0 \\[2mm] \lim\limits_{x \to 0^-} f(x) = \lim\limits_{x \to 0^-} 0 = 0 \end{array}\right\} \implies \lim\limits_{x \to 0} f(x) = 0 = f(0): \quad \text{continuous}$$

Differentiability:

$$f'(0) = \lim_{h \to 0} \frac{f(0+h) - f(0)}{h} = \lim_{h \to 0} \frac{f(h) - 0}{h} = \lim_{h \to 0} \frac{f(h)}{h} = ?$$

$$\left.\begin{array}{l} \lim\limits_{h \to 0^+} \dfrac{f(h)}{h} = \lim\limits_{h \to 0^+} \dfrac{2h}{h} = \lim\limits_{h \to 0^+} 2 = 2 \\[3mm] \lim\limits_{h \to 0^-} \dfrac{f(h)}{h} = \lim\limits_{h \to 0^-} \dfrac{0}{h} = \lim\limits_{h \to 0^-} 0 = 0 \end{array}\right\} \implies \lim\limits_{h \to 0} \dfrac{f(h)}{h} \text{ does not exist}$$

$\implies$ f is not differentiable at 0.

Note! At $x = 0$: f is continuous, but not differentiable.

Example 8: Consider $f(x) = \begin{cases} 0 & \text{if } x < 0 \\ x^2 & \text{if } x \geq 0 \end{cases}$

A sketch of the graph of $y = f(x)$ is:

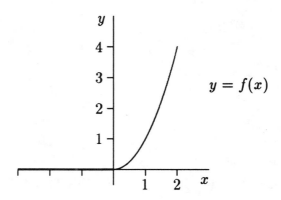

Consider $a = 0$.

Continuity: $f(0) = 0^2 = 0$ $\qquad\qquad \lim_{x \to 0} f(x) = ?$

$\left. \begin{array}{l} \displaystyle\lim_{x \to 0^-} f(x) = \lim_{x \to 0^-} 0 = 0 \\[3mm] \displaystyle\lim_{x \to 0^+} f(x) = \lim_{x \to 0^+} x^2 = 0^2 = 0 \end{array} \right\} \implies \lim_{x \to 0} f(x) = 0 = f(0): \quad \text{continuous}$

Differentiability:

$$f'(0) = \lim_{h \to 0} \frac{f(0+h) - f(0)}{h} = \lim_{h \to 0} \frac{f(h) - 0}{h} = \lim_{h \to 0} \frac{f(h)}{h} = ?$$

$\left. \begin{array}{l} \displaystyle\lim_{h \to 0^-} \frac{f(h)}{h} = \lim_{h \to 0^-} \frac{0}{h} = \lim_{h \to 0^-} 0 = 0 \\[5mm] \displaystyle\lim_{h \to 0^+} \frac{f(h)}{h} = \lim_{h \to 0^+} \frac{h^2}{h} = \lim_{h \to 0^+} h = 0 \end{array} \right\} \implies \lim_{h \to 0} \frac{f(h)}{h} = 0 = f'(0)$

$\implies$ f is differentiable at 0.

Note! At $x = 0$: f is both continuous and differentiable.

2.2   Differentiation Formulas

In this section some general rules will be developed from the definition of the derivative in order to provide efficient techniques for differentiation.

Theorem (2.9): If f is a constant function, $f(x) = c$, then $f'(x) = 0$.

Proof:

$$f'(x) = \lim_{h \to 0} \frac{f(x+h) - f(x)}{h} = \lim_{h \to 0} \frac{c - c}{h} = \lim_{h \to 0} \frac{0}{h} = \lim_{h \to 0} 0 = 0$$

$\qquad\qquad\qquad\qquad\qquad\uparrow\qquad\qquad\qquad\qquad\qquad\qquad\uparrow$

$\qquad (f(x) = c \implies f(x+h) = c)\qquad\qquad\qquad (h \neq 0)$

Remarks:

(R1) In Leibniz notation: $\dfrac{d}{dx} c = 0$

(R2) In words: the derivative of a constant is zero.

(R3) Graphically: $y = c$

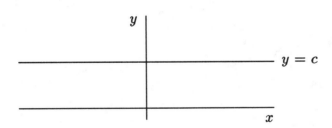

The slope of the tangent line to the graph of $y = c$ is 0 at each x.

The Power Rule (2.10): If n is a positive integer and $f(x) = x^n$, then

$$f'(x) = nx^{n-1}$$

Proof:

$$f'(x) = \lim_{h \to 0} \frac{f(x+h) - f(x)}{h} = \lim_{h \to 0} \frac{(x+h)^n - x^n}{h}$$

$$= \lim_{\substack{h \to 0 \\ \uparrow}} \frac{\left[x^n + nx^{n-1}h + \dfrac{n(n-1)}{2}x^{n-2}h^2 + \cdots + nxh^{n-1} + h^n \right] - x^n}{h}$$

(expanding $(x+h)^n$ via the Binomial Theorem: see Appendix A)

$$= \lim_{\substack{h \to 0 \\ \uparrow}} \frac{nx^{n-1}h + \dfrac{n(n-1)}{2}x^{n-2}h^2 + \cdots + nxh^{n-1} + h^n}{h}$$

(add out non h-terms in numerator)

$$= \lim_{\substack{h \to 0 \\ \uparrow}} \left[nx^{n-1} + \dfrac{n(n-1)}{2}x^{n-2}h + \cdots + nxh^{n-2} + h^{n-1} \right]$$

(divide out original h in denominator)

$$= nx^{n-1} + \dfrac{n(n-1)}{2}x^{n-2}(0) + \cdots + nx(0)^{n-2} + (0)^{n-1} = nx^{n-1}$$
$$\uparrow$$

(n and x: fixed)

Remarks:

(R1) In Leibniz notation: $\dfrac{d}{dx}(x^n) = nx^{n-1}$.

(R2) In words: the derivative (with respect to the variable) of a variable to a positive integer power is the power times the variable to the power minus one.

(R3) Consider $n = 1$: $f(x) = x$

$$f'(x) = \lim_{h \to 0} \frac{f(x+h) - f(x)}{h} = \lim_{h \to 0} \frac{(x+h) - x}{h} = \lim_{h \to 0} \frac{h}{h} = \lim_{h \to 0} 1 = 1 = 1x^0 = 1x^{1-1} = nx^{n-1}$$
$$\uparrow$$

(treating $x^0 = 1$ at each $x \in (-\infty, \infty)$)

Example 1: $f'(x) = ?$ if $f(x) = x^n$,   where;

(A1) $n = 7$.

$$f(x) = x^7 \implies f'(x) = 7x^{7-1} = 7x^6$$

(A2) $n = 23$.

$$f(x) = x^{23} \implies f'(x) = 23x^{23-1} = 23x^{22}$$

Example 2: $\dfrac{d}{dt}(t^5) = ?$

$$\frac{d}{dt}(t^5) = 5t^4 \qquad\qquad \text{(derivative with respect to } t \text{ of } t^5;\ n = 5 \implies n - 1 = 4)$$

Theorem (2.12): Suppose c is a constant and $f'(x)$ and $g'(x)$ exist.

(a) If $F(x) = cf(x)$, then $F'(x)$ exists and $F'(x) = cf'(x)$.

(b) If $G(x) = f(x) + g(x)$, then $G'(x)$ exists and $G'(x) = f'(x) + g'(x)$.

(c) If $H(x) = f(x) - g(x)$, then $H'(x)$ exists and $H'(x) = f'(x) - g'(x)$.

In short:

(a) $(cf)' = cf'$ (b) $(f + g)' = f' + g'$ (c) $(f - g)' = f' - g'$

Proof:

(a) See p. 118 of text.

(b) $G(x) = f(x) + g(x)$

$$G'(x) = \lim_{h \to 0} \frac{G(x + h) - G(x)}{h} = \lim_{h \to 0} \frac{[f(x + h) + g(x + h)] - [f(x) + g(x)]}{h}$$

$$= \lim_{h \to 0} \left[\frac{f(x + h) - f(x)}{h} + \frac{g(x + h) - g(x)}{h} \right]$$

$$= \lim_{h \to 0} \frac{f(x + h) - f(x)}{h} + \lim_{h \to 0} \frac{g(x + h) - g(x)}{h} = f'(x) + g'(x)$$

$$\uparrow$$
$$\text{(given: } f'(x) \text{ and } g'(x) \text{ exist)}$$

(c) $H(x) = f(x) - g(x) = f(x) + [-g(x)]$

$$\Longrightarrow \quad H'(x) = f'(x) + [-g(x)]' = f'(x) + (-1)g'(x) = f'(x) - g'(x)$$

$$\uparrow \qquad\qquad\qquad \uparrow$$

(Theorem (2.12)(b))   (Theorem (2.12)(a)): $c = -1$)

Remarks:

(R1) In Leibniz notation:

(a) $\dfrac{d}{dx}(cf) = c\dfrac{df}{dx}$ 　　　　　　(b) $\dfrac{d}{dx}(f+g) = \dfrac{df}{dx} + \dfrac{dg}{dx}$

(c) $\dfrac{d}{dx}(f-g) = \dfrac{df}{dx} - \dfrac{dg}{dx}$

(R2) In words:

(a) the derivative of a constant times a function is the constant times the derivative of the function;

(b) the derivative of a sum is the sum of the derivatives;

(c) the derivative of a difference is the corresponding difference of the derivatives.

(R3) Theorem (2.12)(b) can be extended to the sum of any number of functions.

Example 3: $f'(x) = ?$ if:

(A1) $f(x) = x^3 - 2x + 4$

$$f'(x) = \frac{d}{dx}(x^3 - 2x + 4) = \frac{d}{dx}(x^3) - 2\frac{d}{dx}(x) + \frac{d}{dx}(5)$$

$$= 3x^2 - 2(1) + 0 = 3x^2 - 2$$

(A2) $f(x) = (x^3 - x)(2x + 1)$

$$f'(x) = \frac{d}{dx}[(x^3 - x)(2x + 1)] = \frac{d}{dx}(2x^4 + x^3 - 2x^2 - x)$$

$$\uparrow$$

(no rule for derivative of products yet)

$$= 2(4x^3) + 3x^2 - 2(2x) - 1 = 8x^3 + 3x^2 - 4x - 1$$

$$\uparrow$$

(Theorem (2.12))

The Product Rule (2.13): If $F(x) = f(x)g(x)$ and $f'(x)$ and $g'(x)$ both exist, then

$$F'(x) = f(x)g'(x) + g(x)f'(x)$$

In short:
$$(fg)' = fg' + gf'$$

Proof:

$$F'(x) = \lim_{h \to 0} \frac{F(x+h) - F(x)}{h} = \lim_{h \to 0} \frac{f(x+h)g(x+h) - f(x)g(x)}{h}$$

$$= \lim_{h \to 0} \frac{f(x+h)g(x+h) - f(x+h)g(x) + f(x+h)g(x) - f(x)g(x)}{h}$$
$\uparrow$
(added 0)

$$= \lim_{h \to 0} \left[f(x+h)\frac{g(x+h) - g(x)}{h} + g(x)\frac{f(x+h) - f(x)}{h} \right] \qquad \text{(rearrange)}$$

$$= \lim_{h \to 0} f(x+h) \lim_{h \to 0} \frac{g(x+h) - g(x)}{h} + \lim_{h \to 0} g(x) \lim_{h \to 0} \frac{f(x+h) - f(x)}{h}$$

$$= f(x)g'(x) + g(x)f'(x)$$
$\uparrow$
$$\left(\begin{array}{l} f'(x) \text{ exists} \implies f \text{ continuous at } x \implies \lim_{h \to 0} f(x+h) = f(x); \\ x \text{ fixed} \implies \lim_{h \to 0} g(x) = g(x) \end{array} \right)$$

Remarks:

(R1) In Leibniz notation:
$$\frac{d}{dx}(fg) = f\frac{dg}{dx} + g\frac{df}{dx}$$

(R2) In words: the derivative of the product of two functions is the first function times the derivative of the second function plus the second function times the derivative of the first.

(R3) $\dfrac{d}{dx}(fg) \neq \dfrac{df}{dx}\dfrac{dg}{dx}$:

the derivative of a product is not the product of the derivatives.

(R4) See Exercise 59(a), Section 2.2, to extend the product rule to the derivative of the product of three functions: If f, g, and h are differentiable functions, then

$$\frac{d}{dx}(fgh) = fg\,\frac{dh}{dx} + f\,\frac{dg}{dx}\,h + \frac{df}{dx}\,gh$$

Note! Consider the pattern of the expression on the right: 3 functions: 3 terms summed; each term consists of the product of 3 functions: 2 of the given functions and the derivative of the remaining function such that the derivative of each given function occurs once.

Example 4: $f'(x) = ?$ if:

(A1) $f(x) = (x^3 - x)(2x + 1)$

$$f'(x) = (x^3 - x)\,\frac{d}{dx}\,(2x + 1) + (2x + 1)\,\frac{d}{dx}\,(x^3 - x)$$

$$= (x^3 - x)(2 + 0) + (2x + 1)(3x^2 - 1) = 2(x^3 - x) + (2x + 1)(3x^2 - 1)$$

$$= 2x^3 - 2x + 6x^3 + 3x^2 - 2x - 1 = 8x^3 + 3x^2 - 4x - 1,$$
↑

(if simplification is desired)

which agrees with the result of Example 3(A2) above (where multiplication was carried out prior to differentiation).

(A2) $f(x) = (x^3 + 5x)^2$

Note! $f(x)$ is not expressed as x to a power but as a different function of x to a power, so The Power Rule (2.10) is not applicable.

$$f(x) = (x^3 + 5x)^2 = \underbrace{(x^3 + 5x)(x^3 + 5x)}$$

(express f as a product)

$$\implies f'(x) = (x^3 + 5x)\,\frac{d}{dx}\,(x^3 + 5x) + (x^3 + 5x)\,\frac{d}{dx}\,(x^3 + 5x)$$

$$= 2(x^3 + 5x)\,\frac{d}{dx}\,(x^3 + 5x) = 2(x^3 + 5x)(3x^2 + 5)$$

Example 5: Find an equation of the tangent line to the graph of $y = (x^2+5x)(x^4+x+2)(1-x)$ at $x = -1$.

$$\frac{dy}{dx} = (x^2 + 5x)(x^4 + x + 2)\frac{d}{dx}(1 - x) + (x^2 + 5x)\left[\frac{d}{dx}(x^4 + x + 2)\right](1 - x)$$

$$+ \left[\frac{d}{dx}(x^2 + 5x)\right](x^4 + x + 2)(1 - x)$$

$$= (x^2 + 5x)(x^4 + x + 2)(-1) + (x^2 + 5x)(4x^3 + 1)(1 - x)$$

$$+ (2x + 5)(x^4 + x + 2)(1 - x)$$

$a = -1:\qquad f(-1) = -16\quad \text{and}\quad f'(-1) = 44$

$\implies \quad \underbrace{y - (-16) = 44[x - (-1)]}_{\text{(point-slope form)}} \quad \implies \quad y = 44x + 28$

The Quotient Rule (2.14): If $F(x) = f(x)/g(x)$ and $f'(x)$ and $g'(x)$ both exist, then $F'(x)$ exists and

$$F'(x) = \frac{g(x)f'(x) - f(x)g'(x)}{[g(x)]^2}$$

In short:

$$\left(\frac{f}{g}\right)' = \frac{gf' - fg'}{g^2}$$

Proof:

$$F'(x) = \lim_{h\to 0}\frac{F(x + h) - F(x)}{h} = \lim_{h\to 0}\frac{\dfrac{f(x + h)}{g(x + h)} - \dfrac{f(x)}{g(x)}}{h}$$

$$= \lim_{h\to 0}\frac{f(x + h)g(x) - f(x)g(x + h)}{hg(x + h)g(x)}$$

$$= \lim_{\substack{h\to 0 \\ \uparrow}}\frac{f(x + h)g(x) - f(x)g(x) + f(x)g(x) - f(x)g(x + h)}{hg(x + h)g(x)}$$

(added 0)

$$= \lim_{\substack{h \to 0 \\ \uparrow}} \left[\frac{g(x)\dfrac{f(x+h)-f(x)}{h} - f(x)\dfrac{g(x+h)-g(x)}{h}}{g(x+h)g(x)} \right]$$

(rearrange)

$$= \frac{\displaystyle \lim_{h\to 0} g(x) \lim_{h\to 0} \frac{f(x+h)-f(x)}{h} - \lim_{h\to 0} f(x) \lim_{h\to 0} \frac{g(x+h)-g(x)}{h}}{\displaystyle \lim_{h\to 0} g(x+h) \lim_{h\to 0} g(x)}$$

$$\underset{\uparrow}{=} \frac{g(x)f'(x) - f(x)g'(x)}{[g(x)]^2}$$

$$\left(\begin{array}{llll} x \text{ fixed} & \Longrightarrow & \lim_{h\to 0} g(x) = g(x); & x \text{ fixed} & \Longrightarrow & \lim_{h\to 0} f(x) = f(x) \\ g'(x) \text{ exists} & \Longrightarrow & g \text{ continuous at } x & \Longrightarrow & \lim_{h\to 0} g(x+h) = g(x) \end{array} \right)$$

Remarks:

(R1) In Leibniz notation:

$$\frac{d}{dx}\left(\frac{f}{g}\right) = \frac{g\,\dfrac{df}{dx} - f\,\dfrac{dg}{dx}}{g^2}$$

(R2) In words: the derivative of a quotient is the denominator times the derivative of the numerator minus the numerator times the derivative of the denominator, all divided by the square of the denominator.

(R3) $\dfrac{d}{dx}\left(\dfrac{f}{g}\right) \neq \dfrac{\dfrac{df}{dx}}{\dfrac{dg}{dx}}$:   the derivative of a quotient is not the quotient of the derivatives.

Example 6: $f'(x) = ?$ if:

(A1) $f(x) = \dfrac{x^2 + 4x}{x^3 - 5}$

$$f'(x) = \frac{(x^3-5)\dfrac{d}{dx}(x^2+4x) - (x^2+4x)\dfrac{d}{dx}(x^3-5)}{(x^3-5)^2}$$

$$= \frac{(x^3 - 5)(2x + 4) - (x^2 + 4)(3x^2)}{(x^3 - 5)^2}$$

(A2) $f(x) = \dfrac{(1 - 3x)^2}{x^2 + 4} \qquad \left(= \dfrac{(1 - 3x)(1 - 3x)}{x^2 + 4} \right)$

$$f'(x) = \frac{(x^2 + 4) \dfrac{d}{dx} \left[(1 - 3x)^2\right] - (1 - 3x)^2 \dfrac{d}{dx} (x^2 + 4)}{(x^2 + 4)^2}$$

$$f'(x) = \frac{(x^2 + 4) \dfrac{d}{dx} \left[(1 - 3x)(1 - 3x)\right] - (1 - 3x)^2 \dfrac{d}{dx} (x^2 + 4)}{(x^2 + 4)^2}$$

$$= \frac{(x^2 + 4)[(1 - 3x)(-3) + (1 - 3x)(-3)] - (1 - 3x)^2(2x)}{(x^2 + 4)^2}$$

$$= \frac{-6(1 - 3x)(x^2 + 4) - (1 - 3x)^2(2x)}{(x^2 + 4)^2}$$

(A3) $f(x) = \dfrac{1}{x^2 + 4}$

$$f'(x) = \frac{(x^2 + 4) \dfrac{d}{dx} (1) - 1 \cdot \dfrac{d}{dx} (x^2 + 4)}{(x^2 + 4)^2} = \frac{(x^2 + 4)(0) - 2x}{(x^2 + 4)^2} = -\frac{2x}{(x^2 + 4)^2}$$

Extend The Power Rule to negative integer powers.

Theorem (2.15): If $f(x) = x^{-n}$, where n is a positive integer, then

$$f'(x) = -nx^{-n-1}$$

Proof:

$$f'(x) = \frac{d}{dx} (x^{-n}) = \frac{d}{dx} \left(\frac{1}{x^n}\right) = \frac{x^n \dfrac{d}{dx} (1) - (1) \dfrac{d}{dx} (x^n)}{(x^n)^2}$$

$$\uparrow$$
$$\text{(Quotient Rule)}$$

$$= \frac{x^n \cdot 0 - nx^{n-1}}{x^{2n}} = -nx^{n-1-2n} = -nx^{-n-1}$$

↑

(n: positive integer, so use Power Rule (2.10))

Example 7: $f'(x) = ?$ if:

(A1) $f(x) = 3x^5 - \dfrac{7}{x^2}$

$f(x) = 3x^5 - 7x^{-2}$

$f'(x) = 3(5x^4) - 7(-2x^{-3}) = 15x^4 + 14x^{-3}$

Note! Could have used The Quotient Rule to determine $\dfrac{d}{dx}\left(\dfrac{1}{x^2}\right)$.

$$f'(x) = 3(5x^4) - 7\left[\frac{x^2 \cdot 0 - 1 \cdot 2x}{(x^2)^2}\right] = 15x^4 - 7\left(-\frac{2x}{x^4}\right) = 15x^4 + \frac{14}{x^3}\quad \text{(as above)}$$

(A3) $f(x) = \dfrac{x^3 + \dfrac{5}{x^3}}{x^2 + 4}$

$f(x) = \dfrac{x^3 + 5x^{-3}}{x^2 + 4}$

$$f'(x) = \frac{(x^2 + 4)[3x^2 + 5(-3x^{-4})] - (x^3 + 5x^{-3})(2x)}{(x^2 + 4)^2}$$

↑

(quotient rule; power rule)

$$= \frac{(x^2 + 4)(3x^2 - 15x^{-4}) - 2x(x^3 + 5x^{-3})}{(x^2 + 4)^2}$$

Generalize The Power Rule to powers which can be any real number.

The Power Rule (General Version) (2.16):

Let n be any real number. If $f(x) = x^n$, then $f'(x) = nx^{n-1}$.

Or, in Leibniz notation, $\dfrac{d}{dx}(x^n) = nx^{n-1}$.

Proof: Given in Section 6.5, p. 368 of text.

Example 8: $g'(t) = ?$ if:

(A1) $g(t) = t^{3/2} + \dfrac{1}{t^{1/2}}$

$g(t) = t^{3/2} + t^{-1/2}$

$g'(t) = \dfrac{3}{2}t^{1/2} + \left(-\dfrac{1}{2}\right)t^{-3/2} = \dfrac{3}{2}t^{1/2} - \dfrac{1}{2t^{3/2}}$

Note! Could have used The Quotient Rule to determine $\dfrac{d}{dt}\left(\dfrac{1}{t^{1/2}}\right)$.

$$g'(t) = \frac{3}{2}t^{1/2} + \frac{t^{1/2}\cdot 0 - 1 \cdot \frac{1}{2}t^{-1/2}}{(t^{1/2})^2} \underset{\uparrow}{=} \frac{3}{2}t^{1/2} - \frac{1}{2t^{3/2}} \qquad \text{(as above)}$$

$$\left((t^{1/2})^2 = t; \quad t \cdot t^{1/2} = t^{3/2}\right)$$

(A2) $g(t) = (t^2 + 3)(t^\pi - 7t^{\sqrt{2}})$

$g'(t) = (t^2 + 3)[\pi t^{\pi-1} - 7(\sqrt{2}t^{\sqrt{2}-1})] + (t^\pi - 7t^{\sqrt{2}})(2t)$

$ = (t^2 + 3)[\pi t^{\pi-1} - 7\sqrt{2}t^{\sqrt{2}-1}] + 2t(t^\pi - 7t^{\sqrt{2}})$

(A3) $g(t) = \dfrac{t^3 + 2}{\sqrt{t}}$

$g(t) = \dfrac{t^3 + 2}{t^{1/2}}$

$$g'(t) = \frac{t^{1/2}(3t^2) - (t^3 + 2)\left(\frac{1}{2}t^{-1/2}\right)}{(t^{1/2})^2}$$

$$= \frac{3t^{5/2} - \dfrac{t^3 + 2}{2t^{1/2}}}{t} = \frac{6t^3 - (t^3 + 2)}{2t^{3/2}} = \frac{5t^3 - 2}{2t^{3/2}}$$

Note! Could have used The Product Rule:

$$g(t) = (t^3 + 2)t^{-1/2}$$

$$g'(t) = (t^3 + 2)\left(-\frac{1}{2}t^{-3/2}\right) + t^{-1/2}(3t^2)$$

$$= -\frac{1}{2}(t^{3/2} + 2t^{-3/2}) + 3t^{3/2} = \frac{5t^3 - 2}{2t^{3/2}} \qquad \text{(as above)}$$

(A4) $g(t) = 3^{-2} + t^{-2}$

$$g'(t) = 0 + (-2t^{-3}) = -2t^{-3}$$
$$\uparrow$$
$$\left(3^{-2} = \frac{1}{3^2} = \frac{1}{9}: \quad \text{a constant}\right)$$

Consider the graph of $y = f(x)$. At the point $P(a, f(a))$, find an equation of:

(C1) the tangent line.

point-slope form: $\left\{ \begin{array}{ll} \text{point} & (a, f(a)) \\ \text{slope} & m = f'(a) \end{array} \right\} \implies y - f(a) = f'(a)(x - a)$

(C2) the normal line, which is perpendicular to the tangent line at the point $P(a, f(a))$.

point-slope form: $\left\{ \begin{array}{ll} \text{point} & (a, f(a)) \\ \text{slope} & m_\perp = -\dfrac{1}{f'(a)} \quad \text{provided } f'(a) \neq 0 \end{array} \right\}$

$$\implies y - f(a) = -\frac{1}{f'(a)}(x - a)$$

Note!

(N1) $m \neq 0 \implies m_\perp = -\dfrac{1}{m}$

(N2) $f'(a) = 0 \implies$ the tangent line is horizontal

$\implies$ the normal line is vertical and has equation $\underbrace{x = a}$

(vertical line through $(a, f(a))$)

Example 9: Consider $y = x^3 + 3x^2 - 7$

(A1) At $x = -1$, find an equation of the tangent line and the normal line to the graph of the curve.

$$y = x^3 + 3x^2 - 7 \implies y' = 3x^2 + 6x$$

At $a = -1:$ $y = (-1)^3 + 3(-1)^2 - 7 = -5$ and $y' = 3(-1)^2 + 6(-1) = -3$

tangent line: $\left\{ \begin{array}{ll} \text{point} & (-1, -5) \\ \text{slope} & m = -3 \end{array} \right\}$

$\implies \underbrace{y - (-5) = -3[x - (-1)]}_{\text{(point-slope form)}} \implies 3x + y + 8 = 0$

normal line: $\left\{ \begin{array}{ll} \text{point} & (-1, -5) \\ \text{slope} & m_\perp = -\dfrac{1}{-3} = \dfrac{1}{3} \end{array} \right\}$

$\implies \underbrace{y - (-5) = \dfrac{1}{3}[x - (-1)]}_{\text{(point-slope form)}} \implies x - 3y - 14 = 0$

(A2) At $x = 0$, find an equation of the tangent line and the normal line to the graph of the curve.

At $a = 0:$ $y = 0^3 + 3(0)^2 - 7 = -7$ and $y' = 3(0)^2 + 6(0) = 0$

tangent line: $\left\{ \begin{array}{ll} \text{point} & (0, -7) \\ \text{slope} & m = 0 \end{array} \right\}$

$$\implies \quad y - (-7) = 0(x - 0) \quad \implies \quad y = -7$$

Note! $m = 0 \quad \implies \quad$ tangent line is horizontal

normal line: $\left\{ \begin{array}{l} \text{point} \ (0, -7) \\ m = 0 \quad \implies \quad \text{normal line is vertical} \end{array} \right\} \quad \implies \quad x = 0$

(A3) Find an equation of the normal line at the point where the tangent line to the graph has slope -3.

$$y' = 3x^2 + 6x \quad \implies \quad 3x^2 + 6x = -3 \quad \implies \quad 3(x^2 + 2x + 1) = 3(x+1)^2 = 0 \quad \implies \quad x = -1$$

$$\uparrow$$
$$(m = -3)$$

$$\implies \quad \underbrace{\text{normal line: } x - 3y - 14 = 0}$$
$$\text{(see (A1) above: } x = -1)$$

2.3   Rates of Change in the Natural and Social Sciences

Recall: If the quantity y depends upon the quantity x in the fashion $y = f(x)$, then:

(R1) the average rate of change of y with respect to x between the values x_1 and x_2 is

$$\frac{\Delta y}{\Delta x} = \frac{f(x_2) - f(x_1)}{x_2 - x_1} : \qquad \text{difference quotient;}$$

(R2) the instantaneous rate of change of y with respect to x at $(x_1, f(x_1))$ is

$$\left. \frac{dy}{dx} \right|_{x=x_1} = \lim_{\Delta x \to 0} \frac{\Delta y}{\Delta x} : \qquad \text{derivative.}$$

Remark: The specific interpretation of the difference quotient and the derivative as a rate of change depends upon the interpretation of $y = f(x)$ in its particular application.

Physics: Mass of a "thin" rod.

(C1) A rod that is homogeneous has a uniform (the same) linear density at each point along the rod.

Notation: $\left\{ \begin{array}{ll} \rho : & \text{density} \\ m : & \text{mass of rod} \\ l : & \text{length of rod} \end{array} \right\} \implies \rho = \dfrac{m}{l}$: linear density equals mass per unit length.

(C2) Consider a rod that is not homogeneous.

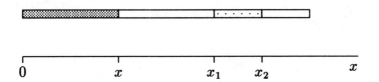

Represent the mass of the rod as a function of x, the distance from one end (left end in drawing above) of the rod: let $m = f(x)$.

The average linear density of the rod between the values x_1 and x_2:

$$\rho_{ave} = \frac{f(x_2) - f(x_1)}{x_2 - x_1} \qquad \left(= \frac{\Delta m}{\Delta x} \right).$$

The linear density of the rod at x_1 (instantaneous):

$$\rho = f'(x) \qquad \left(= \left. \frac{dm}{dx} \right|_{x=x_1} \right).$$

Example 1: A rod of 2 meters has its mass, in kilograms, given by

$$m = 4x - x^2, \qquad \text{where} \qquad 0 \le x \le 2 \qquad (x: \text{meters})$$

(A1) Find the average density of the rod between $x = 1$ and $x = 1.2$.

$$\underset{\uparrow}{\rho_{ave}} = \frac{f(1.2) - f(1)}{1.2 - 1} = \frac{4(1.2) - (1.2)^2 - [4(1) - (1)^2]}{.2} = 1.8 \text{ kg/m}$$

(a choice: $x_1 = 1$, $x_2 = 1.2$)

(A2) Find the density of the rod at $x = 1$.

$$\frac{dm}{dx} = 4 - 2x \implies \rho = \left. \frac{dm}{dx} \right|_{x=1} = 4 - 2x \Big|_{x=1} = 4 - 2(1) = 2 \text{ kg/m}$$

(A3) Find the density of the rod at $x = 1.2$.

$$\rho = \left. \frac{dm}{dx} \right|_{x=1.2} = 4 - 2x \Big|_{x=1.2} = 4 - 2(1.2) = 1.6 \text{ kg/m}$$

Physics: Rectilinear motion of a particle.

If $s = f(t)$ is the position of the particle at time t, then $v = \dfrac{ds}{dt}$ is the velocity (rate of change of position with respect to time) of the particle.

Example 2: (Example 1 on pp. 126, 127 of text): The position of a particle in rectilinear motion is given by
$$s = t^3 - 6t + 9t \qquad (t\text{: seconds; s: meters})$$

(A1) Find the velocity at time t.

$$v(t) = \frac{ds}{dt} = 3t^2 - 12t + 9$$

(A2) What is the velocity after 2 s?

From (A1): $v(2) = 3(2)^2 - 12(2) + 9 = -3$ m/s

(A3) What is the velocity after 4 s?

From (A1): $v(4) = 3(4)^2 - 12(4) + 9 = 9$ m/s

(A4) When is the particle at rest?

At rest: $v(t) = 0 \implies 3t^2 - 12t + 9 = 0 \implies 3(t-1)(t-3) = 0$
$$\uparrow$$
$$(A1)$$

$t = 1,\ 3\quad (> 0) \implies$ after 1 s or after 3 s

(A5) When is the particle moving in the positive direction?

For $\Delta t = t_2 - t_1 > 0$ (i.e., $t_2 > t_1$ and letting $s = f(t)$), want $\Delta s = \underbrace{f(t_2) - f(t_1) > 0}$
(moving in positive direction, so $f(t_2) > f(t_1)$)

$\implies \dfrac{\Delta s}{\Delta t} > 0.$ Intuitively: $v(t) = \dfrac{ds}{dt} = \lim\limits_{\Delta t \to 0} \dfrac{\Delta s}{\Delta t} > 0.$

Note!

(N1) A complete discussion showing that the particle moves in the positive direction of s if $\dfrac{ds}{dt} > 0$ is given in Section 3.3.

(N2) The particle moves in the negative direction of s if $\dfrac{ds}{dt} < 0$.

$$v(t) > 0 \quad \Longrightarrow \quad 3(t-1)(t-3) > 0$$

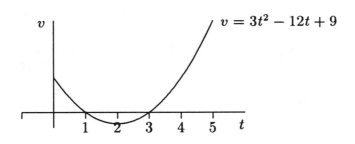

(exclude since $t \geq 0$)

$$\Longrightarrow \quad v(t) > 0 \qquad \text{when} \qquad 0 \leq t < 1 \qquad \text{or} \qquad t > 3$$

Note! Could solve inequality via a sketch in a t, v coordinate system.

$v(t) = 3t^2 - 12t + 9$: a parabola opening upward

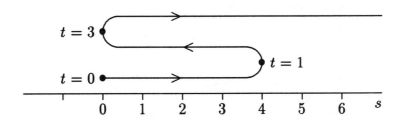

$$\Longrightarrow \quad v(t) > 0 \qquad \text{when} \qquad 0 \leq t < 1 \qquad \text{or} \qquad t > 3$$

(A6) Draw a diagram to represent the motion (position and direction) of the particle.

$A = \pi r^2$ ↘ derivative
$C = 2\pi r$ ↗

Note! Determined diagram from the results of (A5):

$$t = 0: \quad s = 0, \quad \text{then goes to the right}$$
$$t = 1: \quad s = 4, \quad \text{then goes to the left}$$
$$t = 3: \quad s = 0, \quad \text{then goes to the right}$$

(A7) Find the total distance traveled by the particle during the first 5 s.

$$t \in [0,1]: \quad \underbrace{|f(1) - f(0)|} = |4 - 0| = 4 \text{ m to the right}$$
$$\text{(distance} \geq 0; \ s = f(t))$$

$$t \in [1,3]: \quad |f(3) - f(1)| = |0 - 4| = 4 \text{ m to the left}$$

$$t \in [3,5]: \quad |f(5) - f(3)| = |20 - 0| = 20 \text{ m to the right}$$

Total distance $= 4 + 4 + 20 = 28$ m

Economics: Marginal cost.

Let $C = f(x)$ be the cost for producing x units of a commodity.

Increase the number of items produced from x_1 to x_2: the additional cost is
$\Delta C = f(x_2) - f(x_1)$

$\Longrightarrow$ the related average rate of change of the cost is

$$\frac{\Delta C}{\Delta x} = \frac{f(x_2) - f(x_1)}{x_2 - x_1}.$$

Definition: Marginal cost is the instantaneous rate of change of cost with respect to the number of units produced.

Hence, the marginal cost is
$$\frac{dC}{dx} = \lim_{\Delta x \to 0} \frac{\Delta C}{\Delta x}.$$

Since x takes on integer values only, $\Delta x \to 0$ is not possible;

take $\Delta x = 1$ (smallest possible vale for $\Delta x > 0$)

If $\quad x_1 = n \quad$ and $\quad x_2 = n + 1, \quad$ then

$$\frac{\Delta C}{\Delta x} = \frac{f(n+1) - f(n)}{(n+1) - n} = \frac{f(n+1) - f(n)}{1} \approx f'(n)$$

$$\uparrow$$

$$\left(\frac{dC}{dx} = f'(x) \right)$$

$\implies f'(n) \approx f(n+1) - f(n)$, the cost to produce 1 unit beyond the present production level n: the cost to produce the $(n+1)$st unit

Example 3: Consider the cost (in dollars) function $f(x) = \dfrac{x^2 - 1}{x}$ for $x \geq 2$.

(A1) Find the marginal cost function.

$$f'(x) = \frac{x(2x) - (x^2 - 1)(1)}{x^2} = \frac{x^2 + 1}{x}$$

(A2) Find the marginal cost at the production level of 7 units.

$$x = 7: \quad f'(7) = \frac{7^2 + 1}{7^2} = \frac{50}{49} = \$1.02/\text{unit}$$

(A3) Find the cost of producing the 8th unit.

$$f(8) - f(7) = \frac{8^2 - 1}{8} - \frac{7^2 - 1}{7} = \frac{63}{8} - \frac{48}{7} = \frac{57}{56} = \$1.018 \approx \$1.02 = f'(7)$$

Note! Marginal cost at 7 units "closely" approximates the cost of producing the 8th unit.

(A4) Find the marginal cost at the production level of 70 units.

$$x = 70: \quad f'(70) = \frac{(70)^2 + 1}{(70)^2} = \frac{4901}{4900} = \$1/\text{unit}$$

(A5) Approximate the cost of producing the 71st unit.

$$f(71) - f(70) \approx f'(70) = \$1$$

Note! The cost of producing the 71st unit is

$$f(71) - f(70) = \frac{(71)^2 - 1}{71} - \frac{(70)^2 - 1}{70} = \frac{4971}{4970} = \$1$$

2.4 Derivatives of Trigonometric Functions

In this section, limits and derivatives of trigonometric functions are developed: see Appendix B for a review of trigonometric functions.

Theorem (2.19): $\lim\limits_{\theta \to 0} \sin \theta = 0$

Remark: θ is measured in **radians**.

Proof: Consider $0 < \theta < \dfrac{\pi}{2}$ to calculate $\lim\limits_{\theta \to 0^+} \sin \theta$: use a geometric argument.

Let θ be the central angle of a sector of a circle with center O and radius 1.

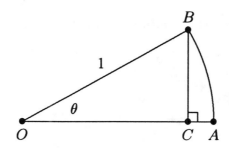

Determine C such that $BC \perp OA$.

$$0 < \sin \theta = \underbrace{\frac{|BC|}{|OB|}}_{\left(\frac{\text{opp}}{\text{hyp}}\right)} = \frac{|BC|}{1} = |BC| < \operatorname{arc} AB = \theta$$

$\quad\ \uparrow$ $\qquad\qquad\qquad\qquad\qquad\qquad\qquad \uparrow$

$\left(0 < \theta < \dfrac{\pi}{2}\right)$ $\qquad\qquad\qquad$ (definition of radian measure)

$$\lim_{\theta \to 0^+} 0 = 0 = \lim_{\theta \to 0^+} \theta \ \implies \ \lim_{\theta \to 0^+} \sin \theta = 0$$

$\qquad\qquad\qquad\qquad \uparrow$

(Squeeze Theorem: $0 < \sin \theta < \theta$)

Consider $-\dfrac{\pi}{2} < \theta < 0$ to calculate $\lim\limits_{\theta \to 0^-} \sin \theta$

$$-\frac{\pi}{2} < \theta < 0 \quad \Longrightarrow \quad \frac{\pi}{2} > -\theta > 0 \quad \text{or, rewritten,} \quad 0 < -\theta < \frac{\pi}{2}$$

$$\Longrightarrow \quad 0 < \sin(-\theta) < -\theta$$
↑

$(-\theta$ replaces θ in preceding discussion for $0 < \theta < \pi/2)$

$$\Longrightarrow \quad 0 < -\sin\theta < -\theta$$
↑

(Identity (10a), Appendix B: $\sin(-\theta) = -\sin\theta$)

$$\Longrightarrow \quad 0 > \sin\theta > \theta \quad \text{or, rewritten,} \quad \theta < \sin\theta < 0$$

$$\lim_{\theta \to 0^-} \theta = 0 = \lim_{\theta \to 0^-} 0 \quad \Longrightarrow \quad \lim_{\theta \to 0^-} \sin\theta = 0$$
↑

(Squeeze Theorem: $\theta < \sin\theta < 0$)

Hence, $\quad \displaystyle\lim_{\theta \to 0^-} \sin\theta = 0 = \lim_{\theta \to 0^+} \sin\theta \quad \Longrightarrow \quad \lim_{\theta \to 0} \sin\theta = 0$

Remark: Consider $f(\theta) = \sin\theta$. Have shown: $\displaystyle\lim_{\theta \to 0} f(\theta) = \lim_{\theta \to 0} \sin\theta = 0 = f(0)$

$\Longrightarrow$ the sine function is continuous at 0.

Corollary (2.21): $\qquad \displaystyle\lim_{\theta \to 0} \cos\theta = 1$

Proof: Need only consider $\quad -\dfrac{\pi}{2} < \theta < \dfrac{\pi}{2}$

$$\Longrightarrow \quad \cos\theta = \sqrt{1 - \sin^2\theta}$$
↑

$\left(\sin^2\theta + \cos^2\theta = 1; \ \cos\theta > 0\right)$

$$\Longrightarrow \quad \lim_{\theta \to 0} \cos\theta = \lim_{\theta \to 0} \sqrt{1 - \sin^2\theta} = \sqrt{1 - 0^2} = 1$$
↑

$\left(\text{limit laws; Theorem (2.19): } \displaystyle\lim_{\theta \to 0} \sin\theta = 0 \right)$

Remark: Consider $f(\theta) = \cos\theta$. Have shown: $\lim\limits_{\theta\to0} f(\theta) = \lim\limits_{\theta\to0}\cos\theta = 1 = f(0)$

$\implies$ the cosine function is continuous at 0.

Theorem (2.22): $\lim\limits_{\theta\to0}\dfrac{\sin\theta}{\theta} = 1$

Remark: $\lim\limits_{\theta\to0}\dfrac{\sin\theta}{\theta} = \dfrac{\sin 0}{0} = \dfrac{0}{0}$: do

Proof: Consider $0 < \theta < \pi/2$. As in proof of Theorem (2.19), let θ be the central angle of a sector of a circle with center O and radius 1.

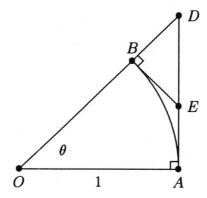

D is determined by the intersection of OB extended with the tangent to the circle at A. E is determined by the intersection of AD with the tangent to the circle at B.

Tangent to circle is perpendicular to radius of circle $\implies m(\angle OAD) = \dfrac{\pi}{2} = m(\angle OBE)$

$\underset{\uparrow}{\implies} m(\angle EBD) = \dfrac{\pi}{2}$

$(m(\angle OBE) + m(\angle EBD) = \pi)$

$\underset{\uparrow}{\theta} = \text{arc } AB < \underset{\uparrow}{|AE| + |EB|} < |AE| + |ED| = |AD|$

(def of radian measure) $\left(\left\{\begin{array}{l} |ED| : \text{ hypotenuse} \\ |EB| : \text{ leg} \end{array}\right\} : \text{of same triangle}\right)$

$$= |OA| \tan \theta = \tan \theta = \frac{\sin \theta}{\cos \theta}$$

$\uparrow$

$$\left(\tan \theta = \frac{\text{opp}}{\text{adj}} \implies \tan \theta = \frac{|AD|}{|OA|} \right)$$

$$\implies \quad \cos \theta < \frac{\sin \theta}{\theta} \qquad \implies \quad \cos \theta < \frac{\sin \theta}{\theta} < 1$$

$\uparrow$ $\qquad\qquad\qquad\qquad\qquad\qquad\qquad \uparrow$

$$\left(0 < \theta < \frac{\pi}{2} \right) \text{(from proof of Theorem (2.19): } 0 < \theta < \pi/2 \implies \sin \theta < \theta)$$

$$\lim_{\theta \to 0^+} \cos \theta = 1 = \lim_{\theta \to 0^+} 1 \implies \lim_{\theta \to 0^+} \frac{\sin \theta}{\theta} = 1$$

$\qquad\qquad\quad \uparrow \qquad\qquad\qquad\qquad \uparrow$

$$\text{(Corollary (2.21))} \qquad \left(\text{Squeeze Theorem: } \cos \theta < \frac{\sin \theta}{\theta} < 1 \right)$$

Consider $-\dfrac{\pi}{2} < \theta < 0$ in order to calculate $\displaystyle\lim_{\theta \to 0^-} \frac{\sin \theta}{\theta}$:

$$-\frac{\pi}{2} < \theta < 0 \implies \frac{\pi}{2} > -\theta > 0 \quad \text{or, rewritten,} \quad 0 < -\theta < \frac{\pi}{2}$$

$$\implies \quad \cos(-\theta) < \frac{\sin(-\theta)}{-\theta} < 1$$

$\uparrow$

$(-\theta$ replaces θ in preceding discussion for $0 < \theta < \pi/2)$

$$\implies \quad \cos \theta < \frac{-\sin \theta}{-\theta} < 1 \implies \cos \theta < \frac{\sin \theta}{\theta} < 1$$

$\uparrow$

(Identities (10b) and (10a), Appendix B: $\cos(-\theta) = \cos \theta$, $\sin(-\theta) = -\sin \theta$)

$$\lim_{\theta \to 0^-} \cos \theta = 1 = \lim_{\theta \to 0^-} 1 \implies \lim_{\theta \to 0^-} \frac{\sin \theta}{\theta} = 1$$

$\qquad\qquad\quad \uparrow \qquad\qquad\qquad\qquad \uparrow$

$$\text{(Corollary (2.21))} \qquad \left(\text{Squeeze Theorem: } \cos \theta < \frac{\sin \theta}{\theta} < 1 \right)$$

Hence, $\displaystyle\lim_{\theta \to 0^+} \frac{\sin \theta}{\theta} = 1 = \lim_{\theta \to 0^-} \frac{\sin \theta}{\theta} \implies \lim_{\theta \to 0} \frac{\sin \theta}{\theta} = 1$

Remark: In words: the limit of the sine of something all divided by that something as that something goes to zero is one.

Example 1: $\lim\limits_{x\to 0}\dfrac{\sin 5x}{2x} = ?$

$$\lim_{x\to 0}\frac{\sin 5x}{2x} = \underset{\uparrow}{\frac{1}{2}}\lim_{x\to 0}\frac{5\sin 5x}{5x} = \frac{5}{2}\lim_{x\to 0}\frac{\sin 5x}{5x} = \frac{5}{2}(1) = \frac{5}{2}$$

$$\left(\frac{0}{0}: \text{ do; } \sin\theta \text{ with } \theta = 5x; \frac{5}{5} = 1\right)\left(x\to 0 \implies 5x\to 0: \text{Theorem (2.22)}\right)$$

Example 2: $\lim\limits_{x\to 0}\dfrac{2x}{\tan 3x} = ?$

$$\lim_{x\to 0}\frac{2x}{\tan 3x} = \underset{\uparrow}{\lim_{x\to 0}}\frac{2x}{\left(\dfrac{\sin 3x}{\cos 3x}\right)} = 2\lim_{x\to 0}\frac{\cos 3x}{\left(\dfrac{\sin 3x}{x}\right)} = \underset{\uparrow}{2\lim_{x\to 0}}\frac{\cos 3x}{\left(\dfrac{3\sin 3x}{3x}\right)}$$

$$\left(\frac{0}{0}: \text{ do}\right) \qquad\qquad \left(\theta = 3x; \frac{3}{3} = 1\right)$$

$$= \frac{2}{3}\lim_{x\to 0}\frac{\cos 3x}{\left(\dfrac{\sin 3x}{3x}\right)} = \underset{\uparrow}{\frac{2(1)}{3(1)}} = \frac{2}{3}$$

$$(x\to 0 \implies 3x\to 0: \text{Corollary (2.21), Theorem (2.22)})$$

Example 3: $\lim\limits_{x\to 0}\dfrac{5x + \sin x}{3x} = ?$

$$\lim_{x\to 0}\frac{5x + \sin x}{3x} = \underset{\uparrow}{\frac{1}{3}}\lim_{x\to 0}\left(\frac{5x + \sin x}{x}\right) = \frac{1}{3}\lim_{x\to 0}\left(5 + \frac{\sin x}{x}\right) = \underset{\uparrow}{\frac{1}{3}}(5 + 1) = 2$$

$$\left(\frac{0}{0}: \text{ do}\right) \qquad\qquad (\theta = x: \text{ Theorem (2.22)})$$

Example 4: $\lim\limits_{x \to 0} \dfrac{\sin 7x}{\sin 4x} = ?$

$$\lim_{x \to 0} \frac{\sin 7x}{\sin 4x} \underset{\underset{\left(\frac{0}{0}\,:\, \text{do}\right)}{\uparrow}}{=} \lim_{x \to 0} \frac{\left(\dfrac{\sin 7x}{x}\right)}{\left(\dfrac{\sin 4x}{x}\right)} \underset{\underset{(\theta = 7x;\ \theta = 4x)}{\uparrow}}{=} \lim_{x \to 0} \frac{7\left(\dfrac{\sin 7x}{7x}\right)}{4\left(\dfrac{\sin 4x}{4x}\right)}$$

$$= \frac{7}{4} \lim_{x \to 0} \frac{\left(\dfrac{\sin 7x}{7x}\right)}{\left(\dfrac{\sin 4x}{4x}\right)} \underset{\underset{(x \to 0 \ \Longrightarrow \ 7x \to 0,\ 4x \to 0:\ \text{Theorem } (2.22))}{\uparrow}}{=} \frac{7\,(1)}{4\,(1)} = \frac{7}{4}$$

Corollary (2.25): $\qquad \lim\limits_{\theta \to 0} \dfrac{\cos \theta - 1}{\theta} = 0$

Proof:

$$\lim_{\theta \to 0} \frac{\cos \theta - 1}{\theta} \underset{\underset{\left(\frac{0}{0}\,:\, \text{do};\ \theta \text{ near } 0 \Longrightarrow \cos \theta + 1 \neq 0\right)}{\uparrow}}{=} \lim_{\theta \to 0} \left(\frac{\cos \theta - 1}{\theta} \cdot \frac{\cos \theta + 1}{\cos \theta + 1}\right) = \lim_{\theta \to 0} \frac{\cos^2 \theta - 1}{\theta(\cos \theta + 1)} \underset{\underset{\left(\sin^2 \theta + \cos^2 \theta = 1\right)}{\uparrow}}{=} \lim_{\theta \to 0} \frac{-\sin^2 \theta}{\theta(\cos \theta + 1)}$$

$$\underset{\underset{\left(\frac{0}{0}\,:\, \text{do}\right)}{\uparrow}}{=} -\lim_{\theta \to 0}\left(\frac{\sin \theta}{\theta} \cdot \frac{\sin \theta}{\cos \theta + 1}\right) \underset{\underset{(\text{Theorem } (2.22);\ \text{Theorem } (2.19);\ \text{Corollary } (2.21))}{\uparrow}}{=} -\,(1)\left(\frac{0}{1+1}\right) = 0$$

Remark: This proof, in order to obtain a trigonometric identity, used a method similar to a previous rationalizing of numerator procedure.

Example 5: $\lim\limits_{x \to 0} \dfrac{1 - \cos x}{\sin x} = ?$

$$\lim_{x \to 0} \frac{1 - \cos x}{\sin x} = \lim_{x \to 0} \left(\frac{1 - \cos x}{\sin x} \cdot \frac{1 + \cos x}{1 + \cos x} \right) = \lim_{x \to 0} \frac{1 - \cos^2 x}{\sin x (1 + \cos x)}$$

$$\uparrow$$
$$\left(\frac{0}{0} : \text{do} \right)$$

$$= \lim_{x \to 0} \frac{\sin^2 x}{\sin x (1 + \cos x)} = \lim_{x \to 0} \frac{\sin x}{1 + \cos x} = \frac{0}{1 + 1} = 0$$

$$\uparrow \qquad\qquad\qquad\qquad \uparrow$$

(x near 0, $x \neq 0$: $\sin x \neq 0$)   (Theorem (2.19); Corollary (2.21))

Note! Another method of solution:

$$\lim_{x \to 0} \frac{1 - \cos x}{\sin x} = \lim_{x \to 0} \frac{\left(\dfrac{1 - \cos x}{x} \right)}{\left(\dfrac{\sin x}{x} \right)} = \lim_{x \to 0} \frac{-\dfrac{\cos x - 1}{x}}{\left(\dfrac{\sin x}{x} \right)} = \frac{-0}{1} = 0$$

$$\uparrow \qquad\qquad\qquad\qquad\qquad\qquad \uparrow$$
$$\left(\frac{0}{0} : \text{do} \right) \qquad\qquad\qquad \text{(Corollary (2.25); Theorem (2.22))}$$

Theorem (2.26): $\qquad \dfrac{d}{dx} \sin x = \cos x$

Proof: Let $f(x) = \sin x$.

$$f'(x) = \lim_{h \to 0} \frac{f(x + h) - f(x)}{h} = \lim_{h \to 0} \frac{\sin(x + h) - \sin x}{h}$$

$$= \lim_{h \to 0} \frac{\sin x \cos h + \cos x \sin h - \sin x}{h} = \lim_{h \to 0} \left[\sin x \left(\frac{\cos h - 1}{h} \right) + \cos x \left(\frac{\sin h}{h} \right) \right]$$

$$\uparrow \qquad\qquad\qquad\qquad\qquad\qquad\qquad \uparrow$$
$$\left(\frac{0}{0} : \text{do}; \begin{array}{l} \text{addition formula (12a)} \\ \text{Appendix B} \end{array} \right) \qquad \left(\frac{0}{0} : \text{do} \right)$$

$$= (\sin x) \lim_{h \to 0} \frac{\cos h - 1}{h} + (\cos x) \lim_{h \to 0} \frac{\sin h}{h} = (\sin x)(0) + (\cos x)(1) = \cos x$$

$$\uparrow \qquad\qquad\qquad\qquad\qquad\qquad\qquad\qquad \uparrow$$

(x fixed; h is variable in limit)   (Corollary (2.25); Theorem (2.22))

Example 6: $\dfrac{d}{dx}\left(\sqrt{x}\sin x\right) = \ ?$

$$\frac{d}{dx}\left(\sqrt{x}\sin x\right) = \frac{d}{dx}\left(x^{1/2}\sin x\right) = x^{1/2}\frac{d}{dx}(\sin x) + (\sin x)\frac{d}{dx}(x^{1/2})$$

$$= x^{1/2}\cos x + (\sin x)\left(\frac{1}{2}x^{-1/2}\right) = \sqrt{x}\cos x + \frac{\sin x}{2\sqrt{x}}$$

↑

(Theorem (2.26))

Example 7: $\dfrac{d}{dx}\left(x^{\pi}\csc x\right) = \ ?$

$$\frac{d}{dx}\left(x^{\pi}\csc x\right) = \frac{d}{dx}\left(\frac{x^{\pi}}{\sin x}\right) = \frac{(\sin x)(\pi x^{\pi-1}) - x^{\pi}\cos x}{\sin^2 x}$$

Table of Derivatives of Trigonometric Functions (2.29):

✳ $\dfrac{d}{dx}(\sin x) = \cos x$ $\dfrac{d}{dx}(\csc x) = -\csc x \cot x$

✳ $\dfrac{d}{dx}(\cos x) = -\sin x$ $\dfrac{d}{dx}(\sec x) = \sec x \tan x$

$\dfrac{d}{dx}(\tan x) = \sec^2 x$ $\dfrac{d}{dx}(\cot x) = -\csc^2 x$

Proof:

(T1) $f(x) = \cos x$

$$f'(x) = \lim_{h\to 0}\frac{\cos(x+h) - \cos x}{h} = \lim_{h\to 0}\frac{\cos x \cos h - \sin x \sin h - \cos x}{h}$$

↑

$\left(\dfrac{0}{0}: \text{ do; addition formula (12b), Appendix B}\right)$

$\sin(A+B) = \sin A \cos B + \sin B \cos A$

$$= \lim_{h \to 0} \left[\cos x \left(\frac{\cos h - 1}{h} \right) - \sin x \left(\frac{\sin h}{h} \right) \right]$$

↑

$$\left(\frac{0}{0} : \text{do} \right)$$

$$= (\cos x) \lim_{h \to 0} \frac{\cos h - 1}{h} - (\sin x) \lim_{h \to 0} \frac{\sin h}{h} = (\cos x)(0) - (\sin x)(1) = -\sin x$$

↑

(Corollary (2.25); Theorem (2.22))

(T2) $f(x) = \tan x \implies f(x) = \dfrac{\sin x}{\cos x}$

$$f'(x) = \frac{(\cos x) \dfrac{d}{dx} (\sin x) - (\sin x) \dfrac{d}{dx} (\cos x)}{\cos^2 x}$$

$$= \frac{(\cos x)(\cos x) - (\sin x)(-\sin x)}{\cos^2 x} = \frac{1}{\cos^2 x} = \sec^2 x$$

↑

$$\left(\sin^2 x + \cos^2 x = 1 \right)$$

(T3) $f(x) = \csc x$:   proceed as in (T2)

(T4) $f(x) = \sec x$:   proceed as in (T2)

(T5) $f(x) = \cot x$:   proceed as in (T2)

Remark: $\underbrace{f(x) = \mathbf{co}\text{trig} \implies f'(x) = - \text{something}}$

(derivative of a "cofunction" has a negative sign in its calculation)

Example 8: $f'(x) = ?$ if $f(x) = \cos^2 x.$

$f(x) = \cos^2 x = (\cos x)(\cos x)$

$$f'(x) = (\cos x) \frac{d}{dx} (\cos x) + (\cos x) \frac{d}{dx} (\cos x)$$

↑

(product rule)

$$= 2(\cos x) \frac{d}{dx} (\cos x) = 2(\cos x)(-\sin x) = -2\cos x \sin x$$

Note!

(N1) Another method:

$$f(x) = \cos^2 x = 1 - \sin^2 x = 1 - (\sin x)(\sin x)$$

$$f'(x) = 0 - 2(\sin x) \frac{d}{dx} (\sin x) = -2(\sin x)(\cos x) = -2\sin x \cos x$$
$$\uparrow$$

(product rule)

(N2) $f'(x) = -2\sin x \cos x = -\sin 2x$
$$\uparrow$$

(Identity (15a), Appendix B)

Example 9: $\dfrac{dy}{dx} = ?$ if $y = \dfrac{\tan x}{\sec x + 1}$.

$$\frac{dy}{dx} = \frac{(\sec x + 1) \dfrac{d}{dx} (\tan x) - (\tan x) \dfrac{d}{dx} (\sec x + 1)}{(\sec x + 1)^2}$$

$$= \frac{(\sec x + 1) \sec^2 x - (\tan x)(\sec x \tan x + 0)}{(\sec x + 1)^2} = \frac{\sec^3 x + \sec^2 x - \tan^2 x \sec x}{(\sec x + 1)^2}$$

Note! The last expression can be simplified:

$$\frac{dy}{dx} = \frac{\sec^3 x + \sec^2 x - (\sec^2 x - 1)\sec x}{(\sec x + 1)^2} = \frac{\sec^3 x + \sec^2 x - \sec^3 x + \sec x}{(\sec x + 1)^2}$$
$$\uparrow$$
$$\left(1 + \tan^2 x = \sec^2 x\right)$$

$$= \frac{\sec x(\sec x + 1)}{(\sec x + 1)^2} = \frac{\sec x}{\sec x + 1}$$

$$f(x) = \cos 2x$$
$$f'(x) = -2\sin 2x$$

Example 10: $y' = ?$ if $y = (x^2 + 3x)\cot x$.

$$y' = (x^2 + 3x)\,\frac{d}{dx}\,(\cot x) + (\cot x)\,\frac{d}{dx}\,(x^2 + 3x)$$

$$= (x^2 + 3x)(-\csc^2 x) + (\cot x)(2x + 3)$$

Example 11: Find an equation of the tangent line to the graph of $y = \csc x$ at $x = \pi/4$.

$$f(x) = \csc x \quad \Longrightarrow \quad f'(x) = -\csc x \cot x$$

$$a = \frac{\pi}{4}: \ f\left(\frac{\pi}{4}\right) = \csc\frac{\pi}{4} = \sqrt{2} \ \text{ and } \ f'\left(\frac{\pi}{4}\right) = -\csc\frac{\pi}{4}\cot\frac{\pi}{4} = -\sqrt{2}(1) = -\sqrt{2}$$

$$\Longrightarrow \quad \underbrace{y - \sqrt{2} = -\sqrt{2}\left(x - \frac{\pi}{4}\right)}$$
$$\text{(point-slope form)}$$

2.5   The Chain Rule

> **The Chain Rule (2.30):** If the derivatives $g'(x)$ and $f'(g(x))$ both exist, and $F = f \circ g$ is the composite function defined by $F(x) = f(g(x))$, then $F'(x)$ exists and is given by the product
> $$F'(x) = f'(g(x))g'(x)$$
> In Leibniz notation, if $y = f(u)$ and $u = g(x)$ are both differentiable functions, then
> $$\frac{dy}{dx} = \frac{dy}{du}\frac{du}{dx}$$

Remarks:

(R1) In words: the derivative of the composite of two functions is the product of the derivative of the "outer" function evaluated at the "inner" function with the derivative of the inner function.

(R2) Another notation:
$$(f \circ g)'(x) = (f' \circ g)(x)g'(x)$$

(R3) Another notation:
$$\frac{d}{dx}\,(f(g(x)) = f'(g(x))g'(x)$$

(R4) In Leibniz notation, if the symbol representing the derivative is treated as a quotient, then du could be canceled:

$$\frac{dy}{du}\frac{du}{dx} = \frac{dy}{1}\frac{1}{dx} = \frac{dy}{dx}$$

(operate as if dy, du, and dx each has been defined with $du \neq 0$, $dx \neq 0$)

(R5) The variable u introduced in Leibniz notation is referred to as an *intermediate* variable.

Proof:

$$\frac{dy}{dx} = \lim_{\Delta x \to 0}\frac{\Delta y}{\Delta x} = \lim_{\Delta x \to 0}\left(\frac{\Delta y}{\Delta x}\cdot\frac{\Delta u}{\Delta u}\right) = \lim_{\Delta x \to 0}\left(\frac{\Delta y}{\Delta u}\cdot\frac{\Delta u}{\Delta x}\right)$$

$\uparrow$

(assumption: $\Delta u \neq 0$ for Δx near 0)

$$= \left(\lim_{\Delta x \to 0}\frac{\Delta y}{\Delta u}\right)\left(\lim_{\Delta x \to 0}\frac{\Delta u}{\Delta x}\right) = \left(\lim_{\Delta x \to 0}\frac{\Delta y}{\Delta u}\right)\frac{du}{dx}$$

Question: Does $\Delta u \to 0$ as $\Delta x \to 0$?

Given: $g'(x)$ exists

$\Longrightarrow$ $u = g(x)$ continuous at x $\Longrightarrow$ $\lim\limits_{\Delta x \to 0} g(x + \Delta x) = g(x)$

$u = g(x)$ $\Longrightarrow$ $\Delta u = g(x + \Delta x) - g(x)$

$\Longrightarrow$ $\lim\limits_{\Delta x \to 0}\Delta u = \lim\limits_{\Delta x \to 0}[g(x + \Delta x) - g(x)]$

$= \lim\limits_{\Delta x \to 0} g(x + \Delta x) - \lim\limits_{\Delta x \to 0} g(x) = g(x) - g(x) = 0$

$\uparrow$

$\left(\lim\limits_{\Delta x \to 0} g(x + \Delta x) = g(x); \left\{\begin{array}{l}\Delta x : \text{ varying} \\ x : \text{ fixed}\end{array}\right\} \Longrightarrow g(x) \text{ constant with respect to } \Delta x\right)$

$\Longrightarrow$ $\Delta u \to 0$ as $\Delta x \to 0$

$\Longrightarrow$ $\dfrac{dy}{dx} = \left(\lim\limits_{\Delta x \to 0}\dfrac{\Delta y}{\Delta u}\right)\dfrac{du}{dx} = \dfrac{dy}{du}\dfrac{du}{dx}$

Remarks:

(R1) The proof given above is not complete due to the assumption $\Delta u \neq 0$ for Δx near 0. There are differentiable functions $u = g(x)$ such that $\Delta u = 0$ for infinitely many values of Δx near 0 (so the assumption does not always hold). A "complete" proof of The Chain Rule is given in the text on pp. 147, 148.

(R2) A procedure when using The Chain Rule to find $(f \circ g)'(x)$ for a given $(f \circ g)(x)$:

  (S1) identify $f(x)$ and $g(x)$

  (S2) find $f'(x)$ and $g'(x)$

  (S3) answer is $f'(g(x))g'(x)$

Example 1: $\dfrac{dy}{dx} = ?$ if $y = (x^3 - 7x)^{-2}$.

Let $y = (f \circ g)(x)$, where $\begin{cases} f(x) = x^{-2} & \implies f'(x) = -2x^{-3} \\ g(x) = x^3 - 7x & \implies g'(x) = 3x^2 - 7 \end{cases}$

$\implies \dfrac{dy}{dx} = \underbrace{-2(x^3 - 7x)^{-3}}_{f'(g(x))} \underbrace{(3x^2 - 7)}_{g'(x)}$

Note! Solving via Leibniz notation:

Let $y = f(u)$ and $u = g(x)$ with $\begin{cases} y = u^{-2} & \implies \dfrac{dy}{du} = -2u^{-3} \\ u = x^3 - 7x & \implies \dfrac{du}{dx} = 3x^2 - 7 \end{cases}$

$\implies \dfrac{dy}{dx} = \dfrac{dy}{du}\dfrac{du}{dx} = -2u^{-3}(3x^2 - 7) = -2(x^3 - 7x)^{-3}(3x^2 - 7)$

$\uparrow$

$\left(\text{express } \dfrac{dy}{du} \text{ in terms of } x \text{ via } u = g(x) \right)$

Example 2: $F'(x) = ?$ if $F(x) = \cos(x^2)$.

Let $F(x) = (f \circ g)(x)$, where $\begin{cases} f(x) = \cos x & \implies f'(x) = -\sin x \\ g(x) = x^2 & \implies g'(x) = 2x \end{cases}$

$$\implies F'(x) = \underbrace{[-\sin(x^2)]}_{f'(g(x))} \underbrace{2x}_{g'(x)} = -2x\sin(x^2)$$

Note! Solving via Leibniz notation:

For $y = F(x)$, let $y = f(u)$ and $u = g(x)$ with $\begin{cases} y = \cos u & \implies \dfrac{dy}{du} = -\sin u \\ u = x^2 & \implies \dfrac{du}{dx} = 2x \end{cases}$

$$\implies \frac{dy}{dx} = \frac{dy}{du}\frac{du}{dx} = (-\sin u)(2x) = -2x\sin(x^2)$$

$$\uparrow$$

$$\left(\text{express } \frac{dy}{du} \text{ in terms of } x \text{ via } u = g(x) \right)$$

Example 3: $F'(x) = ?$ if $F(x) = (\cos x)^2$.

Let $F(x) = (f \circ g)(x)$, where $\begin{cases} f(x) = x^2 & \implies f'(x) = 2x \\ g(x) = \cos x & \implies g'(x) = -\sin x \end{cases}$

$$\implies F'(x) = \underbrace{(2\cos x)}_{f'(g(x))} \underbrace{(-\sin x)}_{g'(x)} = -2\cos x \sin x$$

Note!

(N1) This problem was solved as Example 8, Section 2.4, lecture notes via the product rule.

(N2) Solving via Leibniz notation:

For $y = F(x)$, let $y = f(u)$ and $u = g(x)$ with $\begin{cases} y = u^2 & \implies \dfrac{dy}{du} = 2u \\ u = \cos x & \implies \dfrac{du}{dx} = -\sin x \end{cases}$

let $u =$ inside function

$$\implies \frac{dy}{dx} = \frac{dy}{du}\frac{du}{dx} = (2u)(-\sin x) = 2\cos x(-\sin x) = -2\cos x \sin x$$
$$\uparrow$$
$$\left(\text{express } \frac{dy}{du} \text{ in terms of } x \text{ via } u = g(x)\right)$$

Example 4: Consider $F(x) = \cos\dfrac{\pi}{x}$. At $x = 1$, find an equation for:

(A1) the tangent line to the curve $y = F(x)$.

For $F(x) = (f \circ g)(x)$, let $y = f(u)$ and $u = g(x)$ with

$$\begin{cases} y = \cos u & \implies \dfrac{dy}{du} = -\sin u \\[2mm] u = \dfrac{\pi}{x} = \pi x^{-1} & \implies \dfrac{du}{dx} = -\pi x^{-2} \end{cases}$$

$$\implies \frac{dy}{dx} = \frac{dy}{du}\frac{du}{dx} = (-\sin u)(-\pi x^{-2}) = \frac{\pi\sin(\pi/x)}{x^2}$$
$$\uparrow$$
$$(u = \pi/x)$$

At $a = 1$: $y = \cos\pi = -1$ and $\dfrac{dy}{dx} = \pi\sin\pi = 0$

$$\implies \underbrace{y - (-1) = 0(x - 1)}_{\text{(point-slope form)}} \implies y = -1: \text{ a horizontal line}$$

(A2) the normal line to the curve $y = F(x)$.

From (A1): $m = 0$ (slope of tangent line)

$$\implies \text{ the normal line is a vertical line: } \underbrace{x = 1}_{(x = a)}$$

Example 5: $\dfrac{dy}{dx} = ?$ if $y = \cos^3 \dfrac{\pi}{x}. = \left(\cos \dfrac{\pi}{x} \right)^3$

For $y = F(x)$, let
$$
\begin{cases}
y = u^3 & \implies \dfrac{dy}{du} = 3u^2 \\[2mm]
u = \cos \dfrac{\pi}{x} & \implies \underset{\uparrow}{\dfrac{du}{dx}} = ? \\[2mm]
& \text{(}u\text{ is a composite function)}
\end{cases}
$$

For $u = G(x)$, let
$$
\begin{cases}
u = \cos v & \implies \dfrac{du}{dv} = -\sin v \\[2mm]
v = \dfrac{\pi}{x} = \pi x^{-1} & \implies \dfrac{dv}{dx} = -\pi x^{-2}
\end{cases}
$$

$$
\implies \underset{\uparrow}{\dfrac{du}{dx}} = \dfrac{du}{dv}\dfrac{dv}{dx} = (-\sin v)(-\pi x^{-2}) = \dfrac{\pi \sin(\pi/x)}{x^2}
$$

(Chain Rule: introduced v as the intermediate variable; $v = \pi/x$)

Hence, $\dfrac{dy}{dx} = \dfrac{dy}{du}\dfrac{du}{dx} = 3u^2 \left[\dfrac{\pi \sin(\pi/x)}{x^2} \right] = \underset{\uparrow}{\left(3\cos^2 \dfrac{\pi}{x} \right)} \left[\dfrac{\pi \sin(\pi/x)}{x^2} \right] = \dfrac{3\pi}{x^2} \cos^2 \dfrac{\pi}{x} \sin \dfrac{\pi}{x}$

$(u = \cos(\pi/x))$

The Power Rule Combined with the Chain Rule (2.34): If n is any real number and $u = g(x)$ is differentiable, then

$$
\dfrac{d}{dx}(u^n) = nu^{n-1}\dfrac{du}{dx}
$$

Alternatively,

$$
\dfrac{d}{dx}[g(x)]^n = n[g(x)]^{n-1} \cdot g'(x)
$$

Proof: For $y = [g(x)]^n$, let
$$
\begin{cases}
y = u^n & \implies \dfrac{dy}{du} = nu^{n-1} \\[2mm]
u = g(x) & \implies \dfrac{du}{dx} = g'(x)
\end{cases}
$$

$$
\implies \dfrac{dy}{dx} = \dfrac{dy}{du}\dfrac{du}{dx} = [nu^{n-1}][g'(x)] = n[g(x)]^{n-1}g'(x)
$$

Remark: In words: the derivative of something to a fixed power is the power down times that something to the power minus one times the derivative of that something.

Example 6: $\dfrac{dy}{dx} = ?$ if $y = \left(\dfrac{x+3}{2x-1}\right)^7$.

$$\dfrac{dy}{dx} = 7 \left(\dfrac{x+3}{2x-1}\right)^6 \dfrac{d}{dx}\left(\dfrac{x+3}{2x-1}\right)$$
$\uparrow$

(y = something to a fixed power; $n - 1 = 6$)

$$= 7\left(\dfrac{x+3}{2x-1}\right)^6 \dfrac{(2x-1)\dfrac{d}{dx}(x+3) - (x+3)\dfrac{d}{dx}(2x-1)}{(2x-1)^2}$$

$$= 7\left(\dfrac{x+3}{2x-1}\right)^6 \dfrac{(2x-1)(1) - (x+3)(2)}{(2x-1)^2} = -49 \dfrac{(x+3)^6}{(2x-1)^8}$$

Example 7: $f'(x) = ?$ if $f(x) = \sqrt[3]{x + \sqrt{x}}$.

$$f(x) = \sqrt[3]{x + \sqrt{x}} = \left(x + x^{1/2}\right)^{1/3}$$
$\uparrow$

(rewrite using exponents)

$$f'(x) = \dfrac{1}{3}\left(x + x^{1/2}\right)^{-2/3} \dfrac{d}{dx}\left(x + x^{1/2}\right) = \dfrac{1}{3}\left(x + x^{1/2}\right)^{-2/3}\left(1 + \dfrac{1}{2}x^{-1/2}\right)$$
$\uparrow$

$$\left(\text{Power Rule/Chain Rule:} \begin{cases} n = 1/3 : \text{ fixed} \\ g(x) = x + x^{1/2} \end{cases}\right)$$

Example 8: $\dfrac{dy}{dx} = ?$ if $y = \cos(\tan^2 x)$.

$$\dfrac{dy}{dx} = -\sin(\tan^2 x)\,\dfrac{d}{dx}\,(\tan^2 x)$$
↑

$$\left(\text{Chain Rule: } y = (f \circ g)(x) \text{ with } \left\{ \begin{array}{l} f(x) = \cos x \implies f'(x) = -\sin x \\ g(x) = \tan^2 x \end{array} \right. \right)$$

$$= -\sin(\tan^2 x)2\tan x\sec^2 x$$
↑

$$\left(\text{Power Rule/Chain Rule: } \left\{ \begin{array}{l} n = 2 : \text{ fixed} \\ g(x) = \tan x \end{array} \right. \right)$$

Example 9: $\dfrac{dy}{dx} = ?$ if $y = \cos^2(\tan x)$

$$\dfrac{dy}{dx} = 2\cos(\tan x)\,\dfrac{d}{dx}\,[\cos(\tan x)]$$
↑

$$\left(\text{Power Rule/Chain Rule: } \left\{ \begin{array}{l} n = 2 : \text{ fixed} \\ g(x) = \cos(\tan x) \end{array} \right. \right)$$

$$= 2\cos(\tan x)[-\sin(\tan x)\sec^2 x]$$
↑

$$\left(\text{Chain Rule: } (f \circ g)(x) \text{ with } \left\{ \begin{array}{l} f(x) = \cos x \\ g(x) = \tan x \end{array} \right. \right)$$

Example 10: $F'(x) = ?$ if $F(x) = (x^2 + 7x)^\pi \sin\sqrt{x}$.

$$F(x) = (x^2 + 7x)^\pi \sin\sqrt{x} = (x^2 + 7x)^\pi \sin x^{1/2}$$
↑

(rewrite using exponents)

$$F'(x) = (x^2 + 7x)^\pi(\cos x^{1/2})\left(\tfrac{1}{2}x^{-1/2}\right) + (\sin x^{1/2})[\pi(x^2 + 7x)^{\pi-1}(2x + 7)]$$
↑

(Product Rule; Chain Rule; Power Rule/Chain Rule)

Example 11: $\dfrac{dy}{dx} = ?$ if $y = \csc^5(\cos^3 x). = \left\{\csc\left[(\cos x)^3\right]\right\}^5$

$$\frac{dy}{dx} = 5\csc^4(\cos^3 x)\ \frac{d}{dx}\left(\csc(\cos^3 x)\right)$$
↑
$$\left(\text{Power Rule/Chain Rule:}\begin{cases} n = 5:\ \text{fixed}\\ g(x) = \csc(\cos^3 x)\end{cases}\right)$$

$$= 5\csc^4(\cos^3 x)[-\csc(\cos^3 x)\cot(\cos^3 x)\ \frac{d}{dx}(\cos^3 x)]$$
↑
$$\left(\text{Chain Rule: } (f\circ g)'(x),\ \text{where}\begin{cases} f(x) = \csc x\\ g(x) = \cos^3 x\end{cases}\right)$$

$$= -5\csc^5(\cos^3 x)\cot(\cos^3 x)(3\cos^2 x)(-\sin x)$$
↑
$$\left(\text{Power Rule/Chain Rule:}\begin{cases} n = 3:\ \text{fixed}\\ g(x) = \cos x\end{cases}\right)$$

Example 12: $f'(x) = ?$ if $f(x) = |2 - x|$.

$$f(x) = |2 - x| = \sqrt{(2 - x)^2} = \left[(2 - x)^2\right]^{1/2}$$
↑
(rewrite using exponents)

$$f'(x) = \frac{1}{2}\left[(2 - x)^2\right]^{-1/2}\ \frac{d}{dx}\left((2 - x)^2\right)$$
↑
$$\left(\text{Power Rule/Chain Rule:}\begin{cases} n = 1/2:\ \text{fixed}\\ g(x) = (2 - x)^2\end{cases}\right)$$

$$= \frac{1}{2\sqrt{(2 - x)^2}}2(2 - x)(-1) = -\frac{2 - x}{|2 - x|}$$
↑
$$\left(\text{Power Rule/Chain Rule:}\begin{cases} n = 2:\ \text{fixed}\\ g(x) = 2 - x\end{cases}\right)$$

$$|2 - x| = \begin{cases} 2 - x & \text{if } 2 - x \geq 0 \\ -(2 - x) & \text{if } 2 - x < 0 \end{cases} \implies |2 - x| = \begin{cases} 2 - x & \text{if } x \leq 2 \\ -(2 - x) & \text{if } x > 2 \end{cases}$$

$$\implies f'(x) = \begin{cases} -1 & \text{if } x < 2 \\ 1 & \text{if } x > 2 \\ \underbrace{\text{does not exist}} & \text{if } x = 2 \\ (x = 2 : |2 - x| = 0) \end{cases}$$

Note! Graph of $y = |2 - x|$.

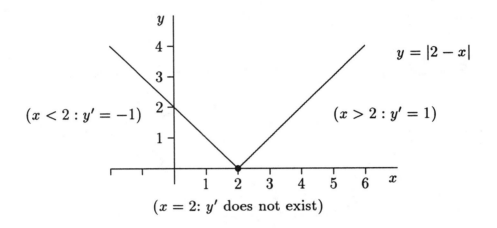

$(x < 2 : y' = -1)$ $(x > 2 : y' = 1)$

$(x = 2 : y'$ does not exist$)$

2.6 Implicit Differentiation

So far the variable y has been expressed explicitly as a function of the variable x: $\underbrace{y = f(x)}$.

$(y$ is solved for in terms of $x)$

Curves in the x, y plane may be represented by a relation involving both variables x and y and it may be difficult to express the relation as an expression in the form $y = f(x)$. The given relation may define y as a function of x or several functions of x.

a relation is made up of fun

Illustration 1: Consider the relation $4x^2 + y^2 = 4$ (an ellipse).

$$y^2 = 4 - 4x^2 \implies y = \pm\sqrt{4 - 4x^2} = \pm 2\sqrt{1 - x^2}$$

$$y = \begin{cases} 2\sqrt{1 - x^2} & (= f(x)) \\ -2\sqrt{1 - x^2} & (= g(x)) \end{cases} \implies y \text{ can be defined as two functions of } x.$$

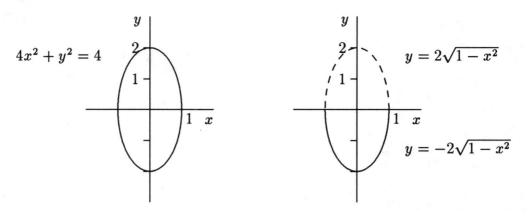

$4x^2 + y^2 = 4$

$y = 2\sqrt{1 - x^2}$

$y = -2\sqrt{1 - x^2}$

Note! The relation $4x^2 + y^2 = 4$ **implicitly** defines y as two functions of x.

Illustration 2: Consider a relation with its related curve sketched as follows:

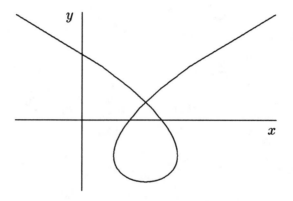

The variable y can be defined as three functions of x:

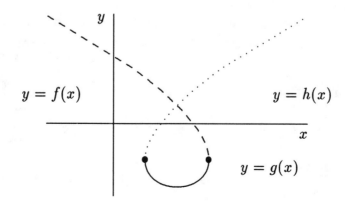

$y = f(x)$

$y = h(x)$

$y = g(x)$

Note! y is implicitly defined as three functions of x.

Question: $y' = \dfrac{dy}{dx} = ?$ if given a relation involving both variables x and y.

A Procedure:

(S1) Treat $y = f(x)$ (i.e., treat y as an expression in terms of x).

(S2) **Differentiate implicitly**: differentiate both sides of the relation with respect to x.

(S3) Solve for $\dfrac{dy}{dx}$: this expression may be in terms of both x and y.

Example 1: $\dfrac{dy}{dx} = ?$ if $x^3 + xy^2 + y^4 = 7$.

$$\frac{d}{dx}\left(x^3 + xy^2 + y^4\right) = \frac{d}{dx}\,(7) \quad \Longrightarrow \quad 3x^2 + \underbrace{x\,\frac{d}{dx}\,(y^2) + y^2\,\frac{d}{dx}\,(x)}_{\text{(Product Rule)}} + \frac{d}{dx}\,(y^4) = 0$$

$$\Longrightarrow \quad 3x^2 + x\left(\underbrace{2y\frac{dy}{dx}}\right) + y^2(1) + \underbrace{4y^3\frac{dy}{dx}} = 0$$

(Power Rule/Chain Rule: y to a fixed power is treated as a function of x to that fixed power)

$$\Longrightarrow \quad \frac{dy}{dx} = -\frac{3x^2 + y^2}{2y\!x + 4y^3} \qquad \text{(in terms of } x \text{ and } y\text{)}$$

Example 2: Consider $x^4 y^2 = x^3 + y^4 + 1$. At the point $(-1, 1)$, find an equation of:

(A1) the tangent line.

$$m = \left.\frac{dy}{dx}\right|_{(-1,1)} = ?$$

$$\frac{d}{dx}\left(x^4 y^2\right) = \frac{d}{dx}\left(x^3 + y^4 + 1\right) \quad \Longrightarrow \quad x^4(2yy') + y^2(4x^3) = 3x^2 + 4y^3 y' + 0$$

$$\Longrightarrow \quad y' = \frac{3x^2 - 4x^3 y^2}{2x^4 y - 4y^3} \qquad \text{(in terms of } x \text{ and } y\text{)}$$

$$y'|_{(-1,1)} = \frac{3(-1)^2 - 4(-1)^3(1)^2}{2(-1)^4(1) - 4(1)^3} = -\frac{7}{2}$$

$$\implies y - 1 = -\frac{7}{2}[x - (-1)] \implies 7x + 2y + 5 = 0$$
↑

$$\left(\text{point-slope form:} \left\{ \begin{array}{l} \text{point } (-1, 1) \\ \text{slope } m = -7/2 \end{array} \right. \right)$$

(A2) the normal line.

$$m_\perp = -\frac{1}{m} = -\frac{1}{\left(-\frac{7}{2}\right)} = \frac{2}{7} \implies y - 1 = \frac{2}{7}[x - (-1)] \implies 2x - 7y + 9 = 0$$
          ↑   ↑
         (A1)

$$\left(\text{point-slope form:} \left\{ \begin{array}{l} \text{point } (-1, 1) \\ \text{slope } m_\perp = 2/7 \end{array} \right. \right)$$

Example 3: $y' = \dfrac{dy}{dx} = ?$ if $(x + y^2)^3 = y - x$.

$$\frac{d}{dx}\left((x + y^2)^3\right) = \frac{d}{dx}(y - x) \implies \underbrace{3(x + y^2)^2 \frac{d}{dx}(x + y^2)}_{\text{(Power Rule/Chain Rule)}} = y' - 1$$

$$\implies 3(x + y^2)^2(1 + \underbrace{2yy'}) = y' - 1 \implies y' = \frac{3(x + y^2)^2 + 1}{1 - 6y(x + y^2)^2}$$
        (Power Rule/Chain Rule)

Example 4: $y' = \dfrac{dy}{dx} = ?$ if $x^2y^3 = \tan(x - y)$.

$$\frac{d}{dx}(x^2y^3) = \frac{d}{dx}(\tan(x - y)) \implies x^2(3y^2y') + y^3(2x) = [\sec^2(x - y)](1 - y')$$

$$\implies y' = \frac{\sec^2(x - y) - 2xy^3}{3x^2y^2 + \sec^2(x - y)}$$

Example 5: If $2x^3 + [f(x)]^3 + 3xf(x) = 0$ and $f(-1) = 2$, find $f'(-1)$.

$$\frac{d}{dx}\left(2x^3 + [f(x)]^3 + 3xf(x)\right) = \frac{d}{dx}(0)$$

$$\Longrightarrow\quad 6x^2 + 3[f(x)]^2 f'(x) + 3[xf'(x) + f(x)(1)] = 0$$

$$\Longrightarrow\quad f'(x) = -\frac{6x^2 + 3f(x)}{3[f(x)]^2 + 3x} \quad\Longrightarrow\quad f'(-1) = -\frac{6(-1)^2 + 3(2)}{3(2)^2 + 3(-1)} = -\frac{4}{3}$$

$$\uparrow$$
$$(f(-1) = 2)$$

2.7   Higher Derivatives

Definition: Let n be a positive integer. If f is a function, then the **nth derivative of f** is the derivative of the $(n-1)$st derivative.

Remarks:

(R1) Notations for the nth derivative of $y = f(x)$ with respect to x:

$$f^{(n)}(x) = y^{(n)} = \frac{d^n y}{dx^n} = \frac{d^n f(x)}{dx^n} = D^n f(x) = D_x^n f(x)$$

(R2) If $y = f(x)$, then

$$f^{(n)}(x) = \frac{d}{dx}\left(f^{(n-1)}(x)\right) \quad\text{or}\quad \frac{d^n y}{dx^n} = \frac{d}{dx}\left(\frac{d^{n-1}y}{dx^{n-1}}\right)$$

(R3) $n = 1$: the first derivative

$$f^{(1)}(x) = f'(x) = \frac{df}{dx} = \frac{df^{(0)}(x)}{dx}$$
$$\uparrow \qquad\qquad \uparrow$$
(using "prime" notation) $\left(\text{notation: } f^{(0)}(x) = f(x)\right)$

(R4) $n = 2$: the second derivative

$$f^{(2)}(x) = f''(x) = \frac{d^2 f}{dx^2} = \frac{d}{dx}\left(f'(x)\right)$$
$$\uparrow$$
(using "double prime" notation)

Example 1: $f(x) = 3x^6 - 7x^3 - 2x$

(A1) $f'(x) = ?$

$$f'(x) = 18x^5 - 21x^2 - 2$$

(A2) $f''(x) = ?$

$$f''(x) = \frac{d}{dx}(18x^5 - 21x^2 - 2) = 90x^4 - 42x$$

(A3) $f'''(x) = ?$

$$f'''(x) = \frac{d}{dx}(90x^4 - 42x) = 360x^3 - 42$$

(A4) $f^{(4)}(x) = ?$

$$f^{(4)}(x) = \frac{d}{dx}(360x^3 - 42) = 1080x^2$$

(A5) $f^{(n)}(x) = ?$ for $n \geq 7$

$$f^{(5)}(x) = 2160x \implies f^{(6)}(x) = 2160 \implies f^{(7)}(x) = 0 \implies f^{(n)}(x) = 0 \quad \text{for} \quad n \geq 7$$
$$\uparrow$$
$$\left(\frac{d}{dx}(0) = 0 \right)$$

Example 2: $f(x) = \cos 2x.$ $\qquad f^{(21)}(x) = ?$

$$\implies f'(x) = \underbrace{(-\sin 2x)(2)}_{\text{(Chain Rule)}} = -2\sin 2x$$

$$\implies f''(x) = -2\underbrace{(\cos 2x)(2)}_{\text{(Chain Rule)}} = -4\cos 2x = -2^2 \cos 2x$$

$$\implies f'''(x) = -2^2(-\sin 2x)(2) = 2^3 \sin 2x$$

$$\implies f^{(4)}(x) = 2^3(\cos 2x)(2) = 2^4 \cos 2x$$

From recognition of "cyclic" pattern: $f^{(21)}(x) = -2^{21} \sin 2x$

pattern		n					
$+2^n \cos 2x$	0	4	8	12	16	20	
$-2^n \sin 2x$	1	5	9	13	17	21	
$-2^n \cos 2x$	2	6	10	14	18		
$+2^n \sin 2x$	3	7	11	15	19		

Example 3: $\dfrac{d^2y}{dx^2} = ?$ if $3x^2 - 2y^4 = 5$.

Differentiate implicitly with respect to x:

$$\frac{d}{dx}\left(3x^2 - 2y^4\right) = \frac{d}{dx}(5) \implies 6x - 8y^3 y' = 0 \implies y' = \frac{6x}{8y^3} = \frac{3x}{4y^3}$$

$$\implies \frac{d^2y}{dx^2} = \frac{d}{dx}\left(\frac{3x}{4y^3}\right) = \frac{3}{4}\frac{d}{dx}\left(\frac{x}{y^3}\right) = \frac{3}{4}\frac{y^3(1) - x(3y^2 y')}{y^6}$$

(Quotient Rule; Power Rule/Chain Rule)

$\frac{3}{4}\,\dfrac{y^3 - 3xy^2 y'}{y^6}$

$\frac{3}{4}\,\dfrac{y - 3xy'}{y^4}$

divide out y^2

$$= \frac{3}{4}\frac{y - 3xy'}{y^4} = \frac{3}{4}\frac{y - 3x\left(\frac{3x}{4y^3}\right)}{y^4} = \frac{3}{16}\frac{4y^4 - 9x^2}{y^7}$$

(replace the symbol y' by its expression in terms of x and y)

Example 4: Consider $2x^3 + y^3 + 3xy = 0$. At the point $(-1,2)$ find the value of y''.

$$\frac{d}{dx}\left(2x^3 + y^3 + 3xy\right) = \frac{d}{dx}(0) \implies 6x^2 + 3y^2 y' + 3[xy' + y(1)] = 0$$

$$\implies y' = -\frac{6x^2 + 3y}{3y^2 + 3x} = -\frac{2x^2 + y}{x + y^2}$$

(simplify if possible)

$$y'' = \frac{d}{dx}(y') = \frac{d}{dx}\left(-\frac{2x^2 + y}{x + y^2}\right) = -\frac{(x + y^2)(4x + y') - (2x^2 + y)(1 + 2yy')}{(x + y^2)^2}$$

$$\implies \quad y''|_{(-1,2)} = -\frac{(-1+2^2)[4(-1)+(-4/3)]-[2(-1)^2+2][1+2(2)(-4/3)]}{(-1+2^2)^2} = -\frac{4}{27}$$

$$\left(\text{from above: } y'|_{(-1,2)} = -\frac{2(-1)^2+2}{-1+2^2} = -\frac{4}{3}\right)$$

Note! If an expression for y'' in terms of x and y is desired:

$$y'' = -\frac{(x+y^2)\left(4x-\dfrac{2x^2+y}{x+y^2}\right)-(2x^2+y)\left[1+2y\left(-\dfrac{2x^2+y}{x+y^2}\right)\right]}{(x+y^2)^2}$$

$$\left(\text{from above: } y' = -\frac{2x^2+y}{x+y^2}\right)$$

Example 5: Find a third degree polynomial P such that $P(-1) = -10$, $P'(-1) = 20$, $P''(-1) = -30$, and $P'''(-1) = 24$.

Let $P(x) = ax^3 + bx^2 + cx + d$, where the constants a, b, c, and d are to be determined.

$$P'(x) = 3ax^2 + 2bx + c \quad \implies \quad P''(x) = 6ax + 2b \quad \implies \quad P'''(x) = 6a$$

$$\left.\begin{array}{lll}
P(-1) = -10 & \implies & -a+b-c+d = -10 \\
P'(-1) = 20 & \implies & 3a-2b+c = 20 \\
P''(-1) = -30 & \implies & -6a+2b = -30 \\
P'''(-1) = 24 & \implies & 6a = 24
\end{array}\right\} : \quad \begin{array}{l}\text{four equations in the} \\ \text{four unknown constants}\end{array}$$

$$6a = 24 \quad \implies \quad a = 4 \quad \implies \quad \underset{\underset{(P''(-1)=-30)}{\uparrow}}{2b = -30 + 6(4) = -6} \quad \implies \quad b = -3$$

$$\implies \quad \underset{\underset{(P'(-1)=20)}{\uparrow}}{c = 20 - 3(4) + 2(-3) = 2} \quad \implies \quad \underset{\underset{(P(-1)=-10)}{\uparrow}}{d = -10 + 4 - (-3) + 2 = -1}$$

$$\implies \quad P(x) = 4x^3 - 3x^2 + 2x - 1$$

Note! Another procedure:

$$P'(x) = 3ax^2 + 2bx + c \implies P''(x) = 6ax + 2b \implies P'''(x) = 6a$$

$$\implies P'''(-1) = 24 \implies 6a = 24 \implies a = 4$$

$$\implies P''(x) = 6(4)x + 2b = 24x + 2b$$

$$P''(-1) = -30 \implies 24(-1) + 2b = -30 \implies b = -3$$

$$\implies P'(x) = 3(4)x^2 + 2(-3)x + c = 12x^2 - 6x + c$$

$$P'(-1) = 20 \implies 12(-1)^2 - 6(-1) + c = 20 \implies c = 2$$

$$\implies P(x) = 4x^3 - 3x^2 + 2x + d$$

$$P(-1) = -10 \implies 4(-1)^3 - 3(-1)^2 + 2(-1) + d = -10 \implies d = -1$$

$$\implies P(x) = 4x^3 - 3x^2 + 2x - 1$$

Interpretation of Second Derivative: Let an object in rectilinear motion have position function $s = f(t)$. The velocity of the object is

$$v(t) = f'(t) = \frac{ds}{dt}$$

(the instantaneous rate of change of distance with respect to time)

The **acceleration** of the object is

$$a(t) = v'(t) = f''(t) = \frac{dv}{dt} = \frac{d^2s}{dt^2}$$

(the instantaneous rate of change of velocity with respect to time)

Remark! Dimensions:

(R1) $s = f(t)$: distance

(R2) $v = f'(t)$: $\dfrac{\text{distance}}{\text{time}}$

(R3) $a = f''(t)$: $\dfrac{\text{distance}}{(\text{time})^2}$

Example 6: Rectilinear motion of an object is described by $s = \dfrac{1}{3}t^3 - 2t^2 + 3t + 4$, where s: feet and t: seconds.

(A1) $a = ?$ when $v = 0$.

$$v = \frac{ds}{dt} = t^2 - 4t + 3 \implies a = \frac{dv}{dt} = 2t - 4$$

$$v = 0 \implies t^2 - 4t + 3 = (t-1)(t-3) = 0 \implies t = 1,\ 3$$

$$a|_{t=1} = 2(1) - 4 = -2 \ \text{ft/s}^2$$

$$a|_{t=3} = 2(3) - 4 = 2 \ \text{ft/s}^2$$

(A2) $s = ?$ and $v = ?$ when $a = 0$.

$$a = 0 \implies \underset{\underset{\text{(A1)}}{\uparrow}}{2t - 4 = 0} \implies t = 2$$

$$s|_{t=2} = \frac{1}{3}(2)^3 - 3(2)^2 + 3(2) + 4 = \frac{14}{3} \ \text{ft}$$

$$v|_{t=2} = \underset{\underset{\text{(A1)}}{\uparrow}}{2^2 - 4(2) + 3 = -1} \ \text{ft/s}$$

2.8   Related Rates

Recall: When the quantity y depends upon time t, say $y = f(t)$, then $\dfrac{dy}{dx} = f'(t)$ represents the instantaneous rate of change of y with respect to t (time).

Terminology: If two or more quantities, which depend on t, are related by an equation, then an expression relating their rates of change with respect to time (known as **related rates**) can be obtained via differentiation of both sides of the equation with respect to t.

Procedure: With slight variations dependent upon the given problem, a method to solve related rate problems is:

(S1) Sketch a figure and enter variables for the quantities which vary.

(S2) Indicate the (instantaneous) rate of change to be determined.

(S3) From the given information, indicate the known rate(s) of change.

(S4) Determine equation(s) relating the quantities whose rates of change are identified in (S2) and (S3).

(S5) Differentiate (with respect to t) both sides of the equation(s) listed in (S4).

(S6) Use the information listed in (S3) to find the value of the unknown rate of change identified in (S2).

Example 1: A pair of opposite sides of a rectangle are becoming longer at the rate of 2 in/s, while the remaining two sides are decreasing in length such that the figure remains a rectangle with area 50 in^2.

(A1) Find the rate of change of the perimeter when the lengthening sides are 5 in.

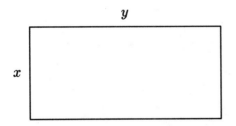

x: lengthening side

$P = 2(x + y)$ (perimeter of rectangle)

Question: $\dfrac{dP}{dt} = ?$ when $x = 5$

Given: $\begin{cases} xy = 50 \quad \text{(area of rectangle)} \\ \dfrac{dx}{dt} = \underbrace{+\,2}_{\text{(lengthening)}} \end{cases}$

$$\frac{dP}{dt} = \frac{d}{dt}\left(2(x+y)\right) = 2\left(\frac{dx}{dt} + \frac{dy}{dt}\right) = 2\left(\underset{\uparrow}{2} + \frac{dy}{dt}\right)$$

$$\left(\text{given: } \frac{dx}{dt} = 2\right)$$

$\dfrac{dy}{dt} = ?$ when $x = 5$.

$$\underbrace{xy = 50}_{(\text{area})} \implies \frac{d}{dt}(xy) = \frac{d}{dt}(50) \implies x\frac{dy}{dt} + y\frac{dx}{dt} = 0; \quad y = ?$$

When $x = 5$: $\underbrace{5y = 50}_{(\text{area})} \implies y = 10 \implies 5\frac{dy}{dt} + (10)(2) = 0 \implies \frac{dy}{dt} = -4$

$$\implies \frac{dP}{dt} = 2[2 + (-4)] = -4 \text{ in/s}$$

Note!

(N1) $\dfrac{dy}{dt} = -4 < 0$: the value of y is decreasing

(N2) $\dfrac{dP}{dt} = -4 < 0$ when $x = 5$: the perimeter is decreasing

(A2) What are the dimensions of the rectangle when the rate of change of the perimeter is zero?

Question: $x = ?$ and $y = ?$ when $\dfrac{dP}{dt} = 0$.

$$\frac{dP}{dt} = 0 \underset{\underset{(A1)}{\uparrow}}{\implies} 2\left(\frac{dx}{dt} + \frac{dy}{dt}\right) = 0 \implies \frac{dx}{dt} + \frac{dy}{dt} = 0 \implies \frac{dy}{dt} = -\frac{dx}{dt} = \underset{\uparrow}{-2}$$

$$\left(\text{given : } \frac{dx}{dt} = 2\right)$$

$$\underbrace{xy = 50}_{(\text{area})} \implies \frac{d}{dt}(xy) = \frac{d}{dt}(50) \implies x\frac{dy}{dt} + y\frac{dx}{dt} = 0$$

$\implies$ when $\dfrac{dP}{dt} = 0$: $x(-2) + y(2) = 0$

$\implies$ $y = x$ $\implies$ $x^2 = 50$ $\implies$ $x = 5\sqrt{2}$ in $\implies$ $y = 5\sqrt{2}$ in

              $\uparrow$ $\uparrow$

    (area: $xy = 50$) (length: $x > 0$)

Note! $\dfrac{dP}{dt} = 0$ when the rectangle becomes a square with side length $5\sqrt{2}$ in.

Example 2: Sand is being poured onto the top of a conical pile at the rate of 10 ft³/min. The coefficient of friction of the sand maintains the cone so its height and base radius are always equal.

(A1) At what rate is the height increasing when the pile is 8 feet high?

    Keywords: conical; onto; ft³/min

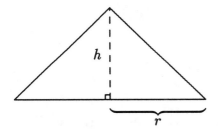

                                         h: height of cone

                                         r: base radius of cone

ft³ $\implies$ volume: $V = \dfrac{1}{3}\pi r^2 h$ (volume of cone)

Question: $\dfrac{dh}{dt} = ?$ when $h = 8$.

Given: $\begin{cases} h = r \\ \dfrac{dV}{dt} = \underbrace{+}\, 10 \text{ ft}^3/\text{min} \\ \qquad\quad\text{(onto, so volume increasing)} \end{cases}$

$V = \dfrac{1}{3}\pi r^2 h = \dfrac{1}{3}\pi h^3$

         $\uparrow$

    (given: $r = h$)

Note! Prefer to relate V and h only, since the given information involves only the rate $\dfrac{dV}{dt}$ and the question pertains to the rate $\dfrac{dh}{dt}$.

$$\frac{dV}{dt} = \frac{d}{dt}\left(\frac{1}{3}\pi h^3\right) = \frac{1}{3}\pi\left(\underbrace{3h^2\frac{dh}{dt}}\right) = \pi h^2\frac{dh}{dt}$$

(Power Rule/Chain Rule)

When $h = 8$: $10 = \pi(8)^2\dfrac{dh}{dt} \implies \dfrac{dh}{dt} = \dfrac{5}{32\pi}$ ft/min

Note! $\dfrac{dh}{dt} = +\dfrac{5}{32\pi} > 0$: the value of h is increasing

(A2) At what rate is the radius increasing when the pile is 5 feet high?

Question: $\dfrac{dr}{dt} = ?$ when $h = 5$

$$V = \frac{1}{3}\pi r^2 h = \frac{1}{3}\pi r^3$$
$$\uparrow$$
(given: $h = r$)

Note! Want to obtain an equation relating the rates $\dfrac{dV}{dt}$, the given, and $\dfrac{dr}{dt}$, the unknown.

$$\frac{dV}{dt} = \frac{d}{dt}\left(\frac{1}{3}\pi r^3\right) = \frac{1}{3}\pi\left(3r^2\frac{dr}{dt}\right) = \pi r^2\frac{dr}{dt}$$

When $h = 5$: $\underbrace{10 = \pi(5)^2\dfrac{dr}{dt}} \implies \dfrac{dr}{dt} = \dfrac{2}{5\pi}$ ft/min

$$\left(\frac{dV}{dt} = 10;\ h = r, h = 5 \implies r = 5\right)$$

Example 3: Repeat (A1) of Example 1 above.

Question: $\dfrac{dP}{dt} = ?$ when $x = 5$.

Given: $\begin{cases} \dfrac{dx}{dt} = 2 \text{ in/s} \\ xy = 50 \end{cases}$

Note! Prefer to relate only the variables P and x since the unknown rate is $\dfrac{dP}{dt}$ and the rate directly given is $\dfrac{dx}{dt}$.

$\begin{cases} P = 2(x + y) \\ xy = 50 \end{cases} \implies P = 2\left(x + \dfrac{50}{x}\right)$ $\left(y = \dfrac{50}{x}\right)$

$\implies \dfrac{dP}{dt} = \dfrac{d}{dt}\left(2\left(x + \dfrac{50}{x}\right)\right) = 2\left[\dfrac{dx}{dt} + 50\left(-x^{-2}\dfrac{dx}{dt}\right)\right] 2\left(1 - \dfrac{50}{x^2}\right)\dfrac{dx}{dt}$

When $x = 5$: $\dfrac{dP}{dt} = 2\left(1 - \dfrac{50}{5^2}\right)(2) = -4 \text{ ft/min}$ (as in Example a (A1) above)

Example 4: A spherical balloon is being deflated at the rate of 16 ft³/min.

(A1) Find the rate of change of the radius of the balloon when its radius is 2 feet.

Keywords: spherical; deflated; ft³/min

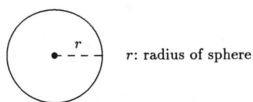

r: radius of sphere

ft³ $\implies$ volume: $V = \dfrac{4}{3}\pi r^3$ (volume of sphere)

Know it's volume because ft³/min and $V = \frac{4}{3}\pi r^3$

Question: $\dfrac{dr}{dt} = ?$ when $r = 2$.

Given: $\dfrac{dV}{dt} = -16 \text{ ft}^3/\text{min}$

↑

(deflated: decreasing)

rate = derivative

Note! Relate r and V.

$$V = \frac{4}{3}\pi r^3 \implies \frac{dV}{dt} = \frac{d}{dt}\left(\frac{4}{3}\pi r^3\right) = \frac{4}{3}\pi\left(3r^2\frac{dr}{dt}\right) = 4\pi r^2\frac{dr}{dt}$$

When $r = 2$: $-16 = 4\pi(2)^2\dfrac{dr}{dt} \implies \dfrac{dr}{dt} = -\dfrac{1}{\pi}$ ft/min

Note! $\dfrac{dr}{dt} = -\dfrac{1}{\pi} < 0$: the value of r is decreasing

(A2) Find the rate of change of the surface area of the balloon when its radius is 2 ft.

Surface area of sphere: $S = 4\pi r^2$

Question: $\dfrac{dS}{dt} = ?$ when $r = 2$.

Given: $\dfrac{dV}{dt} = -16$ ft^3/min

Note! Relate V and S.

$$V = \frac{4}{3}\pi r^3 \implies r = \left(\frac{3V}{4\pi}\right)^{1/3}$$

$$\implies S = 4\pi\left(\frac{3V}{4\pi}\right)^{2/3} = 4\pi\left(\frac{3}{4\pi}\right)^{2/3}V^{2/3} = (36\pi)^{1/3}V^{2/3}$$

$$\implies \frac{dS}{dt} = \frac{d}{dt}\left((36\pi)^{1/3}V^{2/3}\right) = (36\pi)^{1/3}\left(\frac{2}{3}V^{-1/3}\frac{dV}{dt}\right) = \frac{2}{3}(36\pi)^{1/3}\frac{1}{V^{1/3}}\frac{dV}{dt}$$

When $r = 2$: $V = \dfrac{4}{3}\pi(2)^3 = \dfrac{32}{3}\pi$ and

$$\frac{dS}{dt} = \frac{2}{3}(36\pi)^{1/3}\frac{1}{\left(\frac{32\pi}{3}\right)^{1/3}}(-16) = \frac{2}{3}\cdot\frac{3}{2}(-16) = -16 \quad \text{ft}^2/\text{min}$$

Note! Since both S and V are related to r the results of (A1) could be used if the equation involving S and r is differentiated.

$$S = 4\pi r^2 \implies \frac{dS}{dt} = 4\pi \left(2r\frac{dr}{dt}\right) = 8\pi r\frac{dr}{dt}$$

When $r = 2$: $\dfrac{dS}{dt} = 8\pi(2)\underbrace{\left(-\dfrac{1}{\pi}\right)}_{(A1)} = -16 \ \ \text{ft}^2/\text{min}$

Example 5: The base of a triangle is decreasing at the rate of 1 cm/min while the area of the triangle is increasing at the rate of 12 cm²/min. What is the rate of change of the altitude of the triangle when its altitude is 10 cm and its base is 4 cm?

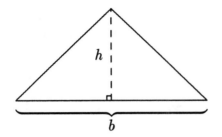

h: altitude of triangle

b: base of triangle

Keywords: triangle; base; area; altitude

area: $A = \dfrac{1}{2}bh$ (area of triangle)

Question: $\dfrac{dh}{dt} = ?$ when $h = 10, \quad b = 4.$

Given:
$$\begin{cases} \dfrac{db}{dt} = -1 \ \text{cm/min} & \text{(decreasing)} & : \dfrac{db}{dt} < 0 \\[3mm] \dfrac{dA}{dt} = +12 \ \text{cm}^2/\text{min} & \text{(increasing)} & : \dfrac{dA}{dt} > 0 \end{cases}$$

Note! Relate h, b, and A.

$$A = \frac{1}{2}bh \implies \frac{dA}{dt} = \frac{d}{dt}\left(\frac{1}{2}bh\right) = \frac{1}{2}\left(b\,\frac{dh}{dt} + h\,\frac{db}{dt}\right)$$

When $h = 10$, $b = 4$: $12 = \dfrac{1}{2}\left[4\,\dfrac{dh}{dt} + 10(-1)\right]$

$$\implies \frac{dh}{dt} = \frac{17}{2} \text{ cm/min}$$

Note! $\dfrac{dh}{dt} = \dfrac{17}{2} > 0$: the value of h is increasing

Example 6: A passenger liner is sailing south at 16 mi/h. A freighter, 32 miles directly south of the liner, is sailing east at 12 mi/h. What is the rate of change of the distance between the ships:

(A1) one hour later?

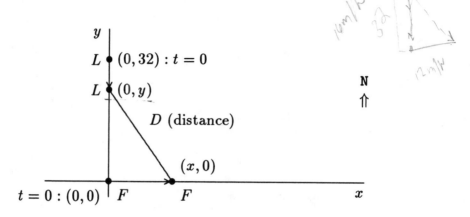

$$t = 0 : \begin{cases} \text{liner (L) at } (0, 32) \\ \text{freighter (F) at } (0, 0) \end{cases} \qquad t > 0 : \begin{cases} \text{liner at } (0, y) & (y < 32) \\ \text{freighter at } (x, 0) & (x > 0) \end{cases}$$

Question: $\dfrac{dD}{dt} = ?$ when $t = 1$

Given: $\begin{cases} \dfrac{dy}{dt} = -16 \text{ mi/h} \quad \text{(decreasing)} \quad : \dfrac{dy}{dt} < 0 \\[3mm] \dfrac{dx}{dt} = +12 \text{ mi/h} \quad \text{(increasing)} \quad : \dfrac{dx}{dt} > 0 \end{cases}$

Note! Relate D, y, and x.

$$D = \sqrt{(x - 0)^2 + (0 - y)^2} = \sqrt{x^2 + y^2}$$

$$\implies \frac{dD}{dt} = \frac{d}{dt}\left(\sqrt{x^2 + y^2}\right) = \frac{d}{dt}\left((x^2 + y^2)^{1/2}\right) = \frac{1}{2}(x^2 + y^2)^{-1/2}\frac{d}{dt}(x^2 + y^2)$$

$$\uparrow \qquad\qquad\qquad\qquad\qquad \uparrow$$

(use exponent notation)   (Power Rule/Chain Rule)

$$= \frac{1}{2}(x^2 + y^2)^{-1/2}\underbrace{\left(2x\frac{dx}{dt} + 2y\frac{dy}{dt}\right)} = \frac{x\dfrac{dx}{dt} + y\dfrac{dy}{dt}}{\sqrt{x^2 + y^2}}$$

(Addition Rule; Power Rule/Chain Rule)

When $t = 1:$ $y = 32 + (-16)(1) = 16,$

$$\uparrow$$

$$\left(\frac{dy}{dt} = -16 : \text{ constant } \implies y = 32 + \left(\frac{dy}{dt}\right)t\right)$$

$$x = 0 + (12)(1) = 12 \quad \text{and}$$

$$\uparrow$$

$$\left(\frac{dx}{dt} = 12 : \text{ constant } \implies x = 0 + \left(\frac{dx}{dt}\right)t\right)$$

$$\frac{dD}{dt} = \frac{12(12) + 16(-16)}{\sqrt{(12)^2 + (16)^2}} = -\frac{28}{5} \text{ mi/h}$$

Note! $\dfrac{dD}{dt} = -\dfrac{28}{5} < 0$, so the distance between the ships is decreasing.

(A2) two hours later?

Proceed as in (A1) above to obtain $\dfrac{dD}{dt} = \dfrac{x\dfrac{dx}{dt} + y\dfrac{dy}{dt}}{\sqrt{x^2 + y^2}}$

When $t = 2:$ $y = 32 + (-16)(2) = 0,$ $x = 0 + (12)(2) = 24$ and

$$\frac{dD}{dt} = \frac{24(12) + 0(-16)}{\sqrt{(24)^2 + 0^2}} = 12 \text{ mi/h}$$

Note! $\dfrac{dD}{dt} = 12 > 0$, so the distance between the ships is increasing.

Example 7: A tank in the shape of a cone (with its vertex downwards) has height 6 feet and base radius 4 feet. If water is leaking from the tank at the rate of .2 ft^3/min, what is the rate of change of the depth of the water when the depth is 3 feet?

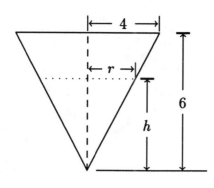

r: base radius of water
h: depth of water

Keywords: cone; leaking; ft^3/min

ft^3 $\Longrightarrow$ volume : $V = \dfrac{1}{3}\pi r^2 h$ (volume of cone of water)

Question: $\dfrac{dh}{dt} = ?$ when $h = 3$.

Given: $\dfrac{dV}{dt} = -.2$ ft^3/min $\left(\text{leaking, so volume is decreasing: } \dfrac{dV}{dt} < 0\right)$

Note! Relate h and V.

For similar triangles (from conical container and cone of water):

$$\dfrac{r}{4} = \dfrac{h}{6} \implies r = \dfrac{2}{3}h \implies V = \dfrac{1}{3}\pi \left(\dfrac{2}{3}h\right)^2 h = \dfrac{4}{27}\pi h^3$$

$$\dfrac{4}{27}\pi h^3 = V \checkmark$$

$$\implies \dfrac{dV}{dt} = \dfrac{d}{dt}\left(\dfrac{4}{27}\pi h^3\right) = \dfrac{4}{27}\pi\left(3h^2\dfrac{dh}{dt}\right) = \dfrac{4}{9}\pi h^2\dfrac{dh}{dt}$$

When $h = 3$: $-.2 = \dfrac{4}{9}\pi(3)^2\dfrac{dh}{dt} \implies \dfrac{dh}{dt} = -\dfrac{.05}{\pi}$ ft/min

Note! $\dfrac{dh}{dt} = -\dfrac{.05}{\pi} < 0$, so the value of h is decreasing.

cone → think - similiar triangles

Example 8: The volume of a contracting cube is changing at the rate of 2 cm³/min. When the volume of the cube is 64 cm³, find the rate of change of:

(A1) an edge.

Keywords: volume; contracting; cube

x: edge of cube   Volume: $V = x^3$ (volume of cube)

Question: $\dfrac{dx}{dt} = ?$ when $V = 64$

Given: $\dfrac{dV}{dt} = -2$ cm³/min $\left(\text{contracting, so volume is decreasing: } \dfrac{dV}{dt} < 0 \right)$

Note! Relate x and V.

$V = x^3 \implies \dfrac{dV}{dt} = \dfrac{d}{dt}(x^3) = 3x^2 \dfrac{dx}{dt}$

When $V = 64$: $64 = x^3 \implies x = 4$; hence $-2 = 3(4)^2 \dfrac{dx}{dt}$

$\implies \dfrac{dx}{dt} = -\dfrac{1}{24}$ cm/min

Note! $\dfrac{dx}{dt} = -\dfrac{1}{24} < 0$, so the value of x is decreasing.

(A2) the total surface area.

Surface area: $S = 6x^2$ (surface area of cube)

Question: $\dfrac{dS}{dt} = ?$ when $V = 64$

Given: $\dfrac{dV}{dt} = -2$ cm³/min

Note! Relate S and V.

$V = x^3 \implies x = V^{1/3} \implies S = 6(V^{1/3})^2 = 6V^{2/3}$

$$\implies \quad \frac{dS}{dt} = \frac{d}{dt}\left(6V^{2/3}\right) = 6\frac{d}{dt}\left(V^{2/3}\right) = 6\left(\frac{2}{3}V^{-1/3}\frac{dV}{dt}\right) = 4\frac{1}{V^{1/3}}\frac{dV}{dt}$$

When $V = 64$: $\dfrac{dS}{dt} = 4\dfrac{1}{(64)^{1/3}}(-2) = -2$ cm^2/min

Note!

(N1) $\dfrac{dS}{dt} = -2 < 0$, so the value of S is decreasing

(N2) Since both S and V are related to x the results of (A1) could be used if the equation involving S and x is differentiated.

$$S = 6x^2 \quad \implies \quad \frac{dS}{dt} = \frac{d}{dt}\left(6x^2\right) = 12x\frac{dx}{dt}$$

When $V = 64$: $64 = x^3 \quad \implies \quad x = 4$;   hence,

$$\frac{dS}{dt} = 12(4)\underbrace{\left(-\frac{1}{24}\right)}_{(A1)} = -2 \text{ cm}^2/\text{min} \quad \text{(as above)}$$

2.9 Differentials and Linear Approximations

Definition (2.46): Let $y = f(x)$, where f is a differentiable function. Then the **differential** dx is an independent variable; that is, dx can be given the value of any real number. The **differential** dy is then defined in terms of dx by the equation

$$dy = f'(x)\,dx$$

Remarks:

(R1) The differential dy depends upon both the variable x (in the domain of f') and the differential dx (which can be assigned any real value).

(R2) From the definition of differentials: if $dx \neq 0$, then

$$f'(x) = \frac{dy}{dx},$$

the ratio of differentials. In Leibniz notation:

$$f'(x) = \frac{dy}{dx},$$

a symbol which does not represent a quotient.

Geometric Interpretation: $dx \neq 0$

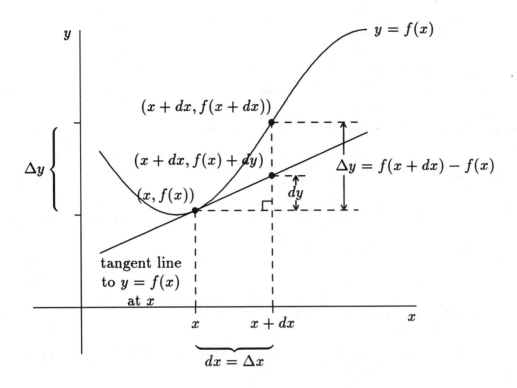

$f'(x)$: slope of tangent line to $y = f(x)$ at x

$$\implies \quad f'(x) = \frac{dy}{dx}: \text{ratio of differentials}$$
$\uparrow$

(rise: dy; run: $dx \neq 0$)

$\implies$ $dy = f'(x) \, dx$ represents the amount the *tangent line* rises/falls when the domain value changes from x by the value $dx \neq 0$.

> *Note!* Δy is the amount the *curve* rises/falls when the domain value changes from x by the amount $dx \neq 0$.

Problem: Given $y = f(x)$, find dy.

Solution:

(S1) Find $f'(x)$.

(S2) Multiply $f'(x)$ by the symbol dx: $\quad dy = f'(x) \, dx$.

Example 1: $dy = ?$ if $y = x^4 + \sqrt{3x - x^{-2}}$.

$$y = x^4 + (3x - x^{-2})^{1/2} \implies \frac{dy}{dx} = 4x^3 + \frac{1}{2}(3x - x^{-2})^{-1/2}(3 + 2x^{-3})$$

$$\implies dy = \left(\frac{dy}{dx}\right)dx = \left(4x^3 + \frac{3 + 2x^{-3}}{2\sqrt{3x - x^{-2}}}\right)dx$$

Note! This technique is equivalent to finding the derivative, using Leibniz notation,

$$\frac{dy}{dx} = \text{expression}$$

and then treating $\dfrac{dy}{dx}$ as a ratio of differentials, provided $dx \neq 0$, to obtain (via multiplication)

$$dy = (\text{expression})\, dx.$$

Example 2: $dy = ?$ if $y = \tan^3(\sin 5x)$.

$$\frac{dy}{dx} = 3\tan^2(\sin 5x)[\sec^2(\sin 5x)])(\cos 5x)5 = 15\tan^2(\sin 5x)\sec^2(\sin 5x)\cos 5x$$

$$\implies dy = 15\tan^2(\sin 5x)\sec^2(\sin 5x)\cos 5x\, dx$$

Example 3: Consider $y = 3 - (x - 1)^2$. At $x = 2$ and $dx = -1$ calculate the:

(A1) value of dy.

$$\frac{dy}{dx} = 0 - 2(x - 1)(1) = -2(x - 1) \implies dy = -2(x - 1)\, dx$$

$$x = 2, \quad dx = -1 \implies dy = -2(2 - 1)(-1) = 2$$

(A2) value of Δy.

$$f(x) = 3 - (x - 1)^2$$

$$\left.\begin{array}{l} x = 2: \quad f(2) = 3 - (2 - 1)^2 = 2 \\ x + dx = 2 + (-1) = 1 : \\ f(1) = 3 - (1 - 1)^2 = 3 \end{array}\right\} \implies \Delta y = \underbrace{f(1) - f(2)}_{(f(x + dx) - f(x))} = 3 - 2 = 1$$

Note! A sketch of the graph of $y = 3 - (x-1)^2$ with dy and Δy included for $x = 2$ and $dx = -1$: find an equation of the tangent line at $x = 2$.

$$\left\{\begin{array}{l} \text{point } (2, f(2)) = (2, 2) \\[2mm] \text{slope } m = \dfrac{dy}{dx}\bigg|_{x=2} = -2(2-1) = -2 \end{array}\right\} \implies \left\{\begin{array}{l} y - 2 = -2(x-2) \\[2mm] \implies y = -2x + 6 \end{array}\right.$$

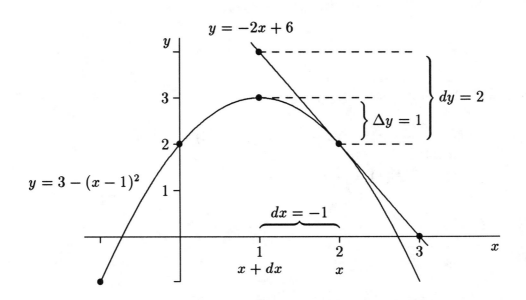

Using Leibniz notation: $\dfrac{dy}{dx} = \lim\limits_{\Delta x \to 0} \dfrac{\Delta y}{\Delta x}$

$\implies$ for Δx close to zero, but $\Delta x \neq 0$: $\dfrac{dy}{dx} \approx \dfrac{\Delta y}{\Delta x}$

$\implies$ letting $dx = \Delta x$: $dy \approx \Delta y$

$\implies$ $dy \approx f(x + \Delta x) - f(x)$

$\implies$ $f(x + \Delta x) \approx \underbrace{f(x) + dy}$
$\qquad\qquad\qquad$ (value of y at $x + \Delta x$ on the tangent line determined at x)

Remarks: For Δx close to zero:

(R1) $\Delta y - dy \approx 0$

(R2) $f(x + \Delta x) \approx f(x) + \underbrace{f'(x)\,dx}_{(dy)}$

Example 4: Consider $y = 2 + 2x - x^2$. Compute Δy, dy, and $\Delta y - dy$ at the value $x = 2$ and the value:

(A1) $\Delta x = 1$.

$$f(x) = 2 + 2x - x^2 \implies f'(x) = 2 - 2x$$

$$\Delta y = f(x + \Delta x) - f(x) = f(2 + 1) - f(2) = f(3) - f(2) = -1 - 2 = -3$$

$$dy = f'(x)\,dx = f'(2)(1) = (-2)(1) = -2$$

$$\Delta y - dy = -3 - (-2) = -1$$

(A2) $\Delta x = 0.5$

$$\Delta y = f(2 + .5) - f(2) = f(2.5) - f(2) = .75 - 2 = -1.25$$

$$dy = f'(2)(.5) = (-2)(.5) = -1$$

$$\Delta y - dy = -1.25 - (-1) = -0.25$$

(A3) $\Delta x = 0.1$

$$\Delta y = f(2 + .1) - f(2) = f(2.1) - f(2) = 1.79 - 2 = -0.21$$

$$dy = f'(2)(.1) = (-2)(.1) = -0.2$$

$$\Delta y - dy = -.21 - (-.2) = -0.01$$

(A4) $\Delta x = 0.05$

$$\Delta y = f(2 + .05) - f(2) = f(2.05) - f(2) = 1.8975 - 2 = -0.1025$$

$$dy = f'(2)(.05) = (-2)(.05) = -0.1$$

$$\Delta y - dy = -.1025 - (-.1) = -0.0025$$

Note!

(N1) As $\Delta x \to 0$, (A1)-(A4) illustrate $\Delta y - dy \to 0$.

(N2) The following graph illustrates $\Delta y - dy$ for $\Delta x = 1$ in (A1).

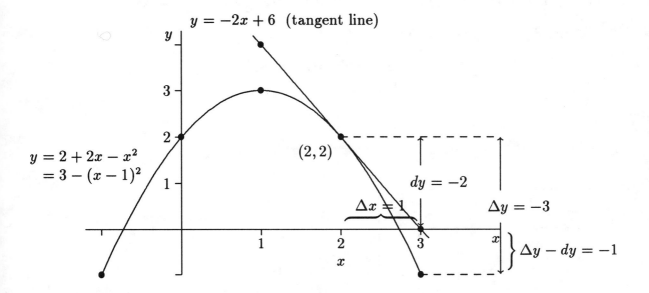

Comment: For $x = 2$, the graph shows,

as $\Delta x = dx \to 0$, $x + \Delta x = 2 + \Delta x \to 2$ and $\Delta y - dy \to 0$.

Problem: Approximate the value of a function using differentials.

A Method:

(S1) Determine, from the given information, both the function (an expression in terms of the variable x) and the value $x + dx$ at which the function is to be approximated.

(S2) Choose x "near" $x + dx$ (hence, dx is "small") such that both $f(x)$ and $f'(x)$ are easily evaluated.

(S3) $\underbrace{f(x + dx)}_{\text{(actual value)}} \approx \underbrace{f(x) + f'(x)\,dx}_{\text{(approximating value)}}$

$$f(x + dx) \approx f(x) + f'(x)\underbrace{[(x + dx) - x]}$$

$$(x + dx: \text{ given, from (S1)}; \ x: \text{ chosen, from (S2)})$$

Note!

(N1) $f(\text{given}) \approx f(\text{chosen}) + f'(\text{chosen})(\text{given} - \text{chosen})$

(N2) $f(x + dx) \approx f(x) + f'(x)\,dx$,   so the error in the approximation is

$$f(x + dx) - [f(x) + f'(x)\,dx] = [f(x + dx) - f(x)] - f'(x)\,dx = \Delta y - dy$$

(N3) "Chosen" close to "given" in order for

$$\Delta x = dx = \text{given} - \text{chosen}$$

to be small.

Illustration: $dx > 0$. Approximate the area of a square of side $x + dx$.

area of square with side x:

$$f(x) = x^2 \implies f'(x) = 2x$$

$$f(x + dx) = (x + dx)^2 = x^2 + 2x\,dx + (dx)^2$$

$$f(x) + f'(x)\,dx = x^2 + 2x\,dx$$

$\implies$ error in approximation: $f(x + dx) - [f(x) + f'(x)\,dx] = (dx)^2$

Note! $dx \to 0 \implies (dx)^2 \to 0$

Example 5: Estimate the volume of a cube with side 4.9 in.

Let $f(x) = x^3$: volume of cube

A choice: $x = 5$ near $x + dx = 4.9$

$\Longrightarrow$ $f(4.9) \approx f(5) + f'(5)(4.9 - 5)$

$\quad\quad f(x) = x^3 \Longrightarrow f'(x) = 3x^2$

$\Longrightarrow$ $f(4.9) \approx 5^3 + 3(5)^2(-.1) = 125 - 7.5 = 117.5$ in^3

Note!

(N1) $f(4.9) = (4.9)^3 = 117.649$ in^3

(N2) The exact volume is $(x + dx)^3 = \underbrace{x^3 + 3x^2\, dx}_{(f(x) + f'(x)\, dx)} + \underbrace{3x(dx)^2 + (dx)^3}_{(\Delta y - dy)}$

$\Longrightarrow$ $117.649 = 117.5 + \underbrace{.149}$

$(x = 5,\ dx = -.1: \quad 3x(dx)^2 + (dx)^3 = .15 - .001 = 0.149)$

Example 6: Approximate $\sqrt{11}$

Let $f(x) = x^{1/2}$: square root.

A choice: $x = 9$ near $x + dx = 11$

$\Longrightarrow$ $f(11) \approx f(9) + f'(9)(11 - 9)$

$\quad\quad f(x) = x^{1/2} \Longrightarrow f'(x) = \frac{1}{2}x^{-1/2} = \frac{1}{2x^{1/2}}$

$\Longrightarrow$ $f(11) \approx 9^{1/2} + \frac{1}{2(9)^{1/2}}(2) = 3 + \frac{1}{3} = \frac{10}{3}$

$\Longrightarrow$ $\sqrt{11} \approx 3.33333333$

Note!

(N1) $\sqrt{11} = 3.31662479$ (via calculator)

$\implies$ error : $3.31662479 - 3.33333333 = -0.016708539$

(N2) Another choice: $x = 16 \implies (x + dx) - x = 11 - 16 = -5$, so
this choice of x is not as near $x + dx = 11$ as the choice $x = 9$ used above.

$$f(11) \approx f(16) + f'(16)(11 - 16)$$

$$\implies \quad f(11) \approx (16)^{1/2} + \frac{1}{2(16)^{1/2}}(-5) = 4 - \frac{5}{8} = \frac{27}{8}$$

$\implies$ $\sqrt{11} \approx 3.375$ $\implies$ error : $3.31662497 - 3.375 = -0.05837521$,
which is greater than the error for the choice $x = 9$.

Example 7: A cube, due to being heated by the sun, is expanding. Approximate the change
in surface area as the side length changes from 5 cm to 5.01 cm.

Let $f(x) = 6x^2$: surface area of cube

Want *change* in surface area: $f(x + dx) - f(x) = \Delta y$

$$\implies \quad f(x + dx) - f(x) \approx dy = f'(x)\,dx = 12x\,dx$$

Given: $\left\{ \begin{array}{l} x = 5 \\ x + dx = 5.01 \end{array} \right\}$ $\implies$ $dx = (x + dx) - x = 5.01 - 5 = 0.01$

$$\implies \quad \Delta y \approx dy = f'(5)(.01) = 12(5)(.01) = 0.6 \quad \text{cm}^2$$

Example 8: The measured edge of a cube is $x = 10 \pm 0.02$ in.

Note! .02 is the largest error when measuring $x = 10$.

(A1) Find the maximum error when using $x = 10$ to calculate the surface area.

Let $S = 6x^2$: surface area of cube

Maximum error when calculating S: $\Delta S = ?$

$$\Delta S \approx dS = S'(x)\,dx = 12x\,dx$$

Given: $\left\{ \begin{array}{l} x = 10 \\ \underline{dx = 0.02} \\ (\text{maximum}) \end{array} \right\}$ $\implies$ $dS = S'(10)(.02) = 12(10)(.02) = 2.4 \text{ in}^2$

$\implies$ maximum error $\approx 2.4 \text{ in}^2$

(A2) Find the **relative error** when using $x = 10$ to calculate the surface area

Note! If y is a quantity, then the relative error $= \dfrac{\Delta y}{y} \left(= \dfrac{\text{error}}{\text{value}} \right)$

For $S = 6x^2$, the relative error $= \dfrac{\Delta S}{S} \approx \dfrac{dS}{S} \implies$ relative error $\approx \dfrac{2.4}{6(10)^2} = 0.004$

$$\uparrow$$
$$((\text{A1}): \; dS = 2.4; \; x = 10)$$

Note!

(N1) "Large" error in S, but "small" relative error.

(N2) The relative error in the measurement of the side length: relative error $= \dfrac{\Delta x}{x}$

$\implies$ relative error $= \dfrac{dx}{x} = \dfrac{.02}{10} = .002$

$$\uparrow$$

(x: independent variable $\implies$ $\Delta x = dx$)

(N3) **Percentage error**: percentage error $= (\text{relative error})(100\%)$

Form (A2): a percentage error of 0.2% in side length produces a 0.4% percentage error in surface area (of a cube).

Consider a curve $y = f(x)$. Fix x_1 in the domain of f'. For x "near" x_1:

$$\Delta y \approx dy \implies f(x) - f(x_1) \approx f'(x_1) \underbrace{(x - x_1)}_{dx}$$

$\implies$ $f(x) \approx \underbrace{f(x_1) + f'(x_1)(x - x_1)}_{\text{(equation of tangent line to } y = f(x) \text{ at } (x_1, f(x_1)))} = y$

Definition: The **linear approximation** of a function f at x_1 is

$$L(x) = f(x_1) + f'(x_1)(x - x_1).$$

Remarks:

(R1) $L(x)$ is the **linearization** of f at x_1.

(R2) Use the value of $L(x)$ to approximate the value of $f(x)$ for x near x_1.

   Note!

   (N1) For x near x_1: $f(x) \approx L(x)$

   (N2) At x_1: $f(x_1) = L(x_1)$

(R3) Graphically:

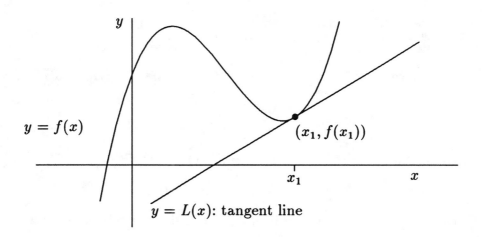

$y = f(x)$

$(x_1, f(x_1))$

x_1

$y = L(x)$: tangent line

$$f(x_1) = L(x_1)$$

Example 9: Find the linearization of $f(x) = \sqrt{x}$ at $x_1 = 9$.

$$f(x) = x^{1/2} \implies f'(x) = \frac{1}{2}x^{-1/2} = \frac{1}{2x^{1/2}}$$

$$x_1 = 9: \quad f(9) = \sqrt{9} = 3, \quad f'(9) = \frac{1}{2\sqrt{9}} = \frac{1}{6}$$

$$\implies L(x) = 3 + \frac{1}{6}(x - 9) = \frac{1}{6}x + \frac{3}{2} \approx \sqrt{x}$$

Note!

(N1) At $x_1 = 9$: $f(9) = 3 = L(9)$

(N2) $f(11) = \sqrt{11} \approx L(11) = \frac{1}{6}(11) + \frac{3}{2} = \frac{10}{3} = 3.33333333$: as in Example 6 above.

(N3) Graphically: sketch of $y = f(x)$ and $L(x)$, where $x_1 = 9$, with $f(11) \approx L(11)$

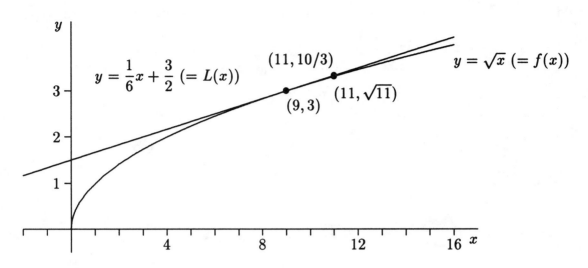

Example 10: Approximate $\cot 43°$.

Let $f(x) = \cot x$, where x is in radians.

A choice: $x = \dfrac{\pi}{4}$ $(= 45°)$ near $x + dx = 43 \underbrace{\left(\dfrac{\pi}{180}\right)}$

$$\underbrace{\text{(convert degrees to radians)}}$$

$$f\left(\frac{43\pi}{180}\right) \approx \underbrace{f\left(\frac{\pi}{4}\right) + f'\left(\frac{\pi}{4}\right)\left(\frac{43\pi}{180} - \frac{\pi}{4}\right)}_{\left(= L\left(\frac{43\pi}{180}\right) \text{ for } x_1 = \frac{\pi}{4}\right)}$$

$$f(x) = \cot x \implies f'(x) = -\csc^2 x$$

$$\underset{\uparrow}{\implies} f\left(\frac{43\pi}{180}\right) \approx 1 + [-(\sqrt{2})^2]\left(-\frac{\pi}{90}\right) = 1 + \frac{\pi}{45} = 1.07 \implies \cot 43° \approx 1.07$$

$$\left(\cot\frac{\pi}{4} = 1; \quad \csc\frac{\pi}{4} = \sqrt{2}; \quad \frac{43\pi}{180} - \frac{\pi}{4} = -\frac{\pi}{90}\right)$$

2.10   Newton's Method

Problem: Given the equation $f(x) = 0$, where f is a differentiable function, solve for x.

Let r be a value in the domain of f such that $\underbrace{f(r) = 0.}$

$$(r \text{ is a } \mathbf{root} \text{ of the equation } f(x) = 0)$$

Consider x_1 in the domain of f. If L_1 denotes the linearization of f at x_1, then

$$f(x) \approx \underbrace{L_1(x) = f(x_1) + f'(x_1)(x - x_1)}$$

$$(\text{tangent line approximation to } f(x) \text{ at } (x_1, f(x_1)))$$

$$\implies \quad f(r) \approx f(x_1) + f'(x_1)(r - x_1)$$

$$\implies \quad 0 \approx f(x_1) + f'(x_1)(r - x_1)$$

$$\uparrow$$

$$(f(r) = 0)$$

$$\implies \quad r \approx x_1 - \frac{f(x_1)}{f'(x_1)}, \quad \text{provided } f'(x_1) \neq 0.$$

Note! $f'(x_1) \neq 0$ requires the tangent line to $y = f(x)$ at $(x_1, f(x_1))$ to be non-horizontal; hence, this tangent line must intersect the x axis for a unique value x satisfying $L_1(x) = 0$.

Let x_2 denote the root of the equation $L_1(x) = 0$.

$$L_1(x_2) = 0 \quad \implies \quad f(x_1) + f'(x_1)(x_2 - x_1) = 0 \quad \implies \quad x_2 = x_1 - \frac{f(x_1)}{f'(x_1)}$$

Note! $f'(x_1) \neq 0$ stipulated earlier.

$$\implies \quad x_2 \approx r \quad (\text{i.e., } x_2 \text{ is the approximation of the value } r \text{ given above})$$

Note! Using the zero of the linearization L_1 to approximate a zero of f.

Now consider the linearization L_2 of f at x_2.

$$L_2(x_3) = 0 \quad \implies \quad x_3 = x_2 - \underbrace{\frac{f(x_2)}{f'(x_2)}}$$

$$(\text{as above: provided } f'(x_2) \neq 0)$$

$$\Longrightarrow \quad x_3 \approx r \qquad \text{(i.e., the zero of } L_2 \text{ is an approximation of the value } r, \text{ a zero of } f)$$

Continue to obtain a sequence of approximations

$$\underbrace{x_{n+1} = x_n - \frac{f(x_n)}{f'(x_n)}}_{}, \quad n = 1, 2, 3, \cdots. \text{ such that } x_{n+1} \approx r$$

(from linearization of L_n of f at x_n such that $L_n(x_{n+1}) = 0$)

Remarks:

(R1) Assumed $f'(x_n) \neq 0$ for each $n = 1, 2, 3, \cdots$

(R2) Graphically:

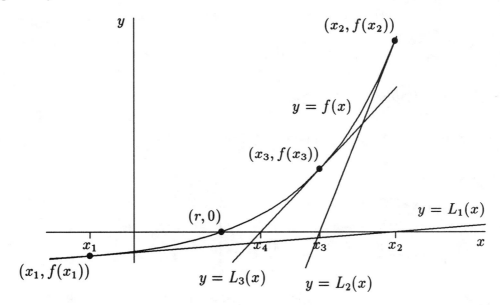

$$x_1 : \text{ chosen; } \quad \underbrace{x_2 : L_1(x_2) = 0; \quad x_3 : L_2(x_3) = 0; \quad x_4 : L_3(x_4) = 0}_{}$$
$$\text{(successive tangent lines intersect } x \text{ axis)}$$

Note! $f'(x_1) \neq 0; \quad f'(x_2) \neq 0; \quad f'(x_3) \neq 0$

(R3) If the $x_n \to r$ as n becomes larger, then the sequence $x_1, x_2, \ldots, x_n, \ldots$ **converges** to r and is denoted by

$$\lim_{n \to \infty} x_n = r$$

(R4) The method for constructing the sequence $x_1, x_2, \ldots$ is called **Newton's method** or the **Newton-Raphson method**.

(R5) Newton's method may fail if $f'(x_n) \approx 0$ for some n. Redo the sequence of approximations by choosing a different initial approximation x_1.

(R6) x_n for some n may not be in the domain of f. Then Newton's method fails (since $f(x_n)$ and, hence, $f'(x_n)$ are not defined). Try a different initial approximation x_1.

(R7) Since the same formula

$$x_{n+1} = x_n - \frac{f(x_n)}{f'(x_n)}$$

is used for each $n = 1, 2, 3, \ldots$, this procedure is called an *iterative* process.

Example 1: Use Newton's method with initial value 1 to find the fourth approximation x_4 to a solution of the equation $x^2 = \cos x$.

Rewrite equation in the form $f(x) = 0 :$ $\cos x - x^2 = 0$

$f(x) = \cos -x^2 \implies f'(x) = -\sin x - 2x$

$\implies f$ differentiable on $(-\infty, \infty)$
$\uparrow$
(domain of $f : (-\infty, \infty)$; domain of $f' : (-\infty, \infty)$)

Note!

$$f(0) = \cos 0 - 0^2 = 1 > 0$$

$$\left.f\left(\frac{\pi}{2}\right) = \cos\frac{\pi}{2} - \left(\frac{\pi}{2}\right)^2 = -\frac{\pi^2}{4} < 0 \right\}$$

$\implies$ there is a value r, where $0 < r < \dfrac{\pi}{2}$, such that $f(r) = 0$
$\uparrow$
(f differentiable $\implies$ f continuous: use Intermediate Value Theorem)

Newton's method: $x_{n+1} = x_n - \underbrace{\dfrac{\cos x_n - x_n^2}{-\sin x_n - 2x_n}}_{\left(x_n - \frac{f(x_n)}{f'(x_n)}\right)} = x_n + \dfrac{\cos x_n - x_n^2}{\sin x_n + 2x_n}$

$x_1 = 1$ (radian): the given initial approximation.

Note! $0 < x_1 < \dfrac{\pi}{2}$

$$x_2 = 1 + \frac{\cos 1 - 1^2}{\sin 1 + 2(1)} = 0.838218409$$

$$x_3 = 0.824241868 \qquad x_4 = 0.824132318 \approx r$$

Note! Graphically:

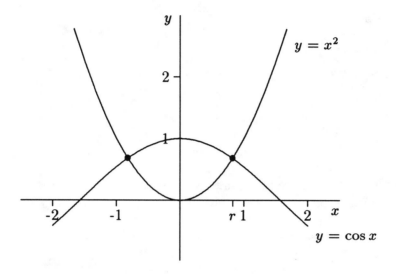

r is the value at which the graphs of $y = x^2$ and $y = \cos x$ intersect. Since both $y = x^2$ and $y = \cos x$ are symmetric with respect to the y axis,

$$r \approx \underbrace{-0.824132319}_{(-x_4)}$$

is the other solution of the equation $x^2 = \cos x$.

Check (via calculator): $\left\{\begin{array}{l} \cos(.824132319) = .679194063 \\ (.824132319)^2 = .679194079 \end{array}\right\}$

$\implies \quad \cos x_4 \approx x_4^2$

Example 2: Use Newton's method with initial value 1 to find the seventh approximation to the number $\sqrt[5]{3}$

Finding $\sqrt[5]{3}$ is equivalent to solving the equation $x^5 - 3 = 0$.

Let $f(x) = x^5 - 3 \implies f'(x) = 5x^4$.

Newton's method: $x_{n+1} = x_n - \dfrac{x_n^5 - 3}{5x_n^4}$

$x_1 = 1$: initial choice
$x_2 = 1.4$
$x_3 = 1.276184923$
$x_4 = 1.247150132$
$x_5 = 1.245734166$
$x_6 = 1.24573094$
$x_7 = 1.24573094 \approx r$

Example 3: Find all roots of the equation $x^4 + x^2 - 2x - 1 = 0$ correct to six decimal places in successive approximations.

Let $f(x) = x^4 + x^2 - 2x - 1 \implies f'(x) = 4x^3 + 2x - 2$

Newton's method: $x_{n+1} = x_n - \dfrac{x_n^4 + x_n^2 - 2x_n - 1}{4x_n^3 + 2x_n - 2}$

Rewrite the given equation and sketch the two related curves in order to determine the number of roots and corresponding initial values for Newton's method:

$$x^4 + x^2 - 2x - 1 = 0 \implies x^4 = -x^2 + 2x + 1$$
$$\uparrow$$
$$\text{(a choice)}$$

Note! Choice yields two basic curves:

$$y = x^4$$

$$y = -x^2 + 2x + 1: \quad \text{parabola, opening down}$$

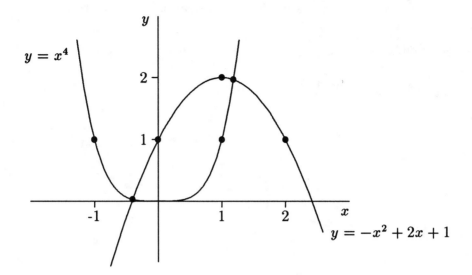

Roots occur in the intervals $[-1, 0]$ and $[1, 2]$.

Note! Using the Intermediate Value Theorem:

$$f(-1) = 3 > 0, \quad f(0) = -1 < 0 \quad \Longrightarrow \quad \text{root in } [-1, 0]$$

$$f(1) = -1 < 0, \quad f(2) = 15 > 0 \quad \Longrightarrow \quad \text{root in } [1, 2]$$

(C1) $[-1, 0]$ $x_1 = 0:$ a choice

                    $x_2 = -.5$

                    $x_3 = -.410714$

                    $x_4 = -.404721$

                    $x_5 = -.404698$

                    $x_6 = -.404698$ agrees with x_5 to 6 decimal places: stop

$$\Longrightarrow \quad r \approx -.404698 \quad (\in [-1, 0])$$

Check: $f(-.404698) = .000000513 \approx 0$

(C2) [1, 2] $x_1 = 1$:   a choice
                $x_2 = 1.25$
                $x_3 = 1.189380$
                $x_4 = 1.184171$
                $x_5 = 1.184135$
                $x_6 = 1.184135$ agrees with x_5 to 6 decimal places: stop

$\implies$ $r \approx 1.184135$ ($\in [1, 2]$)

Check: $f(1.184135) = .0000002386 \approx 0$

3

The Mean Value Theorem and Curve Sketching

3.1 Maximum and Minimum Values

Definition (3.1): A function f has an **absolute maximum** at c if $f(c) \geq f(x)$ for all x in D, where D is the domain of f, and the number $f(c)$ is called the **maximum value** of f on D. Similarly, f has an **absolute minimum** at c if $f(c) \leq f(x)$ for all x in D and the number $f(c)$ is called the **minimum value** of f on D. The maximum and minimum values of f are called the **extreme values** of f.

---

looks at whole thing

always a "y" value

Remarks:

"Global"

(R1) Notation: $\begin{cases} \text{ax:} & \text{absolute maximum} \\ \text{an:} & \text{absolute minimum} \end{cases}$

(R2) **Absolute extrema** refers to both ax and an.

↳ *asks for both max and min*

Definition (3.2): A function f has a **local maximum** (or **relative maximum**) at c if there is an open interval I containing c such that $f(c) \geq f(x)$ for all x in I. Similarly, f has a **local minimum** at c if there is an open interval I containing c such that $f(c) \leq f(x)$ for all x in I.

---

just looks at some values on graph

Remarks:

relative

(R1) Notation: $\begin{cases} \text{lx:} & \text{local maximum} \\ \text{ln:} & \text{local minimum} \end{cases}$

(R2) **Local extrema** refers to both lx and ln.

note: open interval

Illustrations:

(I1)

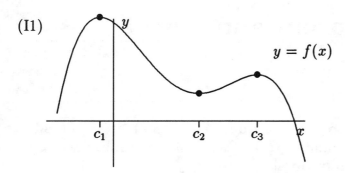

$$y = f(x)$$

lx: $f(c_1)$, $f(c_3)$
ln: $f(c_2)$
ax: $f(c_1)$
an: none

(I2)

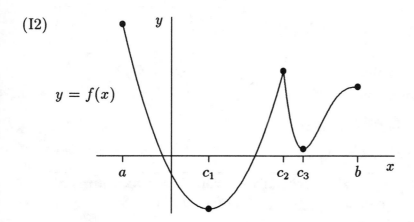

$$y = f(x)$$

lx: $f(c_2)$
ln: $f(c_1)$, $f(c_3)$
ax: $f(a)$
an: $f(c_1)$

Remark: Absolute extrema, if they exist, must occur among the

$$\left\{ \begin{array}{l} \text{local extrema} \\ \text{endpoints of domain} \end{array} \right\};\ \text{if } f(c) \text{ is not } \left\{ \begin{array}{l} \text{a local extreme value} \\ \text{an endpoint value} \end{array} \right\},$$

then for x near c either

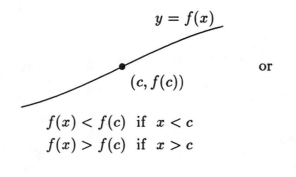

$$y = f(x)$$

$(c, f(c))$

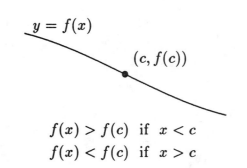

$$y = f(x)$$

$(c, f(c))$

or

$f(x) < f(c)$ if $x < c$
$f(x) > f(c)$ if $x > c$

$f(x) > f(c)$ if $x < c$
$f(x) < f(c)$ if $x > c$

$$\implies \begin{cases} f(x) < f(c): & f(c) \text{ is not an value} \\ f(x) > f(c): & f(c) \text{ is not ax value} \end{cases}$$

The Extreme Value Theorem (3.3): If f is continuous on a closed interval $[a, b]$, then f attains an absolute maximum value $f(c)$ and an absolute minimum value $f(d)$ at some numbers c and d in $[a, b]$.

Remarks:

(R1) Notation: EVT: Extreme Value Theorem

(R2) EVT is an existence theorem:

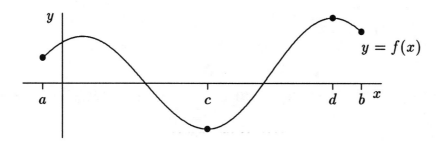

If f is continuous on $[a, b]$, then both

$$\begin{cases} c \text{ such that } a \le c \le b \text{ with } f(c): \text{ an value} \\ \text{and} \\ d \text{ such that } a \le d \le b \text{ with } f(d): \text{ ax value} \end{cases} \quad \text{exist.}$$

(R3) In the hypothesis of the EVT: If both conditions of continuity and closed interval are not satisfied, then the absolute extrema are not guaranteed.

Example 1: $f(x) = \dfrac{1}{x}, \quad 0 < x \leq 1.$ f is not continuous on a closed interval.

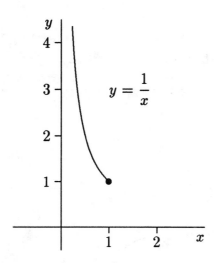

ax: none

an: $f(1) = 1$ (at endpoint)

Example 2: $f(x) = \dfrac{1}{x}, \quad -1 \leq x < 0 \quad$ or $\quad 0 < x \leq 1.$ f is not continuous on a closed interval.

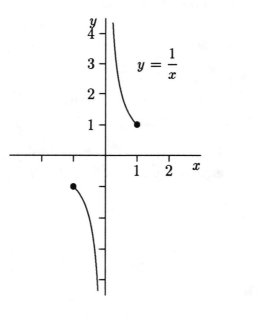

ax: none

an: none

Example 3: $f(x) = \left\{ \begin{array}{ll} 1/2 & \text{if } -1 \le x < 0 \\ x & \text{if } 0 \le x \le 1 \end{array} \right\}$. f is not continuous on a closed interval.

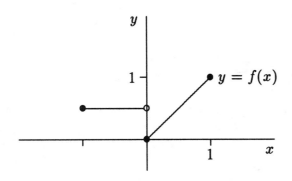

ax: $f(1) = 1$ (an endpoint)

an: $f(0) = 0$ (ln)

y=0 is local min

y=½ is local min and max

Remarks:

(R1) If f satisfies the hypothesis of the EVT, compare the values of $f(x)$

at $\left\{ \begin{array}{l} \text{local extrema} \\ \text{endpoints } a \text{ and } b \end{array} \right\} : \left\{ \begin{array}{l} \text{ax: largest value} \\ \text{an: least value} \end{array} \right\}$

Question: How are local extrema found?   (To be answered.)

(R2) If f does not satisfy the hypothesis of the EVT, sketch the graph of $y = f(x)$: read results from graph.

Note! May also use the graph of $y = f(x)$ when f satisfies the hypothesis of EVT.

Fermat's Theorem (3.4): If f has a local extremum (that is, maximum or minimum) at c, and if $f'(c)$ exists, then $f'(c) = 0$.

Proof:

Assume $f(c)$: lx $\implies$ $f(c) \ge f(x)$ for all x near c.

Let $x = c + h$ such that $h \ne 0$, h near 0 (hence, $x \ne c$ but x near c).

$\implies$ $f(c) \ge f(c + h)$ $\implies$ $f(c + h) - f(c) \le 0$

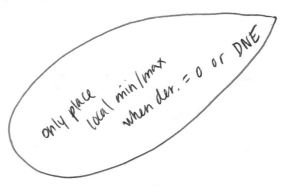

only place local min/max when der. = 0 or DNE

Consider $h > 0$: $\dfrac{f(c+h) - f(c)}{h} \leq 0$

$\Longrightarrow \quad \lim\limits_{h \to 0^+} \dfrac{f(c+h) - f(c)}{h} \leq \lim\limits_{h \to 0^+} 0 = 0$ (Thm (1.9), p. 75)

$\Longrightarrow \quad f'(c) = \lim\limits_{h \to 0} \dfrac{f(c+h) - f(c)}{h} = \lim\limits_{h \to 0^+} \dfrac{f(c+h) - f(c)}{h} \leq 0$

(hypothesis: $f'(c)$ exists; (1.5), p. 67)

Consider $h < 0$: $\dfrac{f(c+h) - f(c)}{h} \geq 0$

$\Longrightarrow \quad \lim\limits_{h \to 0^-} \dfrac{f(c+h) - f(c)}{h} \geq \lim\limits_{h \to 0^-} 0 = 0$ (Thm (1.9), p. 75)

$\Longrightarrow \quad f'(c) = \lim\limits_{h \to 0} \dfrac{f(c+h) - f(c)}{h} = \lim\limits_{h \to 0^-} \dfrac{f(c+h) - f(c)}{h} \geq 0$

(hypothesis: $f'(c)$ exists; (1.5), p. 67)

Hence, $f'(c) \leq 0 \leq f'(c) \implies f'(c) = 0$

Remarks:

(R1) To complete the proof proceed similarly under the assumption $f(c)$: ln.

(R2) Contrapositive of Fermat's Theorem: If $f'(c)$ exists with $f'(c) \neq 0$, then $f(c)$ is not a local extreme value.

(R3) Converse of Fermat's Theorem: If $f'(c) = 0$, then $f(c)$ is a local extreme value. This converse is not true (see Example 4 below).

$f'(c) > 0$ can't be min or max

$(c, f(c))$

$f'(c) < 0$ not min or max

only when derivative is zero or DNE can it possibly be min or max
↳ critical number

Example 4: $f(x) = x^3 + 1$

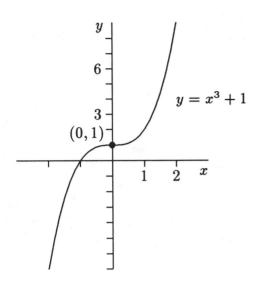

$f'(x) = 3x^2$: exists on $(-\infty, \infty)$

$f'(x) = 0 \implies 3x^2 = 0 \implies x = 0$,

but $f(0) = 1$ is not a local extrema

$$\left(\begin{array}{l} f(x) < f(0) \ \ \text{if} \ \ x < 0 \\ f(x) > f(0) \ \ \text{if} \ \ x > 0 \end{array} \right)$$

Definition (3.6): A **critical number** of a function f is a number c in the domain of f such that either $f'(c) = 0$ or $f'(c)$ does not exist.

Remarks:

(R1) $f'(c) = 0$: horizontal tangent line at $(c, f(c))$.

(R2) $f'(c)$ does not exist: no tangent line or vertical tangent line at $(c, f(c))$.

(R3) c a critical number of $f \implies f(c)$ is a *possible* local extreme value of f.

Example 5: Find the critical numbers of $f(x) = \dfrac{x^4}{4} - \dfrac{x^3}{3} - x^2 + 1$

$f'(x) = x^3 - x^2 - 2x$

$f'(x) = 0$: $x^3 - x^2 - 2x = x(x^2 - x - 2) = x(x + 1)(x - 2) = 0 \implies x = 0, \ -1, \ 2$

$f'(x)$ DNE (does not exist): none

Critical numbers: $x = -1, \ 0, \ 2$

Example 6: Find the critical numbers of $f(x) = x(x-6)^{1/3}$

$$f'(x) = x\left(\frac{1}{3}\right)(x-6)^{-2/3} + (x-6)^{1/3} = \frac{x+3(x-6)}{3(x-6)^{2/3}} = \frac{4x-18}{3(x-6)^{2/3}}$$

$f'(x) = 0: \quad 4x - 18 = 0 \implies x = \dfrac{9}{2}$

$f'(x)$ DNE: $x - 6 = 0 \implies x = 6$

Critical numbers: $x = \dfrac{9}{2}$, 6

Example 7: Find the critical numbers of $f(x) = \dfrac{x^2+1}{x}$

$$f'(x) = \frac{x(2x) - (x^2+1)}{x^2} = \frac{x^2-1}{x^2}$$

$f'(x) = 0: \quad x^2 - 1 = (x-1)(x+1) = 0 \implies x = 1, \; -1$

$f'(x)$ DNE: $x^2 = 0 \implies x = 0$: eliminate since $f(0)$ is not defined

Critical numbers: $x = -1, \; 1$

Example 8: Find the critical numbers of $f(x) = \cos 2x + 2\cos x, \quad 0 < x < 2\pi$

$$f'(x) = -2\sin 2x - 2\sin x = -2(\sin 2x + \sin x)$$

$$= -2(2\sin x \cos x + \sin x) = -2\sin x(2\cos x + 1)$$

$f'(x) = 0: \quad \sin x = 0 \implies x = \pi \in (0, 2\pi)$

$$2\cos x + 1 = 0 \implies \cos x = -\frac{1}{2} \implies x = \pi - \frac{\pi}{3} = \frac{2\pi}{3}, \quad \pi + \frac{\pi}{3} = \frac{4\pi}{3}$$

$$\uparrow \qquad\qquad\qquad \uparrow$$

(reference angle)

$f'(x)$ DNE: none   Critical numbers: $x = \dfrac{2\pi}{3}, \; \pi, \; \dfrac{4\pi}{3}$

Procedure (3.8): To find the *absolute* maximum and minimum values of a continuous function f on a closed interval $[a, b]$:

1. Find the values of f at the critical numbers of f in (a, b).

2. Find the values of $f(a)$ and $f(b)$.

3. The largest of the values from steps 1 and 2 is the absolute maximum value; the smallest of these values is the absolute minimum value.

Remark: From Fermat's Theorem: the critical numbers include all local extrema.

Example 9: Find the absolute extrema of $f(x) = \dfrac{x^4}{4} - \dfrac{x^3}{3} - x^2 + 1$, $-2 \leq x \leq 3$

$f(x)$: polynomial, so continuous everywhere $\implies$ f continuous on $[-2, 3]$.

Critical numbers: $\underbrace{x = -1,\ 0,\ 2}$

(see Example 5 above)

	(endpoints)		(critical # s)		
x	-2	3	-1	0	2
$f(x)$	$11/3$	$13/4$	$7/12$	1	$-5/3$
	ax				an

Example 10: Find the absolute extrema of $f(x) = \cos 2x + 2 \cos x$, $0 \leq x \leq 2\pi$.

$f(x)$: sum of functions continuous everywhere $\implies$ f continuous on $[0, 2\pi]$

Critical numbers: $\underbrace{x = \dfrac{2\pi}{3},\ \pi,\ \dfrac{4\pi}{3}}$

(see Example 8 above)

	(endpoints)		(critical # s)		
x	0	2π	$2\pi/3$	π	$4\pi/3$
$f(x)$	3	3	$-3/2$	-1	$-3/2$
	ax		an		

Note!

(N1) ax occurs at two values of x : $x = 0, \ 2\pi$

(N2) an occurs at two values of x : $\dfrac{2\pi}{3}, \ \dfrac{4\pi}{3}$

Example 11: Find the absolute extrema of $f(x) = \begin{cases} x+1 & \text{if } -1 \le x \le 0 \\ x^2 & \text{if } 0 \le x \le 2 \end{cases}$

$$\left. \begin{array}{l} \lim\limits_{x \to 0^-} f(x) = \lim\limits_{x \to 0^-} (x+1) = 1 \\[2mm] \lim\limits_{x \to 0^+} f(x) = \lim\limits_{x \to 0^+} x^2 = 0 \end{array} \right\} \implies \lim\limits_{x \to 0} f(x) \text{ does not exist}$$

$\implies$ f is not continuous at $x = 0$ $\implies$ f is not continuous on $\underbrace{[-1,2]}$
$$\text{(domain of } f)$$

Sketch graph of $y = f(x)$:

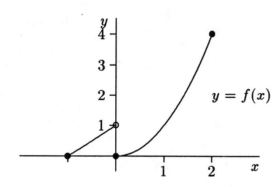

ax: $4 = f(2)$

an: $0 = f(0)$; also $f(-1) = 0$

Example 12: Find the absolute extrema of $f(x) = \dfrac{x^2 + 1}{x}$.

Domain of f is $(-\infty, 0) \cup (0, \infty)$, which is not a closed interval.

Sketch graph of $y = f(x)$.

Take my hand,
and walk with me
and tell me who you are.
Make a wish
if you can see
the first star from above...

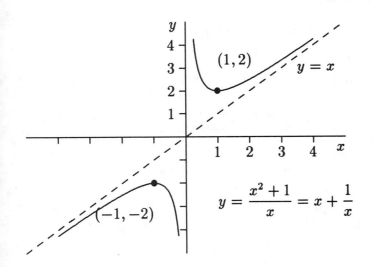

Note! $\underbrace{f'(x) = 0 \text{ at } x = -1, \ 1}$
(see Example 7 above)

ax: none

an: none

Example 13: Find the absolute extrema of $f(x) = \dfrac{x^2 + 1}{x}, \quad -3 \le x \le -\dfrac{1}{2}$

$f(x)$: rational function discontinuous at $x = 0 \implies f$ continuous on $[-3, -1/2]$

Critical number: $\underbrace{x = -1}$ $(\in [-3, -1/2])$
(see Example 7 above)

x	-3	$-1/2$	-1
$f(x)$	$-10/3$	$-5/2$	-2

(endpoints) over -3 and $-1/2$; (critical #) over -1.

an (under -3 / $-10/3$); ax (under -1 / -2)

Note! See graph of $y = \dfrac{x^2 + 1}{x}$ in Example 12 above.

3.2   The Mean Value Theorem

Rolle's Theorem (3.9): Let f be a function that satisfies the following three hypotheses:

1. f is continuous on the closed interval $[a, b]$.

2. f is differentiable on the open interval (a, b).

3. $f(a) = f(b)$

Then there is a number c in (a, b) such that $f'(c) = 0$.

Illustration: $f'(c) = 0 \implies$ horizontal tangent line at $(c, f(c))$

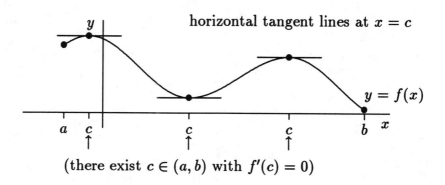

horizontal tangent lines at $x = c$

$y = f(x)$

(there exist $c \in (a, b)$ with $f'(c) = 0$)

Remarks: In hypothesis: f differentiable on $(a, b) \implies$ no *sharp* corners for $x \in (a, b)$.

(R1) Sharp corner such that Rolle's Theorem does not hold:

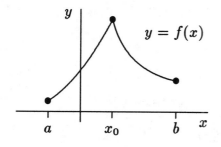

$y = f(x)$

$f'(x_0)$ DNE where $a < x_0 < b$;

no horizontal tangent lines in (a, b)

(R2) Sharp corner such that Rolle's Theorem does hold:

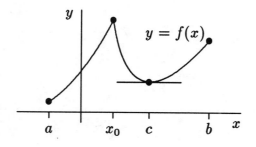

$y = f(x)$

$f'(x_0)$ DNE where $a < x_0 < b$;

$f'(c) = 0$ where $a < c < b$

$\implies$ horizontal tangent line at $(c, f(c))$

Proof (of Rolle's Theorem):

(C1) $f(x) = k$ (constant) $\implies$ $f'(x) = 0$ on (a, b).

$$\left.\begin{cases} f(x) \text{ continuous on } [a, b] \\ f'(x) \text{ exists on } (a, b) \\ f(a) = k = f(b) \end{cases}\right\}$$

$\implies$ hypothesis is satisfied with $f'(c) = 0$ for all c such that $a < c < b$

Illustration:

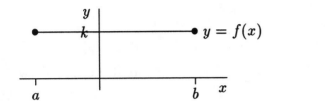

$f'(c) = 0$ for any c

such that $a < c < b$

(C2) $f(x) > f(a)$ for some x in (a, b)

f continuous on $[a, b]$ (hypothesis)

$\implies$ there is a value c in $[a, b]$ such that $\underbrace{f(c)}_{(ax)} \geq f(x)$ for all x in $[a, b]$

↑
(EVT)

$\implies$ $f(c) > f(a) = f(b)$ $\implies$ $c \neq a$ and $c \neq b$ $\implies$ c in (a, b)

↑

$((C2)\colon f(x) > f(a)$ for some x in (a, b); hypothesis: $f(a) = f(b))$

$\implies$ c is a critical number $\implies$ $f'(c) = 0$ or $f'(c)$ DNE $\implies$ $f'(c) = 0$

↑   ↑

$\left(\begin{array}{l} \text{ax on } [a, b] \text{ occurs only at} \\ \text{critical numbers or endpoints} \end{array}\right)$ $(\text{ hypothesis: } f \text{ differentiable on } (a, b))$

Illustration:

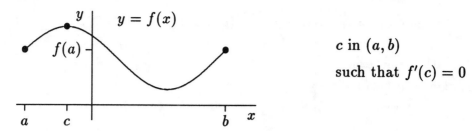

$$c \text{ in } (a, b)$$
$$\text{such that } f'(c) = 0$$

(C3) $f(x) < f(a)$ for some x in (a, b)

Proceed as in (C2) with $f(c)$: an

Example 1: Consider $f(x) = x^3 + 3x^2 - x - 4$, $-3 \le x \le 1$. Find all numbers c which satisfy the conclusion of Rolle's Theorem.

Does f satisfy hypothesis of Rolle's Theorem?

f: polynomial $\implies$ f continuous on $[-3, 1]$

$f'(x) = 3x^2 + 6x - 1$: polynomial $\implies$ f' continuous, hence exists, on $(-3, 1)$

$f(-3) = -1 = f(1)$

f satisfies hypothesis

$f'(x) = 0 \implies 3x^2 + 6x - 1 = 0$

$$x = \frac{-6 \pm \sqrt{48}}{6} = -1 \pm \frac{2}{3}\sqrt{3} \approx -1 \pm \frac{2}{3}(1.732) = -1 \pm 1.155 = 0.155, \ -2.155$$

$\implies \quad c = 0.155, \ -2.155 \quad$ (0.155 and -2.155 in $(-3, 1)$)

Corollary: Let f be a function that satisfies the following four hypotheses:

1. f is continuous on the closed interval $[a, b]$.

2. f is differentiable on the open interval (a, b).

3. $f(a)f(b) < 0$

4. $f'(x) \neq 0$ for all x in (a, b).

Then there is a unique number c in (a, b) such that $f(c) = 0$.

Remarks:

(R1) c: *unique* zero of f on (a, b) $\implies$ $f(x) = 0$ *exactly once* in (a, b).

(R2) $f(a)f(b) < 0$: $f(a)$ and $f(b)$ have opposite signs

$\implies$ there exists c in (a, b) such that $f(c) = 0$. $\hspace{2cm}$ (IVT)

Note! There exists at least one zero of f in (a, b).

(R3) The corollary indicates the existence of a zero and also the uniqueness of the zero.

Illustration: Contrary to hypothesis 4, say $f'(x) = 0$ for some x in (a, b):

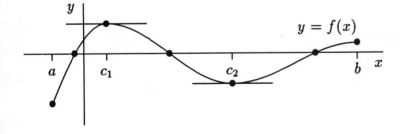

c_1, c_2 in (a, b) such that $f'(c_1) = 0$ and $f'(c_2) = 0$. f has three zeros (hence, not unique) in (a, b).

Note! Hypotheses 1-4 of Corollary hold in:

(N1) (a, c_1). $f'(x) \neq 0$ (no horizontal tangent lines) for all x in (a, c_1) and f has a unique zero in (a, c_1)

(N2) (c_1, c_2). Proceed as in (N1).

(N3) (c_2, b). Proceed as in (N1).

Proof (of Corollary):

$$f(a)f(b) < 0 \implies \text{ there is a number } c \text{ in } (a, b) \text{ such that } f(c) = 0.$$
$$\uparrow$$
(IVT: hypothesis 1 of Corollary and $N = 0$)

By contradiction: assume c in (a, b) is not unique. Say there exist x_1 and x_2 in (a, b) such that $x_1 < x_2$ and $\underline{f(x_1) = 0 = f(x_2)}$.

(f has at least two zeros in (a, b))

$\implies$ there is a number d in (x_1, x_2) such that $f'(d) = 0.$
$\uparrow$

$$\left(\begin{array}{l} \text{apply Rolle's Theorem on } [x_1, x_2] \subset (a, b) \\ \quad 1.\ f \text{ continuous on } [x_1, x_2] \\ \quad 2.\ f \text{ differentiable on } (x_1, x_2) \\ \quad 3.\ f(x_1) = 0 = f(x_2) \end{array} \right)$$

$\implies f'(d) = 0$ for d in (a, b): contradiction to hypothesis 4.
$\uparrow$
$((x_1, x_2) \subset (a, b))$

Example 2: Show $f(x) = x^3 + 3x^2 - x - 3$ has a unique zero in $(-2, 0)$.

Does f satisfy hypothesis of Corollary?

f: polynomial $\implies$ f continuous on $[-2, 0]$

$f'(x) = 3x^2 + 6x - 1$: polynomial $\implies$ f' continuous, hence exists, on $(-2, 0)$

$$\left\{ \begin{array}{l} f(-2) = 3 \\ f(0) = -3 \end{array} \right\} \implies f(-2)f(0) < 0$$

$f'(x) = 0 \implies \underline{x = 0.155,\ -2.155}$
(see Example 1 above)

$\implies f'(x) \neq 0$ for all x in $(-2, 0)$ $(-2.155 < -1 < 0 < 0.155)$

f satisfies hypothesis

$\implies$ there is a unique number c in $(-2, 0)$ such that $f(c) = 0$.

Note!

(N1) $f(-1) = 0 \implies c = -1$ in $(-2, 0)$

(N2) $f(x) = x^3 + 3x^2 - x - 2$ also satisfies hypothesis on $[-2, 0]$. Find unique c in $(-2, 0)$ such that $f(c) = 0$: try Newton's method with $x_1 = -1$; $c \approx -.745902$ $(= x_3)$

Example 3: Show $f(x) = 2x + \cos x$ has a unique zero in $(-1, 2)$

Does f satisfy hypothesis of Corollary?

$f(x) = 2x + \cos x$: continuous everywhere

$f'(x) = 2 - \sin x$: continuous, hence exists, everywhere

$$\left\{ \begin{array}{l} f(-1) = -2 + \cos(-1) \leq -2 + 1 = -1 \\ f(2) = 4 + \cos(2) \geq 4 + (-1) = 3 \end{array} \right\} \implies f(-1)f(2) < 0$$

$f'(x) = 2 - \sin x \geq 2 - 1 = 1$ everywhere $\implies$ $f'(x) \neq 0$ for all x in $(-1, 2)$

f satisfies hypothesis

$\implies$ there is a unique number c in $(-1, 2)$ such that $f(c) = 0$

Note! $c = ?$ Try Newton's method.

The Mean Value Theorem (3.10): Let f be a function that satisfies the following hypotheses:

1. f is continuous on the closed interval $[a, b]$.

2. f is differentiable on the open interval (a, b).

Then there is a number c in (a, b) such that

$$f'(c) = \frac{f(b) - f(a)}{b - a}$$

or, equivalently,

$$f(b) - f(a) = f'(c)(b - a)$$

Remarks:

(R1) Notation: MVT: Mean Value Theorem

(R2) Rolle's Theorem is a special case of MVT: $\underbrace{f(a) = f(b)}$
$$\text{(additional hypothesis in Rolle's Theorem)}$$

$$\implies \underbrace{f'(c) = \frac{f(b) - f(a)}{b - a}}_{\text{(MVT)}} = 0: \text{ conclusion of Rolle's Theorem.}$$

Illustration:

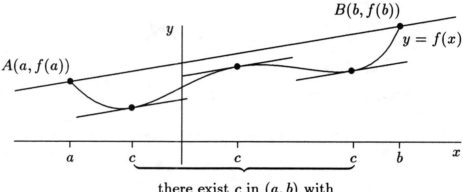

there exist c in (a, b) with

$$f'(c) \quad = \quad \frac{f(b) - f(a)}{b - a}$$

$$\left(\begin{array}{c} \text{slope of tangent line} \\ \text{at point } (c, f(c)) \end{array} \right) = \left(\begin{array}{c} \text{slope of secant line} \\ \text{through the points} \\ A(a, f(a)) \text{ and } B(b, f(b)) \end{array} \right)$$

Note!

(N1) The MVT generalizes Rolle's Theorem by rotating the points $A(a, f(a))$ and $B(b, f(b))$ such that the condition $f(a) = f(b)$ need not hold.

(N2) In hypothesis: f differentiable on $(a, b) \implies$ no sharp corners for $x \in (a, b)$. If sharp corners occur, the conclusion of the MVT is not guaranteed.

Proof (of MVT):

Find an equation of the secant line through points $A(a, f(a))$ and $B(b, f(b))$:

using the point-slope form with $\left\{ \begin{array}{l} \text{point: } (a, f(a)) \\ \text{slope: } \dfrac{f(b) - f(a)}{b - a} \end{array} \right\}$

$$\implies y - f(a) = \frac{f(b) - f(a)}{b - a}(x - a) \implies y = \frac{f(b) - f(a)}{b - a}(x - a) + f(a)$$

For x in $[a, b]$, define the function h via

$$h(x) = f(x) - \underbrace{\left[\frac{f(b) - f(a)}{b - a}(x - a) + f(a) \right]}$$

(the y value along the secant line through points A and B)

$$\implies h(x) = f(x) - f(a) - \frac{f(b) - f(a)}{b - a}(x - a)$$

Note! A geometric interpretation of $h(x)$:

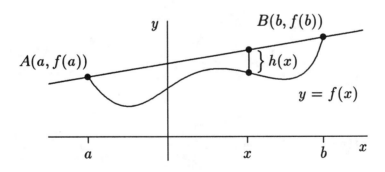

$h(x)$ is the *signed* distance between $y = f(x)$ and the y value along the secant line at each x in $[a, b]$.

Show h satisfies the hypothesis of Rolle's Theorem:

1. $\left\{ \begin{array}{ll} \text{hypothesis 1 of MVT:} & f \text{ continuous on } [a, b] \\ f(a) + \dfrac{f(b) - f(a)}{b - a}(x - a): & \text{polynomial, so continuous on } [a, b] \end{array} \right\}$

$\implies$ h continuous on $[a, b]$ (h: difference of continuous functions)

2. $h'(x) = f'(x) - \dfrac{f(b) - f(a)}{b - a}$

$\implies$ h differentiable on (a, b) (hypothesis 2 of MVT: f differentiable on (a, b))

3. $h(a) = 0 = h(b)$

Since h satisfies the hypothesis of Rolle's Theorem, there exists c in (a, b) such that $h'(c) = 0$

$\implies$ there exists c in (a, b) such that

$$f'(c) - \frac{f(b) - f(a)}{b - a} = 0 \quad \text{or, rewritten,} \quad f'(c) = \frac{f(b) - f(a)}{b - a}$$

Example 4: Consider $f(x) = x^2 - x + 1, \quad -1 \le x \le 4$. Find the numbers c which satisfy the conclusion of the MVT.

Does f satisfy hypothesis of the MVT?

f: polynomial $\implies$ f continuous on $[-1, 4]$

$f'(x) = 2x - 1$: polynomial $\implies$ f' continuous, hence exists, on $(-1, 4)$

f satisfies hypothesis

$$f'(x) = \frac{f(b) - f(a)}{b - a} \quad \implies \quad 2x - 1 = \frac{f(4) - f(-1)}{4 - (-1)} = \frac{13 - 3}{5} = 2$$

$$\implies \quad 2x = 3 \quad \implies \quad x = \frac{3}{2} \quad \implies \quad c = \frac{3}{2} \text{ in } (-1, 4)$$

Note! Demonstrated via graph:

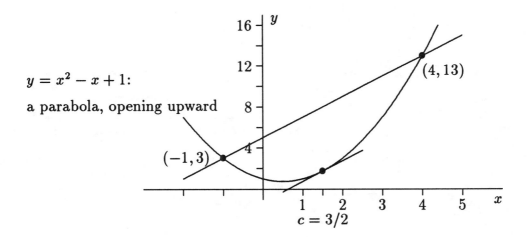

$y = x^2 - x + 1$:

a parabola, opening upward

$(-1, 3)$

$(4, 13)$

$c = 3/2$

The following theorem is a consequence of the MVT.

Theorem (3.15): If $f'(x) = 0$ for all x in an interval (a, b), then f is constant on (a, b).

Recall: Showed, in Section 2.2, that the derivative of a constant function is zero.

Remark: The theorem says: The only function defined on (a, b) such that its derivative is identically zero is a constant function.

Proof:

$f'(x) = 0$ for all x in (a, b)

$\implies$ f: differentiable on (a, b) $\implies$ f: continuous on (a, b)

Choose any x_1, x_2 in (a, b) such that $x_1 < x_2$ $\implies$ $[x_1, x_2] \subset (a, b)$

$\implies \quad \begin{cases} f\text{: continuous on } [x_1, x_2] \\ f\text{: differentiable on } (x_1, x_2) \end{cases}$

$\implies$ there is a number c in (x_1, x_2) such that $f'(c) = \dfrac{f(x_2) - f(x_1)}{x_2 - x_1}$

$\uparrow$

(f satisfies hypothesis of MVT on $[x_1, x_2]$)

$$\implies \quad \frac{f(x_2) - f(x_1)}{x_2 - x_1} = 0 \qquad\qquad (f'(x) = 0 \text{ for all } x \text{ in } (a, b); c \text{ in } (a, b))$$

$$\implies \quad f(x_2) - f(x_1) = 0 \implies f(x_2) = f(x_1)$$

$\implies$ f is constant on (a, b) (since x_1, x_2 in (a, b) are arbitrary)

Corollary (3.17): If $f'(x) = g'(x)$ for all x in an interval (a, b), then $f - g$ is constant on (a, b); that is, $f(x) = g(x) + c$ where c is a constant.

Remark: The corollary says: Two functions with identical derivatives on (a, b) must differ by a constant.

Proof:

Define the function F by $F(x) = f(x) - g(x)$, $a < x < b$

$\implies$ $F'(x) = f'(x) - g'(x) = 0$ for x in (a, b)

$\uparrow$

(hypothesis: $f' = g'$)

$\implies$ $F(x) = c$, where c is a constant   (Theorem (3.15))

$\implies$ $f(x) - g(x) = c$ for x in (a, b) $(F = f - g)$

Illustration: $f'(x) = g'(x)$ for all x in (a, b)

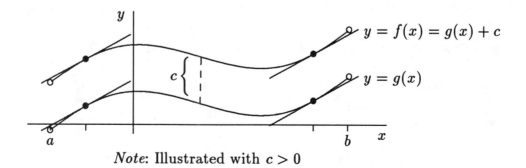

Note: Illustrated with $c > 0$

Example 5: Consider $f(x) = 3x^2 + 7$. Find all functions $F(x)$ such that $F'(x) = f(x)$.

Find one such function: $g(x) = x^3 + 7x$

$$\Longrightarrow \quad F(x) = g(x) + \quad c = x^3 + 7x + c$$
$$\qquad\qquad\quad \uparrow \qquad\qquad \uparrow$$

(Corollary (3.19))   (arbitrary constant)

Example 6: Consider $f(x) = 2x + \cos x$. Find all functions $F(x)$ such that $F'(x) = f(x)$.

Find one such function: $g(x) = x^2 + \sin x$

$$\Longrightarrow \quad F(x) = g(x) + \quad c = x^2 + \sin x + c$$
$$\qquad\qquad\qquad\qquad \uparrow$$

(arbitrary constant)

3.3 Monotonic Functions and the First Derivative Test

> **Definition (3.18)**: A function f is called **increasing** on an interval I if
>
> $$f(x_1) < f(x_2) \qquad \text{whenever } x_1 < x_2 \text{ in } I$$
>
> It is called **decreasing** on I if
>
> $$f(x_1) > f(x_2) \qquad \text{whenever } x_1 < x_2 \text{ in } I$$
>
> It is called **monotonic** on I if it is either increasing or decreasing on I.

Illustration:

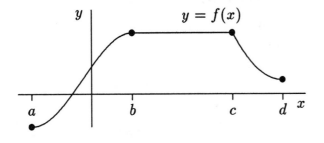

$I_1 = (a, b)$: $f \uparrow$ (increasing)

$I_2 = (c, d)$: $f \downarrow$ (decreasing)

$I_3 = (b, c)$: f neither $\uparrow$ nor $\downarrow$

Remarks!

(R1) $I_1 = [a, b]$, $[a, b)$, $(a, b]$ are also possibilities for interval such that $f \uparrow$.

(R2) f monotonic on I_1 and also on I_2.

The following theorem uses the derivative to determine monotonicity.

Test for Monotonic Functions (3.19): Suppose f is continuous on $[a, b]$ and differentiable on (a, b).

(a) If $f'(x) > 0$ for all x in (a, b), then f is increasing on $[a, b]$.

(b) If $f'(x) < 0$ for all x in (a, b), then f is decreasing on $[a, b]$.

Proof:

(a) Given $f'(x) > 0$ for all x in (a, b).

Choose any x_1 and x_2 in $[a, b]$ such that $x_1 < x_2 \implies [x_1, x_2] \subset [a, b]$

$$\implies \begin{cases} f \text{ is continuous on } [x_1, x_2] \\ f \text{ is differentiable on } (x_1, x_2) \end{cases} \qquad \text{(hypothesis)}$$

$\implies$ there is a number c in (x_1, x_2) such that

$$f'(c) = \frac{f(x_2) - f(x_1)}{x_2 - x_1} \qquad \text{(apply MVT on } [x_1, x_2])$$

$$\implies \frac{f(x_2) - f(x_1)}{x_2 - x_1} > 0 \qquad \text{(given: } f'(x) > 0 \text{ for all } x \text{ in } (a, b); c \text{ in } (a, b))$$

$$\implies f(x_2) - f(x_1) > 0 \implies f(x_2) > f(x_1)$$
$\uparrow$
$$(x_1 < x_2 \implies x_2 - x_1 > 0)$$

$$\implies f \uparrow \text{ on } [a, b] \qquad \text{(since } x_1, x_2 \text{ in } [a, b] \text{ are arbitrary)}$$

(b) Given $f'(x) < 0$ for all x in (a, b).

Proceed as in (a) above.

Remarks:

(R1) If the slope of the tangent line at each point $(x, f(x))$, for x in (a, b), is:

$$\left\{ \begin{array}{l} > 0, \text{ then } f \uparrow \text{ on } [a, b] \\ < 0, \text{ then } f \downarrow \text{ on } [a, b] \end{array} \right\}$$

(R2) The converse is not true: If f is increasing on $[a, b]$, then $f'(x) > 0$ for all x in (a, b).

Note! Similar statement for f decreasing with $f'(x) < 0$ can be made.

Counterexample: Consider $f(x) = x^3$, $-1 \le x \le 1$

$$f'(x) = 3x^2 \implies \left\{ \begin{array}{l} f'(x) > 0 \text{ for } x \text{ in } (-1, 1), \ x \ne 0 \\ f'(0) = 0 \end{array} \right\}$$

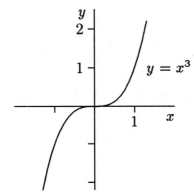

From graph: x_1, x_2 in $[-1, 1]$

$\implies f(x_1) < f(x_2)$ whenever $x_1 < x_2$

$\implies f \uparrow$ on $[-1, 1]$, but $f'(0) = 0$

$\implies f'(x) > 0$ is not satisfied for all x in $(-1, 1)$

(R3) $\left\{ \begin{array}{l} f'(x) > 0 \text{ on } (a, b) \implies f \uparrow \text{ on } [a, b] \\ f'(x) < 0 \text{ on } (a, b) \implies f \downarrow \text{ on } [a, b] \end{array} \right\} \implies$ no local extrema occur in $[a, b]$

Recall: Local extrema occur at points $(x, f(x))$ only when $f'(x) = 0$ or $f'(x)$ DNE (i.e., at critical numbers x).

The First Derivative Test (3.21): Suppose that c is a critical number of a function f that is continuous on $[a, b]$.

(a) If $f'(x) > 0$ for $a < x < c$ and $f'(x) < 0$ for $c < x < b$ (that is, f' changes from positive to negative at c), then f has a local maximum at c.

(b) If $f'(x) < 0$ for $a < x < c$ and $f'(x) > 0$ for $c < x < b$ (that is, f' changes from negative to positive at c), then f has a local minimum at c.

(c) If f' does not change sign at c, then f has no local extremum at c.

Proof:

(a) Consider x in (a, b).

$$a < x < c: \quad f'(x) > 0 \quad \Longrightarrow \quad f \uparrow \text{ on } [a, c] \quad \Longrightarrow \quad f(x) < f(c)$$

$$c < x < b: \quad f'(x) < 0 \quad \Longrightarrow \quad f \downarrow \text{ on } [c, b] \quad \Longrightarrow \quad f(c) > f(x)$$

Hence, for all x in (a, b), $f(c) \geq f(x) \quad \Longrightarrow \quad f$ has a local maximum at c.

(b), (c): See Exercise 48, p. 196

Illustrations: Visualize The First Derivative Test. Let f be continuous on $[a, b]$. If c is a critical number of f with:

$$\text{(I1)} \quad \begin{cases} f'(x) > 0 \text{ for all } x \text{ in } (a, c) \implies f \uparrow \text{ on } [a, c] \\ f'(x) < 0 \text{ for all } x \text{ in } (c, b) \implies f \downarrow \text{ on } [c, b] \end{cases}$$

(C1) $f'(c) = 0$

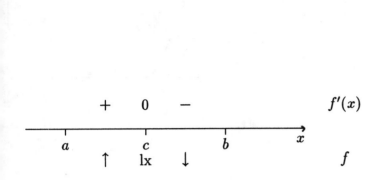
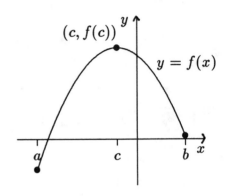

(C2) $f'(c)$ DNE

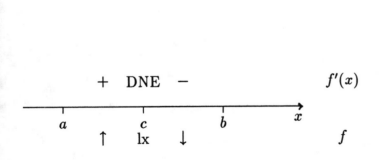
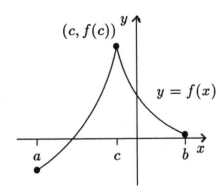

$$(I2) \quad \left\{ \begin{array}{l} f'(x) < 0 \text{ for all } x \text{ in } (a, c) \quad \Longrightarrow \quad f \downarrow \text{ on } [a, c] \\ f'(x) > 0 \text{ for all } x \text{ in } (c, b) \quad \Longrightarrow \quad f \uparrow \text{ on } [c, b] \end{array} \right\}$$

(C1) $f'(c) = 0$

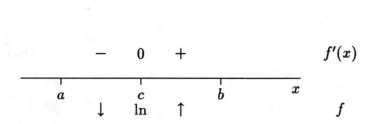 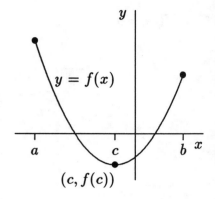

(C2) $f'(c)$ DNE

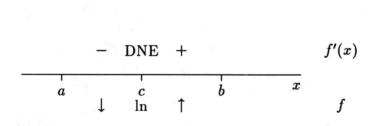 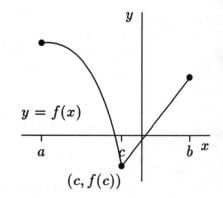

(I3) $\left\{ \begin{array}{l} f'(x) > 0 \text{ for all } x \text{ in } (a,c) \implies f \uparrow \text{ on } [a,c] \\ f'(x) > 0 \text{ for all } x \text{ in } (c,b) \implies f \uparrow \text{ on } [c,b] \end{array} \right\}$

(C1) $f'(c) = 0$

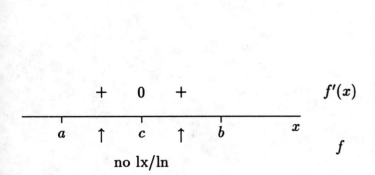

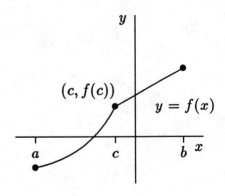

(C2) $f'(c)$ DNE

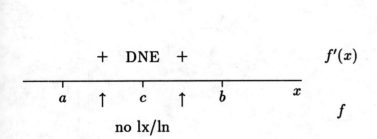

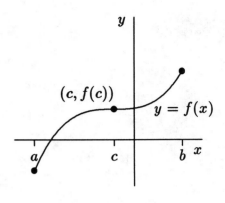

(I4) $\left\{ \begin{array}{l} f'(x) < 0 \text{ for all } x \text{ in } (a,c) \implies f \downarrow \text{ on } [a,c] \\ f'(x) < 0 \text{ for all } x \text{ in } (c,b) \implies f \downarrow \text{ on } [c,b] \end{array} \right\}$

(C1) $f'(c) = 0$

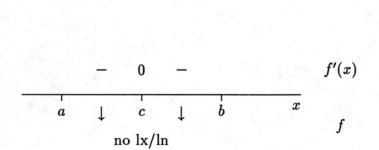

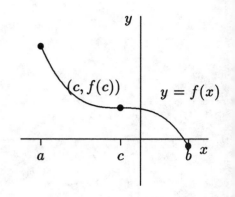

(C2) $f'(c)$ DNE

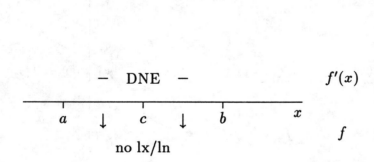

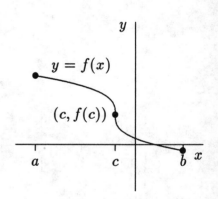

Example 1: Consider $f(x) = \dfrac{x^4}{4} - \dfrac{x^3}{3} - x^2 + 1$.

(A1) Find the interval(s) on which f is increasing or f is decreasing and find the local extreme value(s) of f.

Note! f is continuous on $(-\infty, \infty)$.

$f'(x) = x^3 - x^2 - 2x$

Find critical numbers:

$f'(x) = 0: \; x^3 - x^2 - 2x = x(x^2 - x - 2) = x(x - 2)(x + 1) = 0 \implies x = 0, \; 2, \; -1$

$f'(x)$ DNE: none

$$
\begin{array}{cccccccc}
- & 0 & + & 0 & - & & 0 & + \qquad f'(x)\\
\end{array}
$$

	-1		0		2		x
$\downarrow$		$\uparrow$		$\downarrow$		$\uparrow$	f
	ln		lx		ln		

Note! To determine the sign of f', evaluate $f'(x)$ for a value of x different from a critical number.

(N1) $f'(-2) < 0 \implies f \downarrow$ on $(-\infty, -1)$

(N2) $f'\left(-\dfrac{1}{2}\right) > 0 \implies f \uparrow$ on $(-1, 0)$

(N3) $f'(1) < 0 \implies f \downarrow$ on $(0, 2)$

(N4) $f'(3) > 0 \implies f \uparrow$ on $(2, \infty)$

$f \uparrow$: $[-1, 0]$, $[2, \infty)$; $f \downarrow$: $(-\infty, -1]$, $[0, 2]$

lx: $f(0) = 1$;   ln: $f(-1) = \dfrac{7}{12}$, $f(2) = -\dfrac{5}{3}$

(A2) Use the results of (A1) to sketch the graph of $y = f(x)$

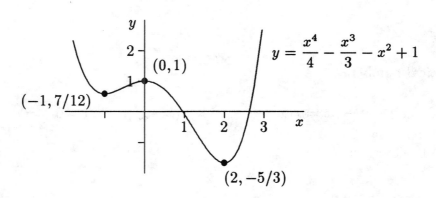

Example 2: Consider $f(x) = x^3 + 3x^2 + 2$.

(A1) Find the interval(s) on which f is increasing or f is decreasing and find the local extreme value(s) of f.

Note! f is continuous on $(-\infty, \infty)$.

$f'(x) = 3x^2 + 6x$

Find critical numbers:

$f'(x) = 0$: $3x^2 + 6x = 3x(x + 2) = 0$ $\implies$ $x = 0, \ -2$

$f'(x)$ DNE: none

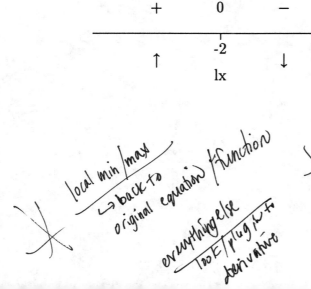

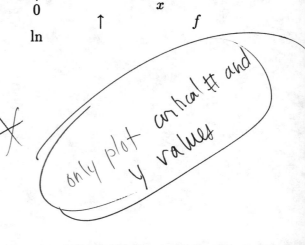

Note! Use factored form of $f'(x)$ to determine the sign of f'. $f'(x) = 3x(x+2)$

(N1) $f'(-3)$: $(-)(-) = +$

(N2) $f'(-1)$: $(-)(+) = -$

(N3) $f'(1)$: $(+)(+) = +$

$f \uparrow$: $(=\infty, -2]$, $[0, \infty)$; $f \downarrow$: $[-2, 0]$

lx: $f(-2) = 6$;   ln: $f(0) = 2$

(A2) Use the results of (A1) to sketch the graph of $y = f(x)$.

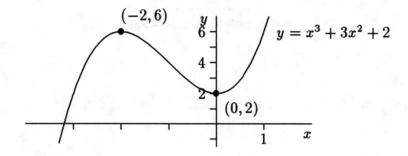

Example 3: Consider $f(x) = \dfrac{x^2 + 1}{x}$.

(A1) Find the interval(s) on which f is increasing or f is decreasing and find the local extreme value(s) of f.

Note! Domain of f is $(-\infty, 0) \cup (0, \infty)$.

$$f'(x) = \frac{x(2x) - (x^2 + 1)}{x^2} = \frac{x^2 - 1}{x^2}$$

Find critical numbers:

$f'(x) = 0$: $x^2 - 1 = (x-1)(x+1) = 0 \implies x = 1, -1$

$f'(x)$ DNE: $x = 0$, which is not in domain of $f \implies$ none

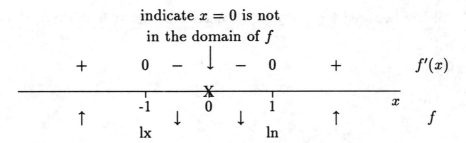

indicate $x = 0$ is not
in the domain of f

Note! Check sign of f' on both sides of $x = 0$, which is not in the domain of f' (since f' could change sign at $x = 0$).

$$f \uparrow: \; (-\infty, -1], \; [1, \infty); \qquad f \downarrow: \; [-1, 0), \; (0, 1]$$
$$(\; x = 0 \text{ is not in the domain of } f)$$

lx: $f(-1) = -2$;   ln: $f(1) = 2$

(A2) Use the results of (A1) to sketch the graph of $y = f(x)$.

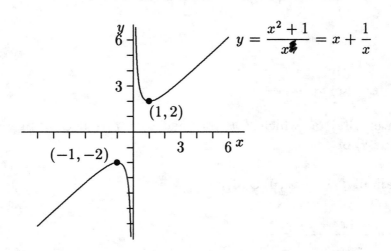

$$y = \frac{x^2 + 1}{x} = x + \frac{1}{x}$$

Example 4: Consider $f(x) = x(x - 6)^{1/3}$.

(A1) Find the interval(s) on which f is increasing or f is decreasing and find the local extreme value(s) of f.

Note! f is continuous on $(-\infty, \infty)$.

$$f(x) = x\left(\frac{1}{3}\right)(x-6)^{-2/3} + (x-6)^{1/3} = \frac{x + 3(x-6)}{3(x-6)^{2/3}} = \frac{4x - 18}{3(x-6)^{2/3}} = \frac{2(2x-9)}{3(x-6)^{2/3}}$$

Find critical numbers:

$$f'(x) = 0: \quad 2(2x-9) = 0 \quad \Longrightarrow \quad x = \frac{9}{2}$$

$$f'(x) \text{ DNE}: \quad x - 6 = 0 \quad \Longrightarrow \quad x = 6$$

$f \uparrow$: $[9/2, \infty)$

$f \downarrow$: $(-\infty, 9/2]$

lx: none

$$\text{ln}: f\left(\frac{9}{2}\right) = -\frac{9}{2}\left(\frac{3}{2}\right)^{1/3}$$

$-$	0	$+$	DNE	$+$	$f'(x)$

$$\downarrow \quad \begin{array}{c} 9/2 \\ \text{ln} \end{array} \quad \uparrow \quad \begin{array}{c} 6 \\ \text{no lx/ln} \end{array} \quad \uparrow \quad f$$

(A2) Use the results of (A1) to sketch the graph of $y = f(x)$.

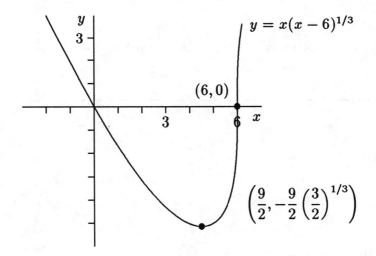

$y = x(x-6)^{1/3}$

$(6, 0)$

$$\left(\frac{9}{2}, -\frac{9}{2}\left(\frac{3}{2}\right)^{1/3}\right)$$

Note!

(N1) Used: $\left(\frac{3}{2}\right)^{1/3} \approx 1.14 \quad \Longrightarrow \quad f\left(\frac{9}{2}\right) \approx -5.13$

(N2) Plotted the point $(6,0)$ since $x = 6$ is a critical number (although a local extreme value does not occur there).

(N3) At point $(6,0)$: $f'(6)$ DNE

(A3) Find the absolute extreme value(s) of f.

Note! The domain of f is not a closed interval, so determine via the sketch in (A2).

ax: none;   an: $f\left(\dfrac{9}{2}\right) = -\dfrac{9}{2}\left(\dfrac{3}{2}\right)^{1/3} \underbrace{\approx -5.13}_{\text{((A2) above)}}$

Example 5: Consider $f(x) = \cos 2x + 2\cos x, \ 0 \le x \le 2\pi$.

(A1) Find the local extreme value(s) of f.

$$f'(x) = -2\sin 2x - 2\sin x = -2(2\sin x \cos x) - 2\sin x = -2\sin x(2\cos x + 1)$$

Find critical numbers:

$$f'(x) = 0: \ \sin x = 0 \implies x = \pi \qquad (x = 0, \ 2\pi \text{ are endpoints})$$

$$2\cos x + 1 = 0 \implies \cos x = -\frac{1}{2} \implies x = \frac{2\pi}{3}, \ \frac{4\pi}{3}$$

$f'(x)$ DNE: none

XXX: indicates values excluded from the domain of f

$$
\begin{array}{ccccccccc}
 & - & & 0 & + & 0 & - & 0 & + & & f'(x)\\
\end{array}
$$

XXX ————————————————————————— XXX x

$$
\begin{array}{ccccccc}
0 & \downarrow & \dfrac{2\pi}{3} & \uparrow & \pi & \downarrow & \dfrac{4\pi}{3} & \uparrow & 2\pi & f\\
 & & \text{ln} & & \text{lx} & & \text{ln} & &
\end{array}
$$

lx: $f(\pi) = -1$;   ln : $f\left(\dfrac{2\pi}{3}\right) = -\dfrac{3}{2} = f\left(\dfrac{4\pi}{3}\right)$

(A2) Find the absolute extreme value(s) of f.

f continuous on its domain $[0, 2\pi]$: a closed interval

	(endpoints)		(critical # s)		
x	0	2π	$2\pi/3$	π	$4\pi/3$
$f(x)$	3	3	$-3/2$	-1	$-3/2$

$\underbrace{\qquad}_{\text{ax}}$ $\qquad$ $\underbrace{\qquad}_{\text{an}}$

(A3) Use the results of (A1) and (A2) to sketch the graph of $y = f(x)$.

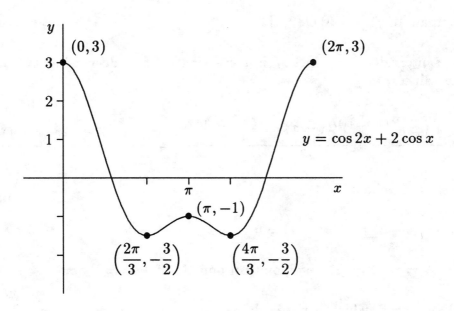

Note!

(N1) One-sided derivatives: $\begin{cases} f'_+(0) = 0 \\ f'_-(2\pi) = 0 \end{cases}$

(N2) Able to use the graph to find absolute extrema if only the results of (A1) are used to sketch the graph.

Example 6: Consider $f(x) = \dfrac{(1 - x^2)^{3/2}}{x}$.

(A1) Find the domain of f.

$\sqrt{1 - x^2}$ requires $1 - x^2 \geq 0 \implies x^2 - 1 \leq 0 \implies (x - 1)(x + 1) \leq 0$

$(x - 1)(x + 1) = 0 \implies x = 1, \ -1$

$$(x-1)(x+1) \leq 0 \ \text{if} \ -1 \leq x \leq 1 \ \implies \ 1 - x^2 \geq 0 \ \text{if} \ -1 \leq x \leq 1$$

$$\implies \ \text{domain of } f: [-1,0) \cup (0,1] \qquad\qquad (x \neq 0 \text{ in denominator})$$

(A2) Find the interval(s) on which f is increasing or f is decreasing and find the local extreme value(s) of f.

$$f'(x) = \frac{x\left(\dfrac{3}{2}\right)(1-x^2)^{1/2}(-2x) - (1-x^2)^{3/2}(1)}{x^2} = -\frac{(1-x^2)^{1/2}(2x^2+1)}{x^2}.$$

Find critical numbers:

$$f'(x) = 0: \ (1-x^2)^{1/2}(2x^2+1) = 0 \ \implies \ 1 - x^2 = 0$$

$$\implies \ x = \pm 1 : \text{endpoints of domain of } f \ \implies \ \text{none}$$

$f'(x)$ DNE: $x = 0$, which is not in domain of $f \ \implies \ \text{none}$

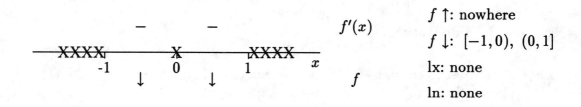

$f \uparrow$: nowhere

$f \downarrow$: $[-1,0)$, $(0,1]$

lx: none

ln: none

Note! $x = 0$ is not in domain of f', so check sign of f' on each side of $x = 0$.

(A3) Use the results of (A1) to sketch the graph of $y = f(x)$ and find the absolute extreme value(s) of f.

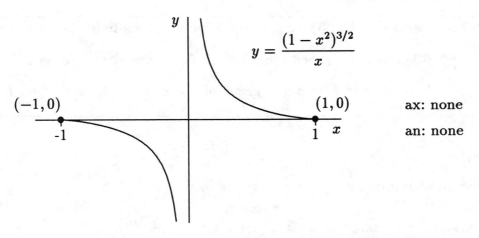

$$y = \frac{(1 - x^2)^{3/2}}{x}$$

$(-1, 0)$

$(1, 0)$ ax: none

-1

1 x an: none

Note! One-sided derivative: $f'_+(-1) = 0,$ $f'_-(1) = 0$

3.4   Concavity and Points of Inflection

Definition (3.22): If the graph of f lies above all of its tangents on an interval I, then it is called **concave upward** on I. If the graph of f lies below all of these tangents, it is called **concave downward** on I.

Notation: CU: concave upward;   CD: concave downward

Illustration:

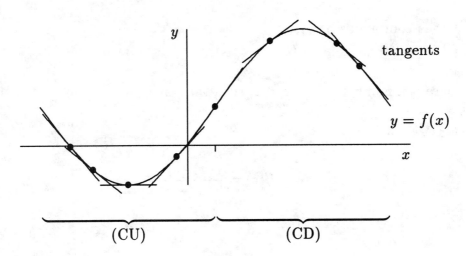

tangents

$y = f(x)$

x

(CU)   (CD)

Use the second derivative of f to determine concavity of f.

The Test for Concavity (3.23): Suppose f is twice differentiable on an interval I.

(a) If $f''(x) > 0$ for all x in I, then the graph of f is concave upward on I.

(b) If $f''(x) < 0$ for all x in I, then the graph of f is concave downward on I.

Proof of (a):

Fix the number a in I.

An equation of tangent line at $(a, f(a))$ is $y = f(a) + f'(a)(x - a)$.

To be concave upward: show $\underbrace{f(x) > f(a) + f'(a)(x - a)}$ at each $x \in I$, $x \neq a$.
(graph of f lies above the tangent line)

Given: $f''(x) > 0$ for all $x \in I \implies f'(x)$ exists for all $x \in I$
$\uparrow$
(contrapositive: $f'(x)$ DNE $\implies f''(x)$ DNE)

$\implies f$ is continuous on I

(C1) Fix $x \in I$ such that $x > a \implies [a, x] \subset I$

$\implies \begin{cases} f \text{ is continuous on } [a, x] \\ f \text{ is differentiable on } (a, x) \end{cases}$

$\implies$ there is a number c, $a < c < x$, such that
$\uparrow$
(apply MVT on $[a, x]$)

$$f'(c) = \frac{f(x) - f(a)}{x - a}$$

Illustration:

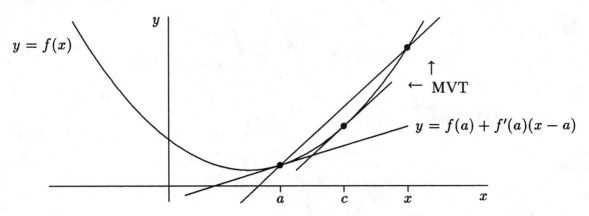

$$\implies \quad f(x) = f(a) + f'(c)(x - a)$$

Given: $f'' > 0$ on $I \implies f'$ increasing on $I \implies f'(a) < f'(c)$

$$\uparrow$$
$$(a < c)$$

$$\implies \quad f'(a)(x - a) < f'(c)(x - a) \qquad\qquad (a < x \implies x - a > 0)$$

$$\implies \quad f(a) + f'(a)(x - a) < f(a) + f'(c)(x - a) = f(x)$$

(C2) Fix $x \in I$ such that $x < a$. Proceed as in (C1).

Combining (C1) and (C2): since a, x in I are arbitrary, f lies above all of its tangents on I.

Definition (3.27): A point P on a curve is called a **point of inflection** if the curve changes from concave upward to concave downward or from concave downward to concave upward at P.

Remarks:

(R1) If $f''(x)$ exists such that $f''(x) \neq 0$, then the concavity is known (depending upon the sign of $f''(x)$) and f can not change concavity at $(x, f(x))$.

Note! $f''(x)$ can only change sign when $f''(x) = 0$ or $f''(x)$ DNE.

(R2) If the curve $y = f(x)$ has a tangent at a point of inflection, then the curve crosses its tangent line at that point.

Illustration:

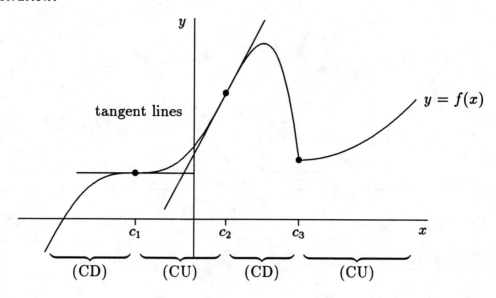

Points of inflection at $\underbrace{(c_1, f(c_1)),\ (c_2, f(c_2))}_{\text{(curve crosses tangent line)}},\ \underbrace{(c_3, f(c_3))}_{\text{(no tangent line)}}$

A Procedure: To determine points of inflection:

(S1) find nominees: those x in the domain of f such that $f''(x) = 0$ or $f''(x)$ DNE

(S2) screen nominees: check for change in sign of f'' on each side of nominee

    (C1) If change in sign occurs, then $(x, f(x))$ is a point of inflection.

    (C2) If a change in sign does not occur, then $(x, f(x))$ is not a point of inflection.

Example 1: Consider $f(x) = x^2 + 1$.

(A1) Find the interval(s) in which f is increasing or f is decreasing and find the local extreme value(s) of f.

$f'(x) = 2x$

Critical numbers:

$f'(x) = 0 :\ 2x = 0 \implies x = 0 \qquad f'(x)$ DNE: none

$$\begin{array}{ccccc} - & 0 & + & & f'(x) \\ \hline & \overset{\displaystyle 0}{} & & x & \\ \downarrow & \ln & \uparrow & & f \end{array}$$

$f \uparrow: [0, \infty)$

$f \downarrow: (-\infty, 0]$

lx: none

ln: $f(0) = 1$

(A2) Find the interval(s) on which f is concave upward or f is concave downward and find all points of inflection.

$f''(x) = 2$

Nominees:

$f''(x) = 0$: none $f''(x)$ DNE: none

$$\begin{array}{ccc} + & & f''(x) \\ \hline & x & \\ \text{CU} & & f \end{array}$$

f CU: $(-\infty, \infty)$

f CD: nowhere

points of inflection: none

(A3) Use the results of (A1) and (A2) to sketch the graph of $y = f(x)$.

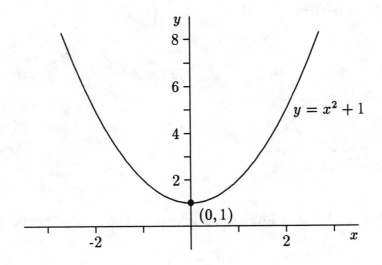

$y = x^2 + 1$

$(0, 1)$

Example 2: Consider $f(x) = x^3 + 1$.

(A1) Find the interval(s) in which f is increasing or f is decreasing and find the local extreme value(s) of f.

$$f'(x) = 3x^2$$

Critical numbers:

$f'(x) = 0: \ 3x^2 = 0 \ \Longrightarrow \ x = 0 \qquad\qquad f'(x)$ DNE: none

$$\begin{array}{ccccc} + & \quad 0 \quad & + & \quad f'(x) & \qquad f \uparrow: \ (-\infty, \infty) \\ \hline & 0 & & x & \qquad f \downarrow: \ \text{nowhere} \\ \uparrow & & \uparrow & f & \qquad \text{lx, ln: none} \\ & \text{no lx/ln} & & & \end{array}$$

(A2) Find the interval(s) on which f is concave up or f is concave down and find all points of inflection.

$$f''(x) = 6x$$

Nominees:

$f''(x) = 0: \ 6x = 0 \ \Longrightarrow \ x = 0 \qquad\qquad f''(x)$ DNE: none

$$\begin{array}{ccccc} - & \quad 0 \quad & + & \quad f''(x) & \qquad f \ \text{CU}: \ (0, \infty) \\ \hline & 0 & & x & \qquad f \ \text{CD}: \ (-\infty, 0) \\ \text{CD} & & \text{CU} & f & \qquad \text{IP}: \ (0, f(0)) = (0, 1) \\ & \underbrace{\text{IP}} & & & \\ & \text{(point of inflection)} & & & \end{array}$$

(A3) Use the results of (A1) and (A2) to sketch the graph of $y = f(x)$.

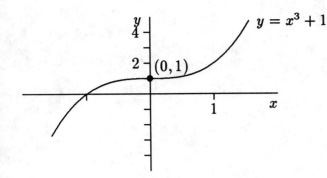

$y = x^3 + 1$

Note! The tangent line at the IP
$(0,1)$ has slope zero

Example 3: Consider $f(x) = x^4 + 1$.

(A1) Find the interval(s) in which f is increasing or f is decreasing and find the local extreme value(s) of f.

$$f'(x) = 4x^3$$

Critical numbers:

$$f'(x) = 0: \ 4x^3 = 0 \implies x = 0 \qquad f'(x) \text{ DNE: none}$$

$f \uparrow$: $[0, \infty)$

$f \downarrow$: $(-\infty, 0]$

lx: none

ln: $f(0) = 1$

(A2) Find the interval(s) on which f is concave up or f is concave down and find all points of inflection.

$$f''(x) = 12x^2$$

Nominees:

$$f''(x) = 0: \ 12x^2 = 0 \implies x = 0 \qquad f''(x) \text{ DNE: none}$$

$$+ \qquad 0 \qquad + \qquad\qquad f''(x)$$

$$\underline{}$$

CU   0   CU x

f CU: $(-\infty, \infty)$

f CD: nowhere

IP: none

(A3) Use the results of (A1) and (A2) to sketch the graph of $y = f(x)$.

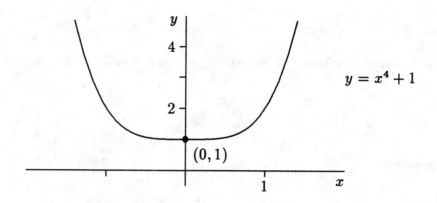

$y = x^4 + 1$

$(0, 1)$

Example 4: Consider $f(x) = (x+1)^2(x-2)$.

(A1) Find the interval(s) in which f is increasing or f is decreasing and find the local extreme value(s) of f.

$$f'(x) = (x+1)^2 + (x-2)(2)(x+1) = 3(x+1)(x-1)$$

Critical numbers:

$$f'(x) = 0: \quad 3(x+1)(x-1) = 0 \implies x = -1, \ 1 \qquad\qquad f'(x) \text{ DNE: none}$$

$$+ \qquad 0 \qquad - \qquad 0 \qquad + \qquad\qquad f'(x)$$

$$\underline{}$$

$\uparrow$ $\downarrow$ $\uparrow$ f

-1   1

lx   ln

$f \uparrow$: $(-\infty, -1]$, $[1, \infty)$

$f \downarrow$: $[-1, 1]$

lx: $f(-1) = 0$

ln: $f(1) = -4$

(A2) Find the interval(s) on which f is concave up or f is concave down and find all points of inflection.

$$f''(x) = 3[(x+1) + (x-1)] = 6x$$

Nominees:

$f''(x) = 0:\ 6x = 0 \implies x = 0$ $f''(x)$ DNE: none

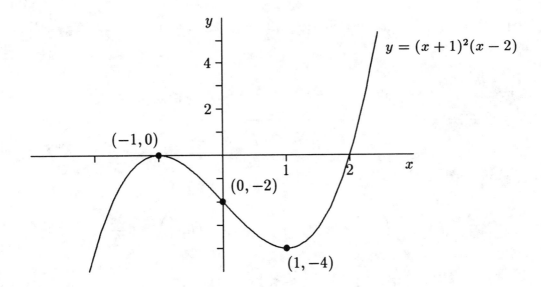

	−	0	+	$f''(x)$

CD	0	CU	x
	IP		f

f CU: $(0, \infty)$

f CD: $(-\infty, 0)$

IP: $(0, f(0)) = (0, -2)$

(A3) Use the results of (A1) and (A2) to sketch the graph of $y = f(x)$.

$$y = (x+1)^2(x-2)$$

$(-1, 0)$

$(0, -2)$

$(1, -4)$

Example 5: Consider $f(x) = x + \cos x,\quad 0 \le x \le 2\pi$.

(A1) Find the interval(s) in which f is increasing or f is decreasing and find the local extreme value(s) of f.

$$f'(x) = 1 - \sin x$$

Critical numbers:

$$f'(x) = 0: \quad 1 - \sin x = 0 \implies \sin x = 1 \implies x = \frac{\pi}{2} \qquad f'(x) \text{ DNE: none}$$

$$\begin{array}{lll} & + \quad 0 & \qquad\qquad + & \qquad\qquad f'(x) & \qquad f \uparrow: [0, 2\pi] \\ \text{XXXX} & & \text{XXXX} & \qquad f \downarrow: \text{nowhere} \\ 0 \quad \uparrow \quad \pi/2 & \qquad \uparrow & 2\pi \qquad f & \qquad \text{lx, ln: none} \\ \text{no lx/ln} \end{array}$$

(A2) Find the interval(s) on which f is concave up or f is concave down and find all points of inflection.

$$f''(x) = -\cos x$$

Nominees:

$$f''(x) = 0: \quad -\cos x = 0 \implies \cos x = 0 \implies x = \frac{\pi}{2}, \frac{3\pi}{2} \qquad f''(x) \text{ DNE: none}$$

$$\begin{array}{llll} & - \quad 0 & + & \quad 0 \quad - & \quad f''(x) \\ \text{XXXX} & & & \text{XXXX} \\ 0 \quad \text{CD} \quad \pi/2 & \quad \text{CU} & \quad 3\pi/2 \ \text{CD} & 2\pi \qquad f \\ \quad \text{IP} & & \quad \text{IP} \end{array}$$

$$f \text{ CU}: \left(\frac{\pi}{2}, \frac{3\pi}{2}\right); \qquad f \text{ CD}: \left(0, \frac{\pi}{2}\right), \left(\frac{3\pi}{2}, 2\pi\right)$$

$$\text{IP}: \left(\frac{\pi}{2}, f\left(\frac{\pi}{2}\right)\right) = \left(\frac{\pi}{2}, \frac{\pi}{2}\right), \quad \left(\frac{3\pi}{2}, f\left(\frac{3\pi}{2}\right)\right) = \left(\frac{3\pi}{2}, \frac{3\pi}{2}\right)$$

(A3) Use the results of (A1) and (A2) to sketch the graph of $y = f(x)$.

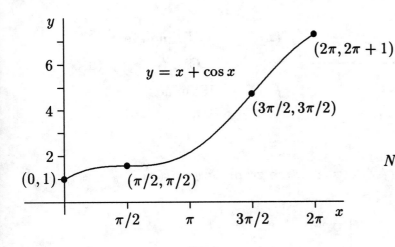

Note! The tangent line at the IP $(\pi/2, \pi/2)$ has slope zero.

Example 6: Consider $f(x) = x^{2/3} + 1$.

(A1) Find the interval(s) in which f is increasing or f is decreasing and find the local extreme value(s) of f.

$$f'(x) = \frac{2}{3}x^{-1/3} = \frac{2}{3x^{1/3}}$$

Critical numbers:

$f'(x) = 0$: none $f'(x)$ DNE: $x^{1/3} = 0 \implies x = 0$

$$\begin{array}{ccccc} & - & \text{DNE} & + & f'(x) \\ \hline & & 0 & & x \\ & \downarrow & & \uparrow & f \\ & & \text{ln} & & \end{array}$$

$f \uparrow$: $[0, \infty)$
$f \downarrow$: $(-\infty, 0]$
lx: none
ln: $f(0) = 1$

(A2) Find the interval(s) on which f is concave up or f is concave down and find all points of inflection.

$$f''(x) = -\frac{2}{9}x^{-4/3} = -\frac{2}{9x^{4/3}}$$

Nominees:

$f''(x) = 0$: none $f''(x)$ DNE: $x^{4/3} = 0 \implies x = 0$

−	DNE	−	$f''(x)$	f CU: nowhere
				f CD: $(-\infty, 0)$, $(0, \infty)$
CD	0	CD	f	IP: none

(A3) Use the results of (A1) and (A2) to sketch the graph of $y = f(x)$.

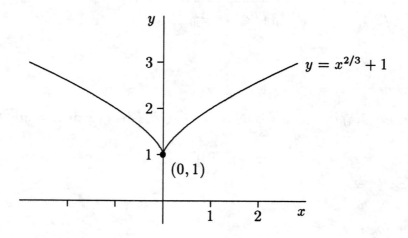

The second derivative can be used to classify local extrema for those critical numbers where the tangent line has slope zero.

The Second Derivative Test (3.28): Suppose f'' is continuous on an open interval that contains c.

(a) If $f'(c) = 0$ and $f''(c) > 0$, then f has a local minimum at c.

(b) If $f'(c) = 0$ and $f''(c) < 0$, then f has a local maximum at c.

Proof of (a):

Given: $f'(c) = 0$

$\implies$ an equation of the tangent line at the point $P(c, f(c))$ is

$$y - f(c) = 0(x - c) \qquad \text{or, rewritten,} \qquad y = f(c)$$

Given: $\begin{cases} f''(c) > 0 \\ f'' \text{ continuous on open interval containing } c \end{cases}$

$\implies$ $f''(x) > 0$ on some open interval I containing c

$\implies$ f is concave upward on I

$\implies$ graph of $y = f(x)$ for all $x \in I$ lies above tangent at P

$\implies$ $f(x) \ge \underbrace{f(c)}$ for all $x \in I$

$(y = f(c)$: equation of tangent line at $P)$

$\implies$ f has a local minimum at the critical number c.

Illustration:

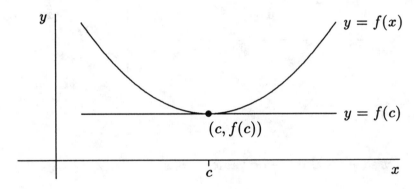

Note! $f'(c) = 0$: slope of tangent line at $(c, f(c))$.

Proof of (b): Proceed similarly to proof of (a)

Remarks: If $f'(c) = 0$ and f'' is continuous "about" c such that:

(R1) $f''(c) > 0$; the graph of f is concave upward about c and $f(c)$ is a local minimum value;

(R2) $f''(c) < 0$; the graph of f is concave downward about c and $f(c)$ is a local maximum value;

(R3) $f''(c) = 0$; the Second Derivative Test gives no information. Use the First Derivative Test to determine if $f(c)$ is either a local maximum value, local minimum value, or neither.

Example 7: Classify each local extreme value of $f(x) = x^3 - 3x^2 + 5$.

$$f'(x) = 3x^2 - 6x = 3x(x - 2)$$

Critical numbers:

$f'(x) = 0: \ 3x(x - 2) = 0 \implies x = 0, \ 2$ \qquad $f'(x)$ DNE: none

$f''(x) = 6x - 6 = 6(x - 1)$: polynomial, so continuous everywhere

$x = 0: \ f''(0) = -6 < 0 \implies$ lx : $f(0) = 5$

$x = 2: \ f''(2) = \ \ 6 > 0 \implies$ ln : $f(2) = 1$

Note! Could use First Derivative Test to classify the local extrema.

$$
\begin{array}{ccccccc}
+ & 0 & - & 0 & + & & f'(x) \\
\hline
 & 0 & & 2 & & x \\
\uparrow & & \downarrow & & \uparrow & & f \\
 & \text{lx} & & \text{ln} & & &
\end{array}
$$

Example 8: Classify each local extreme value of $f(x) = x^3 + 2$.

$$f'(x) = 3x^2$$

Critical numbers:

$f'(x) = 0: \ 3x^2 = 0 \implies x = 0$ \qquad $f'(x)$ DNE: none

$f''(x) = 6x$: polynomial, so continuous everywhere

$x = 0: \ f''(0) = 0 \implies$ no conclusion via the Second Derivative Test;

use First Derivative Test:

$$\begin{array}{cccc} + & 0 & + & f'(x) \end{array}$$

```
_____
            |                    x
            0
    ↑               ↑            f
        no lx/ln
```

Note! $f''(x) = 6x$

Nominees for points of inflection:

$$f''(x) = 0 : \ 6x = 0 \implies x = 0 \qquad\qquad f''(x) = 0 \text{ DNE: none}$$

```
    −        0        +        f''(x)
_____
            |                  x
            0
    CD               CU          f
            IP
```

IP: $(0, f(0)) = (0, 2)$

Example 9: Classify each local extreme value of $f(x) = x^4 + 2$.

$$f'(x) = 4x^3$$

Critical numbers:

$$f'(x) = 0 : \ 4x^3 = 0 \implies x = 0 \qquad\qquad f'(x) \text{ DNE: none}$$

$f''(x) = 12x^2$: polynomial, so continuous everywhere

$x = 0 : \ f''(0) = 0 \implies$ no conclusion via Second Derivative Test;

use First Derivative Test:

```
        −        0        +        f'(x)
_____
                |                  x
                0
        ↓                ↑          f
                ln
```

ln: $f(0) = 2$

Note! $f''(x) = 12x^2$

Nominees for points of inflection:

$f''(x) = 0 : 12x^2 = 0 \implies x = 0$ $\qquad$ $f''(x)$ DNE: none

$$
\begin{array}{ccccc}
+ & 0 & + & & f''(x) \\
\hline
& 0 & & x & \\
\text{CU} & & \text{CU} & & f
\end{array}
$$

$(0, f(0)) = (0, 2)$:

not a point of inflection

Comments:

(C1) Let a be in the domain of f. $f'(a)$ DNE $\implies$ $f''(a)$ DNE; to classify the critical number a, use the First Derivative Test.

(C2) The Second Derivative Test is limited to those critical numbers a such that $f'(a) = 0$. The First Derivative Test can be used for all critical numbers.

3.5   Limits at Infinity; Horizontal Asymptotes

> **Definition (3.29)**: Let f be a function defined on some interval (a, ∞). Then
>
> $$\lim_{x \to \infty} f(x) = L$$
>
> means that the values of $f(x)$ can be made arbitrarily close to L by taking x sufficiently large.

Remarks:

(R1) Alternative notation: $f(x) \to L$ as $\underbrace{x \to \infty}_{\text{(increases without bound)}}$

(R2) L: constant; ∞: not a number, but indicates a "direction" in which x is changing

Example 1: $\displaystyle\lim_{x\to\infty} \frac{2x-1}{x} = ?$

$$\underbrace{\lim_{x\to\infty} \frac{2x-1}{x}}_{\substack{f(x) \\ (x\neq 0)}}\ \underset{\uparrow}{=}\ \lim_{x\to\infty}\left(2 - \frac{1}{x}\right) = 2 - 0 = \underbrace{2}_{L}$$

Illustration: For $x > 0$:

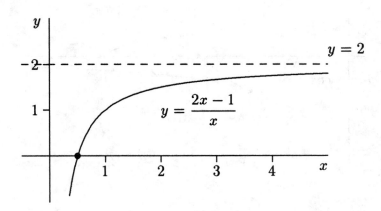

Note! For $x > 0$: $\ y = \dfrac{2x-1}{x} = 2 - \underset{\underset{(1/x>0)}{\uparrow}}{\dfrac{1}{x}} < 2$

Example 2: $\displaystyle\lim_{x\to\infty} \frac{\cos x}{x} = ?$

$$f(x) = \frac{\cos x}{x} = \frac{1}{x}\cos x \qquad\qquad |\cos x| \leq 1 \ \Longrightarrow\ -1 \leq \cos x \leq 1$$

$x \to \infty$: need only consider $x > \dfrac{\pi}{2}\ (>0)\ \Longrightarrow\ \dfrac{1}{x} > 0$

$$\Longrightarrow\quad -\frac{1}{x} \leq \frac{1}{x}\cos x \leq \frac{1}{x}$$

Note!

(N1) $\ -\dfrac{1}{x} \leq \dfrac{1}{x}\cos x$: equality holds when $\cos x = -1\ \Longrightarrow\ x = (2n-1)\pi,\ n = 1, 2, \ldots$

(N2) $\dfrac{1}{x}\cos x \leq \dfrac{1}{x}$: equality holds when $\cos x = 1$ $\implies$ $x = 2n\pi,\ n = 1, 2, \ldots$

Illustration: For $x > \dfrac{\pi}{2}$:

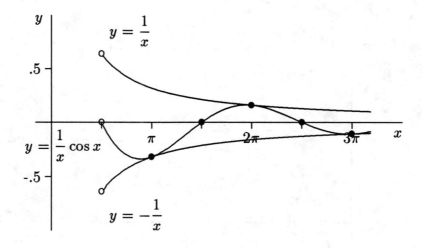

$$\lim_{x \to \infty} \underbrace{\frac{\cos x}{x}}_{f(x)} = \underbrace{0}_{L}$$

Example 3: $\displaystyle\lim_{x \to \infty} \cos x = ?$

$\displaystyle\lim_{x \to \infty} \cos x$ does not exist

Note! $y = \cos x$ does not approach a particular value as $x \to \infty$:

(N1) $\cos x = 1$ for $x = 2n\pi,\ n = 1, 2, \ldots$

(N2) $\cos x = -1$ for $x = (2n - 1)\pi,\ n = 1, 2, \ldots$

Definition (3.30): Let f be a function defined on some interval $(-\infty, a)$. Then

$$\lim_{x \to -\infty} f(x) = L$$

means that the values of $f(x)$ can be made arbitrarily close to L by taking x sufficiently large negative.

Remarks:

(R1) Alternative notation: $f(x) \to L$ as $\underbrace{x \to -\infty}_{\text{(decreases without bound)}}$

(R2) $-\infty$: not a number, but indicates a direction in which x is changing

Example 4: $\lim\limits_{x \to -\infty} \dfrac{x+1}{x} = ?$

$$\lim_{x \to -\infty} \underbrace{\frac{x+1}{x}}_{\substack{f(x) \\ (x \neq 0)}} = \lim_{x \to -\infty} \left(1 + \frac{1}{x}\right) = 1 + 0 = \underbrace{1}_{L}$$

Illustration: For $x < 0$:

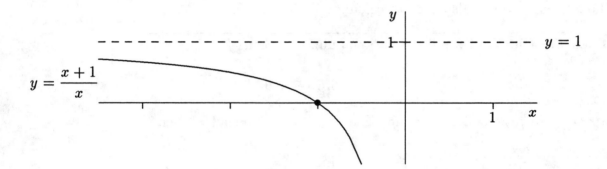

$$y = \frac{x+1}{x} \qquad y = 1$$

Note! For $x < 0$: $\quad y = \dfrac{x+1}{x} = 1 + \underset{\uparrow}{\dfrac{1}{x}} < 1$

$$(1/x < 0)$$

The following eight properties of limits, which can be applied when calculating limits, are identical to those given in Section 1.3.

Limit Laws: Let a represent a finite number, the symbol ∞, or the symbol $-\infty$. Suppose c is a constant and the limits $\lim\limits_{x \to a} f(x)$ and $\lim\limits_{x \to a} g(x)$ exist, then:

1. $\lim\limits_{x \to a}[f(x) + g(x)] = \lim\limits_{x \to a} f(x) + \lim\limits_{x \to a} g(x)$

2. $\lim\limits_{x \to a}[f(x) - g(x)] = \lim\limits_{x \to a} f(x) - \lim\limits_{x \to a} g(x)$

3. $\lim\limits_{x \to a} cf(x) = c \lim\limits_{x \to a} f(x)$

4. $\lim\limits_{x \to a}[f(x)g(x)] = \lim\limits_{x \to a} f(x) \cdot \lim\limits_{x \to a} g(x)$

5. $\lim\limits_{x \to a} \dfrac{f(x)}{g(x)} = \dfrac{\lim\limits_{x \to a} f(x)}{\lim\limits_{x \to a} g(x)}$ if $\lim\limits_{x \to a} g(x) \neq 0$

6. $\lim\limits_{x \to a}[f(x)]^n = \left[\lim\limits_{x \to a} f(x)\right]^n$ where n is a positive integer.

7. $\lim\limits_{x \to a} c = c$

8. $\lim\limits_{x \to a} \sqrt[n]{f(x)} = \sqrt[n]{\lim\limits_{x \to a} f(x)}$ where n is a positive integer.

   (If n is even, we assume that $\lim\limits_{x \to a} f(x) > 0$.)

The following rule is also used to calculate limits at infinity.

Theorem (3.32): If $r > 0$ is a rational number, then

$$\lim_{x \to \infty} \frac{1}{x^r} = 0$$

If $r > 0$ is a rational number such that x^r is defined for all x, then

$$\lim_{x \to -\infty} \frac{1}{x^r} = 0$$

Example 5: $r = 1$ (> 0)

   (A1) $\lim\limits_{x \to \infty} \dfrac{1}{x} = 0$

   (A2) $\lim\limits_{x \to -\infty} \dfrac{1}{x} = 0$

   $\left.\right\}$ see graph of $y = \dfrac{1}{x}$:

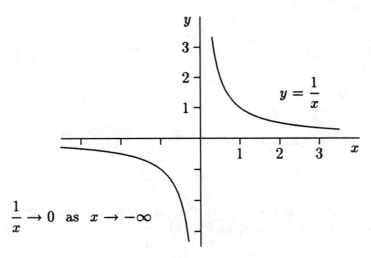

$$\frac{1}{x} \to 0 \quad \text{as} \quad x \to \infty$$

$$\frac{1}{x} \to 0 \quad \text{as} \quad x \to -\infty$$

Example 6: $r = 2 \quad (> 0)$

$$\left.\begin{array}{ll} \text{(A1)} & \lim\limits_{x \to \infty} \dfrac{1}{x^2} = 0 \\[3mm] \text{(A2)} & \lim\limits_{x \to -\infty} \dfrac{1}{x^2} = 0 \end{array}\right\} \quad \text{see graph of } y = \dfrac{1}{x^2}:$$

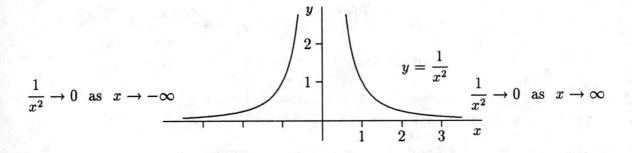

$$\frac{1}{x^2} \to 0 \quad \text{as} \quad x \to -\infty$$

$$\frac{1}{x^2} \to 0 \quad \text{as} \quad x \to \infty$$

Example 7: $r = \dfrac{3}{2} \quad (> 0)$

(A1) $\lim\limits_{x \to \infty} \dfrac{1}{x^{3/2}} = 0$

(A2) $\lim\limits_{x \to -\infty} \dfrac{1}{x^{3/2}} : \quad x^{3/2}$ not defined for $x < 0$

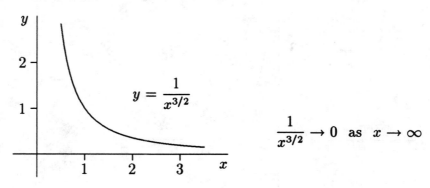

$$\frac{1}{x^{3/2}} \to 0 \quad \text{as} \quad x \to \infty$$

Question: Let P and Q be polynomials. If the degree of Q is greater than or equal to one and the degree of P is less than or equal to the degree of Q, find

$$\lim_{x \to \infty} \frac{P(x)}{Q(x)} \quad \text{or} \quad \lim_{x \to -\infty} \frac{P(x)}{Q(x)}.$$

A Procedure:

(S1) Let n (≥ 1) be the degree of Q.

(S2) Divide $P(x)$ and $Q(x)$ by x^n to obtain the expression $\dfrac{f(x)}{g(x)}$, where

$$f(x) = \frac{P(x)}{x^n} \quad \text{and} \quad g(x) = \frac{Q(x)}{x^n}.$$

   Note! $x \to \pm\infty \implies |x|$ is "large" $\implies x \neq 0 \implies x^n \neq 0$

(S3) $f(x)$ and $g(x)$ each consists of a sum of terms of the form $\dfrac{c}{x^m}$, where c is a real number and m is a nonnegative integer.

(S4) Use Limit Laws 1-3,7 and Theorem (3.32) to determine

$$\lim_{x \to \pm\infty} f(x) = L_1 \quad \text{and} \quad \lim_{x \to \pm\infty} g(x) = L_2$$

   Apply Limit Law 5:

   (C1) Degree of P < degree of Q : $L_1 = 0$, $L_2 \neq 0$, and $\displaystyle\lim_{x \to \pm\infty} \frac{P(x)}{Q(x)} = 0 \quad \left(= \frac{L_1}{L_2}\right)$

   (C2) Degree of P = degree of Q : $L_1 \neq 0$, $L_2 \neq 0$, and $\displaystyle\lim_{x \to \pm\infty} \frac{P(x)}{Q(x)} = \frac{L_1}{L_2} \quad (\neq 0)$

Example 8: $\displaystyle\lim_{x\to-\infty} \frac{3x^2 + 8x - 11}{2x^2 - 3x - 2} = ?$

$$\underbrace{\lim_{x\to-\infty} \frac{3x^2 + 8x - 11}{2x^2 - 3x - 2}}_{\left(\begin{array}{l} \text{rational function:} \\ \deg(P) = \deg(Q) = 2 \end{array}\right)} = \lim_{x\to-\infty} \frac{\left(\dfrac{3x^2 + 8x - 11}{x^2}\right)}{\left(\dfrac{2x^2 - 3x - 2}{x^2}\right)} = \underbrace{\lim_{x\to-\infty} \frac{3 + \dfrac{8}{x} - \dfrac{11}{x^2}}{2 - \dfrac{3}{x} - \dfrac{2}{x^2}}}_{(f(x)/g(x))}$$

$$= \frac{3 + 0 - 0}{2 - 0 - 0} = \frac{3}{2} \quad (\neq 0)$$

Example 9: $\displaystyle\lim_{x\to\infty} \frac{1 - x}{2x^2 - 3x - 2} = ?$

$$\underbrace{\lim_{x\to\infty} \frac{1 - x}{2x^2 - 3x - 2}}_{\left(\begin{array}{l} \text{rational function:} \\ \deg(P) = 1 < 2 = \deg(Q) \end{array}\right)} = \lim_{x\to\infty} \frac{\left(\dfrac{1 - x}{x^2}\right)}{\left(\dfrac{2x^2 - 3x - 2}{x^2}\right)} = \underbrace{\lim_{x\to\infty} \frac{\dfrac{1}{x^2} - \dfrac{1}{x}}{2 - \dfrac{3}{x} - \dfrac{2}{x^2}}}_{(f(x)/g(x))}$$

$$= \frac{0 - 0}{2 - 0 - 0} = \frac{0}{2} = 0$$

Definition (3.31): The line $y = L$ is called a **horizontal asymptote** of the curve $y = f(x)$ if either

$$\lim_{x\to\infty} f(x) = L \qquad \text{or} \qquad \lim_{x\to-\infty} f(x) = L$$

Remark: Geometrically, as the point $P(x, f(x))$ on the graph of $y = f(x)$ moves to the far right ($x \to \infty$) or to the far left ($x \to -\infty$), the point P approaches the horizontal line $y = L$ in the sense that the vertical distance between P and the line $y = L$ tends to zero.

Illustration:

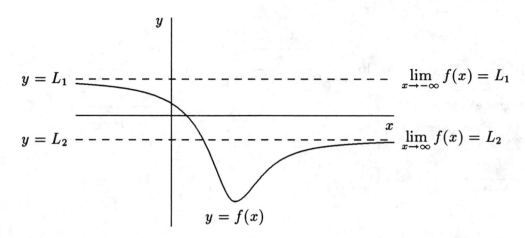

$$y = f(x)$$

Example 10: Consider $f(x) = \dfrac{1 - x^2}{1 + x^2}$.

(A1) Find the interval(s) in which f is increasing or f is decreasing and find the local extreme value(s) of f.

$$f'(x) = \frac{(1 + x^2)(-2x) - (1 - x^2)(2x)}{(1 + x^2)^2} = \frac{-4x}{(1 + x^2)^2}$$

Critical numbers:

$$f'(x) = 0: \ -4x = 0 \implies x = 0 \qquad f'(x) \text{ DNE}: 1 + x^2 = 0: \text{ never} \implies \text{none}$$

		+	0	−		$f'(x)$

		↑	0	↓	x	f
			lx			

$f \uparrow$: $(-\infty, 0]$
$f \downarrow$: $[0, \infty)$
lx: $f(0) = 1$
ln: none

(A2) Find the interval(s) on which f is concave up or f is concave down and find all points of inflection.

$$f''(x) = \frac{(1 + x^2)^2(-4) - (-4x)(2)(1 + x^2)(2x)}{(1 + x^2)^4} = \frac{-4 - 4x^2 + 16x^2}{(1 + x^2)^3} = \frac{4(3x^2 - 1)}{(1 + x^2)^3}$$

has asymptote if $\underset{\textstyle \le}{\text{numerator power}} \ \text{or} = \underset{\textstyle \text{denominator power}}{}$

Nominees:

$$f''(x) = 0 : \quad 4(3x^2 - 1) = 0 \quad \Longrightarrow \quad 3x^2 = 1 = 0 \quad \Longrightarrow \quad x = \pm\frac{1}{\sqrt{3}}$$

$f''(x)$ DNE: $1 + x^2 = 0$: never $\quad \Longrightarrow \quad$ none

	+		0		−		0		+		$f''(x)$
	CU	$-1/\sqrt{3}$	CD		$1/\sqrt{3}$		CU		x		f
		IP			IP						

f CU: $\left(-\infty, -\frac{1}{\sqrt{3}}\right)$, $\left(\frac{1}{\sqrt{3}}, \infty\right)$

f CD: $\left(-\frac{1}{\sqrt{3}}, \frac{1}{\sqrt{3}}\right)$

IP: $\left(-\frac{1}{\sqrt{3}}, f\left(-\frac{1}{\sqrt{3}}\right)\right) = \left(-\frac{1}{\sqrt{3}}, \frac{1}{2}\right)$, $\quad \left(\frac{1}{\sqrt{3}}, f\left(\frac{1}{\sqrt{3}}\right)\right) = \left(\frac{1}{\sqrt{3}}, \frac{1}{2}\right)$

(A3) Find the horizontal asymptote(s) of the graph of $y = f(x)$

$$\lim_{x \to \infty} \frac{1 - x^2}{1 + x^2} = \lim_{x \to \infty} \frac{\left(\dfrac{1 - x^2}{x^2}\right)}{\left(\dfrac{1 + x^2}{x^2}\right)} = \lim_{x \to \infty} \frac{\dfrac{1}{x^2} - 1}{\dfrac{1}{x^2} + 1} = \frac{0 - 1}{0 + 1} = -1$$

$\Longrightarrow \quad y = -1$: horizontal asymptote as $x \to \infty$

$$\lim_{x \to -\infty} \frac{1 - x^2}{1 + x^2} = \lim_{x \to -\infty} \frac{\dfrac{1}{x^2} - 1}{\dfrac{1}{x^2} + 1} = -1$$

$\Longrightarrow \quad y = -1$: horizontal asymptote as $x \to -\infty$

(A4) Use the results of (A1)-(A3) to sketch the graph of $y = f(x)$.

zeros of $f(x)$: $1 - x^2 = 0 \implies x = \pm 1$

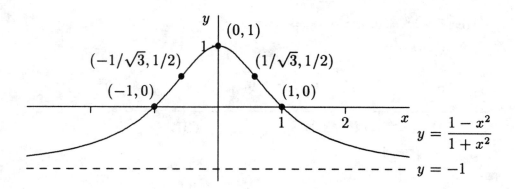

Note! Does graph intersect its horizontal asymptote?

$$\frac{1 - x^2}{1 + x^2} = -1 \implies 1 - x^2 = -(1 + x^2) \implies 1 = -1: \text{never}$$

$\implies$ the point $(x, f(x))$ approaches $y = -1$ only from above as $x \to \pm\infty$

Example 11: $\displaystyle\lim_{x \to \infty} \frac{3 - 2x}{\sqrt{x^2 + 1}} = ?$

$$\lim_{x \to \infty} \frac{3 - 2x}{\sqrt{x^2 + 1}} = \lim_{x \to \infty} \frac{3 - 2x}{\sqrt{x^2\left(1 + \dfrac{1}{x^2}\right)}} = \lim_{x \to \infty} \frac{3 - 2x}{|x|\sqrt{1 + \dfrac{1}{x^2}}} \uparrow = \lim_{x \to \infty} \frac{3 - 2x}{x\sqrt{1 + \dfrac{1}{x^2}}}$$

$$(x \to \infty \implies x > 0 \implies |x| = x)$$

$$= \lim_{x \to \infty} \frac{\dfrac{3 - 2x}{x}}{\sqrt{1 + \dfrac{1}{x^2}}} = \lim_{x \to \infty} \frac{\dfrac{3}{x} - 2}{\sqrt{1 + \dfrac{1}{x^2}}} = \frac{0 - 2}{\sqrt{1 + 0}} = -2$$

Note! $y = -2$ is a horizontal asymptote as $x \to \infty$

Example 12: $\displaystyle\lim_{x\to-\infty}\frac{3-2x}{\sqrt{x^2+1}} = ?$ (same $f(x)$ as in Example 11 above)

$$\lim_{x\to-\infty}\frac{3-2x}{\sqrt{x^2+1}} = \lim_{x\to-\infty}\frac{3-2x}{\sqrt{x^2\left(1+\dfrac{1}{x^2}\right)}} = \lim_{x\to-\infty}\frac{3-2x}{|x|\sqrt{1+\dfrac{1}{x^2}}}\uparrow = \lim_{x\to-\infty}\frac{3-2x}{-x\sqrt{1+\dfrac{1}{x^2}}}$$

$$(x\to-\infty \implies x<0 \implies |x|=-x)$$

$$= -\lim_{x\to-\infty}\frac{\dfrac{3-2x}{x}}{\sqrt{1+\dfrac{1}{x^2}}} = -\lim_{x\to-\infty}\frac{\dfrac{3}{x}-2}{\sqrt{1+\dfrac{1}{x^2}}} = -\frac{0-2}{\sqrt{1+0}} = 2$$

Note! $y=2$ is a horizontal asymptote as $x\to-\infty$, which differs from the horizontal asymptote $y=-2$ as $x\to\infty$.

Example 13: $\displaystyle\lim_{x\to-\infty}\frac{3-2x}{\sqrt[3]{x^3+1}} = ?$

$$\lim_{x\to-\infty}\frac{3-2x}{\sqrt[3]{x^3+1}} = \lim_{x\to-\infty}\frac{3-2x}{\sqrt[3]{x^3\left(1+\dfrac{1}{x^3}\right)}}\uparrow = \lim_{x\to-\infty}\frac{3-2x}{x\sqrt[3]{1+\dfrac{1}{x^3}}} = \frac{0-2}{\sqrt[3]{1+0}} = -2$$

(odd integer root: no need for absolute values)

Note! $\displaystyle\lim_{x\to\infty}\frac{3-2x}{\sqrt[3]{x^3+1}} = -2$ also.

Example 14: $\displaystyle\lim_{x\to\infty}\left(x-\sqrt{x^2+1}\right) = ?$

$$\lim_{x\to\infty}\left(x-\sqrt{x^2+1}\right) = \lim_{x\to\infty}\left[\left(x-\sqrt{x^2+1}\right)\frac{x+\sqrt{x^2+1}}{x+\sqrt{x^2+1}}\right] = \lim_{x\to\infty}\frac{x^2-(x^2+1)}{x+\sqrt{x^2+1}}$$

$$\uparrow$$
(introduce conjugate)

$$= \lim_{x\to\infty}\frac{-1}{x+\sqrt{x^2+1}} = \lim_{x\to\infty}\frac{-1}{\underbrace{x+|x|\sqrt{1+\dfrac{1}{x^2}}}}$$

(as in Example 11 above)

$$= \lim_{x \to \infty} \frac{-\dfrac{1}{x}}{1 + \sqrt{1 + \dfrac{1}{x^2}}} = \frac{0}{1+1} = 0$$

$$(x \to \infty \implies x > 0 \implies |x| = x)$$

Example 15: $\displaystyle \lim_{x \to -\infty} \frac{x}{x - \sqrt{x^2 + 1}} = ?$

$$\lim_{x \to -\infty} \frac{x}{x - \sqrt{x^2 + 1}} = \lim_{x \to -\infty} \frac{x}{x - \sqrt{x^2\left(1 + \dfrac{1}{x^2}\right)}} = \lim_{x \to -\infty} \frac{x}{x - |x|\sqrt{1 + \dfrac{1}{x^2}}}$$

$$= \lim_{x \to -\infty} \frac{x}{x - (-x)\sqrt{1 + \dfrac{1}{x^2}}} = \lim_{x \to -\infty} \frac{1}{1 + \sqrt{1 + \dfrac{1}{x^2}}} = \frac{1}{1+1} = \frac{1}{2}$$

$$(x \to -\infty \implies x < 0 \implies |x| = -x)$$

The Squeeze Theorem:

(a) If $f(x) \le g(x) \le h(x)$ for all x in some interval (a, ∞) and

$$\lim_{x \to \infty} f(x) = \lim_{x \to \infty} h(x) = L, \quad \text{then} \quad \lim_{x \to \infty} g(x) = L$$

(b) If $f(x) \le g(x) \le h(x)$ for all x in some interval $(-\infty, a)$ and

$$\lim_{x \to -\infty} f(x) = \lim_{x \to -\infty} h(x) = L, \quad \text{then} \quad \lim_{x \to -\infty} g(x) = L$$

Example 16: $\displaystyle\lim_{x\to\infty} \frac{1}{x+\sqrt{x^2+1}} = ?$

For $x \in (0,\infty)$: $0 < \underset{\uparrow}{\dfrac{1}{x+\sqrt{x^2+1}}} < \dfrac{1}{2x}$

$$\left(\sqrt{x^2+1} = |x|\underset{\underset{(x>0)}{\uparrow}}{\sqrt{1+\dfrac{1}{x^2}}} = x\underbrace{\sqrt{1+\dfrac{1}{x^2}}}_{(>1)} > x \implies x+\sqrt{x^2+1} > x+x = 2x \right)$$

$$\left.\begin{array}{l} \displaystyle\lim_{x\to\infty} 0 = 0 \\[2mm] \displaystyle\lim_{x\to\infty}\frac{1}{2x} = \frac{1}{2}\lim_{x\to\infty}\frac{1}{x} = \frac{1}{2}(0) = 0 \end{array}\right\} \underset{\uparrow}{\implies} \lim_{x\to\infty}\frac{1}{x+\sqrt{x^2+1}} = 0$$

(Squeeze Theorem)

Note! Another method:

$$\lim_{x\to\infty}\underset{\uparrow}{\frac{1}{x+\sqrt{x^2+1}}} = \lim_{x\to\infty}\frac{1}{x+x\sqrt{1+\dfrac{1}{x^2}}} = \lim_{x\to\infty}\frac{\dfrac{1}{x}}{1+\sqrt{1+\dfrac{1}{x^2}}} = \frac{0}{1+1} = \frac{0}{2} = 0$$

$(x \to \infty \implies x > 0 \implies |x| = x)$

Example 17: $\displaystyle\lim_{x\to-\infty}\frac{\cos x}{x} = ?$

$$\left.\begin{array}{l} |\cos x| \le 1 \implies -1 \le \cos x \le 1 \\[2mm] x \to -\infty : \text{ consider } x \in (-\infty,0) \implies \dfrac{1}{x} < 0 \end{array}\right\} \implies -\frac{1}{x} \ge \frac{1}{x}\cos x \ge \frac{1}{x}$$

$$\left.\begin{array}{l} \displaystyle\lim_{x\to-\infty}\frac{1}{x} = 0 \\[2mm] \displaystyle\lim_{x\to-\infty}-\frac{1}{x} = -\lim_{x\to-\infty}\frac{1}{x} = 0 \end{array}\right\} \underset{\uparrow}{\implies} \lim_{x\to-\infty}\frac{1}{x}\cos x = 0$$

(Squeeze Theorem)

A precise definition of $\displaystyle\lim_{x\to\infty} f(x) = L$ is:

$\dfrac{Large}{Large}$ = do something

Infinity − infinity = do something

$\dfrac{finite}{infinity} = 0$

Definition (3.33): Let f be a function defined on some interval (a, ∞). Then

$$\lim_{x \to \infty} f(x) = L$$

means that for every $\epsilon > 0$ there is a corresponding number N such that

$$|f(x) - L| < \epsilon \qquad \text{whenever} \qquad x > N$$

Remarks:

(R1) $|f(x) - L| < \epsilon \implies -\epsilon < f(x) - L < \epsilon \implies L - \epsilon < f(x) < L + \epsilon$:
$f(x)$ is in the open interval $(L - \epsilon, L + \epsilon)$

(R2) The definition, in words: $f(x)$ can be made as close as we please to L by taking x large enough.

(R3) Geometrically: choose arbitrary $\epsilon > 0$; can one find a number N such that
$L - \epsilon < f(x) < L + \epsilon$ whenever $x > N$?

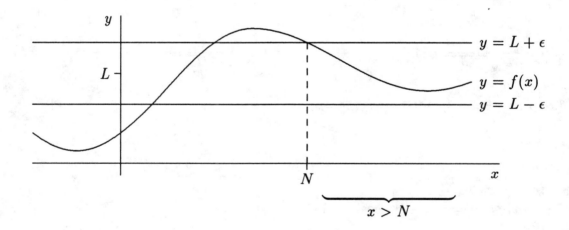

Note!

(N1) The graph of $y = f(x)$ lies between the lines $y = L + \epsilon$ and $y = L - \epsilon$ for all $x > N$.

(N2) Any larger N will suffice for the chosen ϵ

(N3) Usually $N = g(\epsilon)$ (N depends on ϵ) in a way that N must get larger as ϵ gets smaller.

(R4) This definition can be used to prove the Limit Laws, the Squeeze Theorem and Theorem (3.32) given in this section.

Example 18: Using the ϵ, N definition, show $\lim\limits_{x \to \infty} \dfrac{1}{x^2} = 0$.

Note! $f(x) = \dfrac{1}{x^2}$; $L = 0$

Let $\epsilon > 0$ be arbitrary.

$$|f(x) - L| = \left|\dfrac{1}{x^2} - 0\right| = \dfrac{1}{x^2} < \epsilon \implies x^2 > \dfrac{1}{\epsilon}$$

$$\implies \underset{\uparrow}{x} > \dfrac{1}{\sqrt{\epsilon}}; \quad \text{a choice: } N = \dfrac{1}{\sqrt{\epsilon}}$$

$(x \to \infty : \ x > 0)$

Note! $\epsilon \downarrow \implies N \uparrow$

Show $N = \dfrac{1}{\sqrt{\epsilon}}$ works:

$$x > \dfrac{1}{\sqrt{\epsilon}} \implies x^2 > \dfrac{1}{\epsilon} \implies \dfrac{1}{x^2} < \epsilon \implies \left|\dfrac{1}{x^2}\right| < \epsilon$$

$$\implies |f(x) - L| = \left|\dfrac{1}{x^2} - 0\right| < \epsilon$$

Example 19: Using the ϵ, N definition, show $\lim\limits_{x \to \infty} \dfrac{x}{2x - 1} = \dfrac{1}{2}$.

Note! $f(x) = \dfrac{x}{2x - 1}$; $L = \dfrac{1}{2}$

Let $\epsilon > 0$ be arbitrary.

$$|f(x) - L| = \left|\dfrac{x}{2x - 1} - \dfrac{1}{2}\right| = \left|\dfrac{2x - (2x - 1)}{2(2x - 1)}\right| = \dfrac{1}{2|2x - 1|} \underset{\uparrow}{=} \dfrac{1}{2(2x - 1)} < \epsilon$$

$$\left(\begin{array}{l} x \to \infty : \ \text{agree to restrict } x, \text{ say} \\ x \in \underbrace{(1/2, \infty)}_{\text{a choice}} \implies 2x - 1 > 0 \end{array}\right)$$

$$\implies \quad \frac{1}{2x-1} < 2\epsilon \implies 2x - 1 > \frac{1}{2\epsilon} \implies 2x > \frac{1}{2\epsilon} + 1$$

$$\implies \quad x > \frac{1}{2}\left(\frac{1}{2\epsilon} + 1\right); \quad \text{a choice: } N = \max\left\{\frac{1}{2}, \frac{1}{2}\left(\frac{1}{2\epsilon} + 1\right)\right\} = \frac{1}{2}\left(\frac{1}{2\epsilon} + 1\right)$$

$$\left(\epsilon > 0 \implies \frac{1}{2\epsilon} + 1 > 1 \implies \frac{1}{2}\left(\frac{1}{2\epsilon} + 1\right) > \frac{1}{2}\right)$$

A precise definition of $\displaystyle\lim_{x \to -\infty} f(x) = L$ is:

Definition (3.34): Let f be a function defined on some interval $(-\infty, a)$. Then

$$\lim_{x \to -\infty} f(x) = L$$

means that for every $\epsilon > 0$ there is a corresponding number N such that

$$|f(x) - L| < \epsilon \quad \text{whenever} \quad x < N$$

Remarks:

(R1) The definition, in words: $f(x)$ can be made as close as we please to L by taking x large enough negative ($x < 0$ such that $|x|$ is large).

(R2) Geometrically: choose arbitrary $\epsilon > 0$; can one find a number N such that
$L - \epsilon < f(x) < L + \epsilon$ whenever $x < N$?

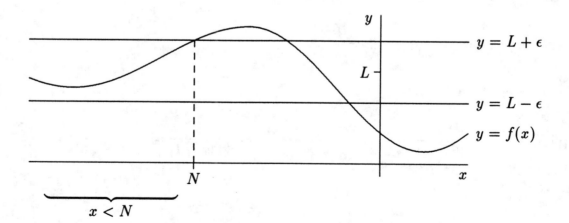

Example 20: Using the ϵ, N definition, show $\displaystyle \lim_{x \to -\infty} \frac{1}{x^2} = 0$.

Note! $f(x) = \dfrac{1}{x^2}$; $L = 0$

Let $\epsilon > 0$ be arbitrary.

$$|f(x) - L| = \left| \frac{1}{x^2} - 0 \right| = \frac{1}{x^2} > \epsilon \implies x^2 > \frac{1}{\epsilon} \implies x > \frac{1}{\sqrt{\epsilon}} \quad \text{or} \quad x < -\frac{1}{\sqrt{\epsilon}}$$

$$\implies x < -\frac{1}{\sqrt{\epsilon}}; \text{ a choice: } N = -\frac{1}{\sqrt{\epsilon}}$$

$\uparrow$

$(x \to -\infty : \ x < 0)$

Note!

(N1) $\epsilon \downarrow \implies N \downarrow$ (in the sense, N gets more negative)

(N2) $\epsilon = \dfrac{1}{100} \implies x < N = -\dfrac{1}{\sqrt{1/100}} = -10$

(N3) $\epsilon = \dfrac{1}{10000} \implies x < N = -\dfrac{1}{\sqrt{1/10000}} = -100$

(N4) Geometrically: $L = 0$

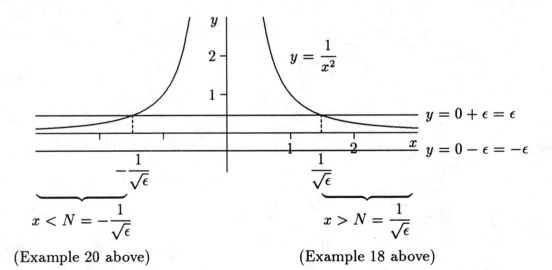

(Example 20 above)   (Example 18 above)

$y = 0$: horizontal asymptote as $x \to \infty$ and as $x \to -\infty$

3.6   Infinite Limits: Vertical Asymptotes

Definition (3.35): Let f be a function defined on both sides of a, except possibly at a itself. Then

$$\lim_{x \to a} f(x) = \infty$$

means that the values of $f(x)$ can be made arbitrarily large (as large as we please) by taking x sufficiently close to a $(x \neq a)$.

Remarks:

(R1) Alternative notation: $f(x) \to \infty$ (increases without bound) as $x \to a$, but $x \neq a$

(R2) The notation $\lim_{x \to a} f(x) = \infty$ indicates that the limit of $f(x)$ as x approaches a does not exist and it designates the functional behavior of $f(x)$ as x approaches a.

Illustration:

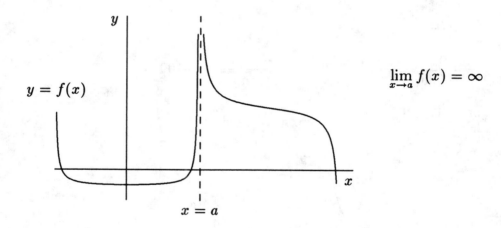

$$\lim_{x \to a} f(x) = \infty$$

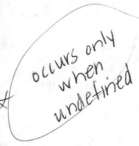

occurs only when undefined

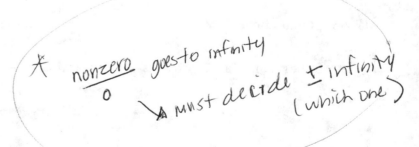

$\ast$ $\dfrac{\text{nonzero}}{0}$ goes to infinity

must decide $\pm$ infinity (which one)

Example 1: $\lim\limits_{x \to 0} \dfrac{1}{x^2} = ?$

Consider $y = x^2$: $x^2 \geq 0$

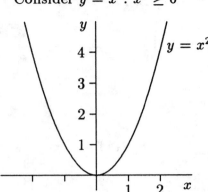

Consider $y = \dfrac{1}{x^2}$: $\dfrac{1}{x^2} \geq 0$ for all $x \neq 0$

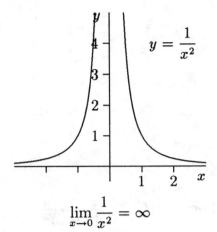

$$\lim_{x \to 0} \frac{1}{x^2} = \infty$$

Definition (3.36): Let f be a function defined on both sides of a, except possibly at a itself. Then

$$\lim_{x \to a} f(x) = -\infty$$

means that the values of $f(x)$ can be made arbitrarily large negative by taking x sufficiently close to a ($x \neq a$).

Remarks:

(R1) Alternative notation: $f(x) \to -\infty$ (decreases without bound) as $x \to a$, but $x \neq a$

(R2) The notation $\lim\limits_{x \to a} f(x) = -\infty$ indicates that the limit of $f(x)$ as x approaches a does not exist and it designates the functional behavior of $f(x)$ as x approaches a.

Illustration:

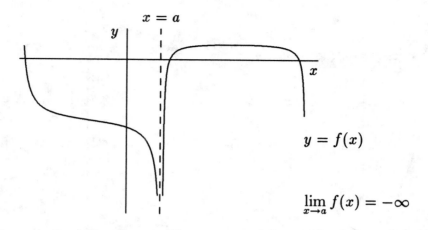

$$x = a$$

$$y = f(x)$$

$$\lim_{x \to a} f(x) = -\infty$$

Example 2: Consider $f(x) = \dfrac{1 - x^2}{(2 - x)^2}$.

(A1) Find the domain of f.

$$2 - x = 0 \implies x = 2;$$

f is a rational function continuous on its domain $(-\infty, 2) \cup (2, \infty)$

(A2) Determine the behavior of f as $x \to 2$.

$$\lim_{x \to 2} \frac{1 - x^2}{(2 - x)^2} = -\infty$$
$$\uparrow$$

$$\left(\begin{array}{l} \text{direct substitution yields: } \dfrac{-3}{0}, \text{ a form of infinity;} \\[2mm] 1 - x^2 < 0 \text{ and } (2 - x)^2 > 0 \text{ for } x \text{ near } 2, x \neq 2 \implies \dfrac{-}{+} = - \end{array} \right)$$

(A3) Find the horizontal asymptotes of the graph of $y = f(x)$.

$$\lim_{x \to \infty} \frac{1 - x^2}{(2 - x)^2} = \lim_{x \to \infty} \frac{\dfrac{1}{x^2} - 1}{\dfrac{4}{x^2} - \dfrac{4}{x} + 1} = \frac{0 - 1}{0 - 0 + 1} = -1$$

$\implies \quad y = -1$: horizontal asymptote as $x \to \infty$

$$\lim_{x \to -\infty} \frac{1 - x^2}{(2 - x)^2} = \lim_{x \to -\infty} \frac{\dfrac{1}{x^2} - 1}{\dfrac{4}{x^2} - \dfrac{4}{x} + 1} = -1$$

$\implies$ $y = -1$: horizontal asymptote as $x \to -\infty$

(A4) Find the interval(s) in which f is increasing or f is decreasing and find the local extreme value(s) of f.

$$f'(x) = \frac{(2 - x)^2(-2x) - (1 - x^2)(2)(2 - x)(-1)}{(2 - x)^4} = \frac{2(1 - 2x)}{(2 - x)^3}$$

Critical numbers:

$$f'(x) = 0: \quad 2(1 - 2x) = 0 \implies 1 - 2x = 0 \implies x = \frac{1}{2}$$

$$f'(x) \text{ DNE}: \quad 2 - x = 0 \implies x = 2: \text{ not in domain of } f$$

$+$	0	$-$		$+$	$f'(x)$
$\uparrow$	$1/2$	$\downarrow$	2	$\uparrow$	x
	lx				f

$f \uparrow: \left(-\infty, \dfrac{1}{2}\right]$, $(2, \infty)$

$f \downarrow: \left[\dfrac{1}{2}, 2\right)$

lx: $f\left(\dfrac{1}{2}\right) = \dfrac{1}{3}$

ln: none

(A5) Use the results of (A1)-(A4) to sketch the graph of $y = f(x)$.

zeros of $f(x): \quad 1 - x^2 = 0 \implies x = \pm 1$

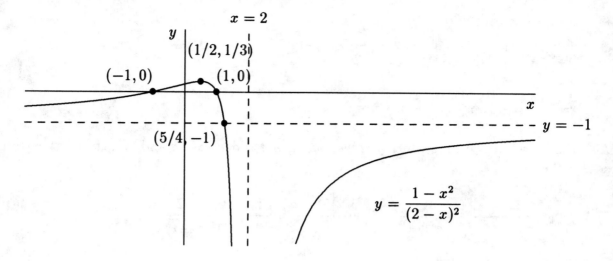

Note! Does the graph intersect its horizontal asymptote?

$$\frac{1-x^2}{(2-x)^2} = -1 \implies 1-x^2 = -(4-4x+x^2) \implies 4x = 5 \implies x = \frac{5}{4}$$

One-sided infinite limits can also be defined.

Illustrations:

(I1) $\displaystyle\lim_{x\to a^+} f(x) = \infty$ (I2) $\displaystyle\lim_{x\to a^-} f(x) = \infty$

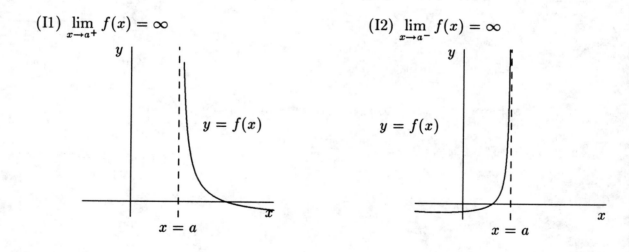

(I3) $\displaystyle\lim_{x\to a^+} f(x) = -\infty$

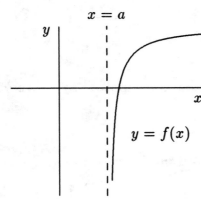

$y = f(x)$

(I4) $\displaystyle\lim_{x\to a^-} f(x) = -\infty$

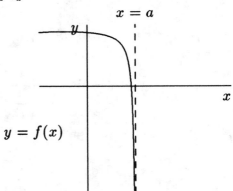

$y = f(x)$

Remarks:

(R1) Right-hand limit $\displaystyle\lim_{x\to a^+} f(x)$: examines behavior of $f(x)$ near a but only for $x > a$.

(R2) Left-hand limit $\displaystyle\lim_{x\to a^-} f(x)$: examines behavior of $f(x)$ near a but only for $x < a$.

(R3) $\displaystyle\lim_{x\to a} f(x) = \infty$ if and only if $\displaystyle\lim_{x\to a^-} f(x) = \infty$ and $\displaystyle\lim_{x\to a^+} f(x) = \infty$.

(R4) $\displaystyle\lim_{x\to a} f(x) = -\infty$ if and only if $\displaystyle\lim_{x\to a^-} f(x) = -\infty$ and $\displaystyle\lim_{x\to a^+} f(x) = -\infty$.

Definition (3.37): The line $x = a$ is called a **vertical asymptote** of the curve $y = f(x)$ if at least one of the following statements is true:

$$\lim_{x\to a} f(x) = \infty \qquad \lim_{x\to a^-} f(x) = \infty \qquad \lim_{x\to a^+} f(x) = \infty$$

$$\lim_{x\to a} f(x) = -\infty \qquad \lim_{x\to a^-} f(x) = -\infty \qquad \lim_{x\to a^+} f(x) = -\infty$$

Remarks:

(R1) From Example 1 above: $x = 0$ is a vertical asymptote of $y = \dfrac{1}{x^2}$.

(R2) From Example 2 above: $x = 2$ is a vertical asymptote of $y = \dfrac{1 - x^2}{(2 - x)^2}$.

Example 3: Consider $f(x) = \dfrac{x^2}{x^2 - 1}$.

(A1) Find the horizontal asymptotes of the graph of $y = f(x)$.

$$\lim_{x \to \infty} \frac{x^2}{x^2 - 1} = \lim_{x \to \infty} \frac{1}{1 - \dfrac{1}{x^2}} = 1 \implies y = 1: \text{ horizontal asymptote as } x \to \infty.$$

$$\lim_{x \to -\infty} \frac{x^2}{x^2 - 1} = \lim_{x \to -\infty} \frac{1}{1 - \dfrac{1}{x^2}} = 1 \implies y = 1: \text{ horizontal asymptote as } x \to -\infty.$$

Note! Does graph intersect its horizontal asymptote?

$$\frac{x^2}{x^2 - 1} = 1 \implies x^2 = x^2 - 1 \implies 0 = -1: \text{ never}$$

(A2) Find the vertical asymptotes of the graph of $y = f(x)$.

Domain of f: $(-\infty, -1) \cup (-1, 1) \cup (1, \infty)$ $(x^2 - 1 = 0 \implies x = \pm 1)$

f: rational function $\implies$ f is continuous on its domain

$\implies$ $x = -1, 1$: only possible vertical asymptotes

$$x = -1: \quad \lim_{x \to -1} \frac{x^2}{x^2 - 1} : \frac{1}{0} \implies \text{check sides}$$
$$\underset{\text{(direct substitution)}}{\uparrow} \qquad \underset{\left(\frac{\neq 0}{0}\right)}{\uparrow}$$

$$\lim_{x \to -1^-} \frac{x^2}{(x - 1)(x + 1)} = +\infty \implies x = -1: \text{ vertical asymptote}$$
$$\uparrow$$

$$\left(\text{sign of factors for } x < -1, x \text{ near } -1: \frac{+}{(-)(-)} = + \right)$$

$$\lim_{x \to -1^+} \frac{x^2}{(x - 1)(x + 1)} = -\infty \qquad \left(x > -1, x \text{ near } -1: \frac{+}{(-)(+)} = - \right)$$

$$x = 1: \lim_{x \to 1} \frac{x^2}{x^2 - 1} : \frac{1}{0} \implies \text{check sides}$$

$$\lim_{x \to 1^-} \frac{x^2}{(x-1)(x+1)} = -\infty \implies x = 1: \text{vertical asymptote}$$
$$\uparrow$$

$$\left(x < 1, x \text{ near } 1: \frac{+}{(-)(+)} = - \right)$$

$$\lim_{x \to 1^+} \frac{x^2}{(x-1)(x+1)} = +\infty \qquad\qquad \left(x > 1, x \text{ near } 1: \frac{+}{(+)(+)} = + \right)$$

(A3) Use the results of (A1) and (A2) to sketch the graph of $y = f(x)$.

$$\text{zeros of } f: \frac{x^2}{x^2 - 1} = 0 \implies x^2 = 0 \implies x = 0$$

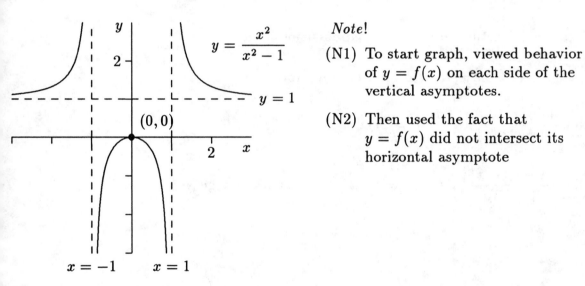

Note!

(N1) To start graph, viewed behavior of $y = f(x)$ on each side of the vertical asymptotes.

(N2) Then used the fact that $y = f(x)$ did not intersect its horizontal asymptote

Example 4: Consider $f(x) = \dfrac{4 + 3x - x^2}{x^2 + 4x + 4}$.

(A1) Find the horizontal asymptotes of the graph of $= f(x)$.

$$\lim_{x \to \infty} \frac{4 + 3x - x^2}{x^2 + 4x + 4} = \lim_{x \to \infty} \frac{\dfrac{4}{x^2} + \dfrac{3}{x} - 1}{1 + \dfrac{4}{x} + \dfrac{4}{x^2}} = -1$$

$\Longrightarrow$ $y = -1$: horizontal asymptote as $x \to \infty$

$$\lim_{x \to -\infty} \frac{4 + 3x - x^2}{x^2 + 4x + 4} = \lim_{x \to -\infty} \frac{\dfrac{4}{x^2} + \dfrac{3}{x} - 1}{1 + \dfrac{4}{x} + \dfrac{4}{x^2}} = -1$$

$\Longrightarrow$ $y = -1$: horizontal asymptote as $x \to -\infty$

Note! Does graph intersect its horizontal asymptote?

$$\frac{4 + 3x - x^2}{x^2 + 4x + 4} = -1 \Longrightarrow 4 + 3x - x^2 = -(x^2 + 4x + 4) \Longrightarrow 7x = -8 \Longrightarrow x = -\frac{8}{7}$$

(A2) Find the vertical asymptotes of the graph of $y = f(x)$.

Domain of f: $(-\infty, -2) \cup (-2, \infty)$ $(x^2 + 4x + 4 = (x + 2)^2 = 0 \implies x = -2)$

f: rational function $\implies$ f is continuous on its domain

$\implies$ $x = -2$: only possible vertical asymptote

$$\lim_{x \to -2} \frac{4 + 3x - x^2}{(x + 2)^2} : \frac{-6}{0} \implies \text{check sides}$$

$$\lim_{x \to -2-} \frac{4 + 3x - x^2}{(x + 2)^2} = -\infty \implies x = -2: \text{vertical asymptote}$$
$$\uparrow$$
$$\left(x < -2, \ x \text{ near } -2: \frac{-}{+} = - \right)$$

$$\lim_{x \to -2+} \frac{4 + 3x - x^2}{(x + 2)^2} = -\infty \qquad\qquad \left(x > -2, \ x \text{ near } -2: \frac{-}{+} = - \right)$$

Note! $\displaystyle \lim_{x \to -2} \frac{4 + 3x - x^2}{(x + 2)^2} = -\infty$

(A3) Use the results of (A1) and (A2) to sketch the graph of $y = f(x)$.

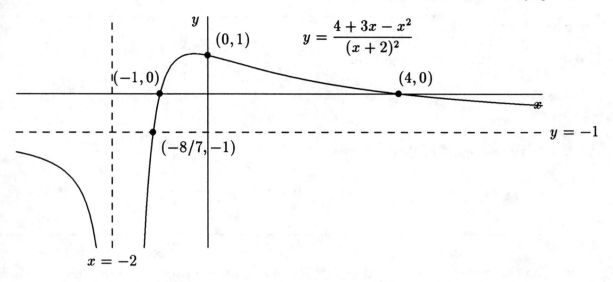

Note! $f'(x) = \dfrac{-(7x+2)}{(x+2)^3}$

Critical numbers:

$f'(x) = 0: \ -(7x+2) = - \implies 7x+2 = 0 \implies x = -\dfrac{2}{7}$

$f'(x)$ DNE: $x+2 = 0 \implies x = -2$: not in the domain of $f \implies$ none

```
         −             +       0          −        f'(x)
  ───────────────✗──────────┬─────────────────────── x
                 −2          −2/7
         ↓            ↑                  ↓           f
                            lx
```

Theorem (3.38):

(a) If n is a positive even integer, then

$$\lim_{x \to a} \frac{1}{(x-a)^n} = \infty$$

(b) If n is a positive odd integer, then

$$\lim_{x \to a^+} \frac{1}{(x-a)^n} = \infty \qquad \text{and} \qquad \lim_{x \to a^-} \frac{1}{(x-a)^n} = -\infty$$

Remark: The theorem can be proved via a precise definition of infinite limits but it can be demonstrated intuitively via the arguments used in the preceding examples.

Example 5: Consider $f(x) = \dfrac{2+x}{\sqrt{x^2-1}}$.

(A1) Find the domain of f.

$\sqrt{x^2-1}$ requires $x^2 - 1 \geq 0 \implies (x-1)(x+1) \geq 0$

$(x-1)(x+1) = 0 \implies x = 1, -1$

$$
\begin{array}{ccccccc}
 & + & 0 & - & 0 & + & (x-1)(x+1) \\
\hline
 & & -1 & & 1 & & x
\end{array}
$$

$(x-1)(x+1) \geq 0$ if

$x \leq -1$ or $x \geq 1$

$\implies$ domain of f: $(-\infty, -1) \cup (1, \infty)$ ($x^2 - 1$ in denominator $\implies x \neq -1, 1$)

(A2) Find the horizontal asymptotes of the graph of $y = f(x)$.

$$\lim_{x \to \infty} \frac{2+x}{\sqrt{x^2-1}} = \lim_{x \to \infty} \frac{2+x}{\sqrt{x^2\left(1 - \frac{1}{x^2}\right)}} = \lim_{x \to \infty} \frac{2+x}{|x|\sqrt{1 - \frac{1}{x^2}}} \uparrow = \lim_{x \to \infty} \frac{\frac{2}{x} + 1}{\sqrt{1 - \frac{1}{x^2}}} = 1$$

$(x \to \infty \implies x > 0 \implies |x| = x)$

$\Longrightarrow$ $y = 1$: horizontal asymptote as $x \to \infty$

$$\lim_{x \to -\infty} \frac{2 + x}{\sqrt{x^2 - 1}} = \lim_{x \to -\infty} \frac{2 + x}{|x|\sqrt{1 - \dfrac{1}{x^2}}} \uparrow = \lim_{x \to -\infty} \frac{\dfrac{2}{x} + 1}{-\sqrt{1 - \dfrac{1}{x^2}}} = -1$$

$$(x \to -\infty \implies x < 0 \implies |x| = -x)$$

$\Longrightarrow$ $y = -1$: horizontal asymptote as $x \to -\infty$

Note!

(N1) Does graph intersect its horizontal asymptote $y = 1$?

$$\frac{2 + x}{\sqrt{x^2 - 1}} = 1 \implies 2 + x = \sqrt{x^2 - 1} \implies (2 + x)^2 = x^2 - 1$$

$$\implies 4 + 4x + x^2 = x^2 - 1 \implies 4x = -5 \implies x = -\frac{5}{4}$$

Check in given equation (due to squaring of terms): $\dfrac{2 + \left(-\dfrac{5}{4}\right)}{\sqrt{\left(-\dfrac{5}{4}\right)^2 - 1}} = 1$

(N2) Does the graph intersect its horizontal asymptote $y = -1$?

$$\frac{2 + x}{\sqrt{x^2 - 1}} = -1 \implies 2 + x = -\sqrt{x^2 - 1} \implies 4 + 4x + x^2 = x^2 - 1 \implies x = -\frac{5}{4}$$

Check in given equation: $\dfrac{2 + \left(-\dfrac{5}{4}\right)}{\sqrt{\left(-\dfrac{5}{4}\right)^2 - 1}} \neq -1$ (extraneous)

(A3) Find the vertical asymptotes of the graph of $y = f(x)$.

Domain of f: $(-\infty, -1) \cup (1, \infty)$.

$x = -1$: $\displaystyle\lim_{x \to -1^-} \frac{2 + x}{\sqrt{x^2 - 1}} = +\infty$ $\left(\text{substitution} : \dfrac{1}{0}; x < -1, x \text{ near } -1 : \dfrac{+}{+} = +\right)$

$\implies$ $x = -1$: vertical asymptote

Note! Since the domain of f does not contain an interval of the form $(-1, a)$, $\lim_{x \to -1^+} f(x)$ is not considered.

$$x = 1: \quad \lim_{x \to 1^+} \frac{2 + x}{\sqrt{x^2 - 1}} = +\infty \qquad \left(\text{substitution}: \frac{3}{0}; x > 1, x \text{ near } 1 : \frac{+}{+} = + \right)$$

(A4) Use the results of (A1)-(A3) to sketch the graph of $y = f(x)$.

zeros of f: $\dfrac{2 + x}{\sqrt{x^2 - 1}} = 0 \implies 2 + x = 0 \implies x = -2$ (in the domain of f)

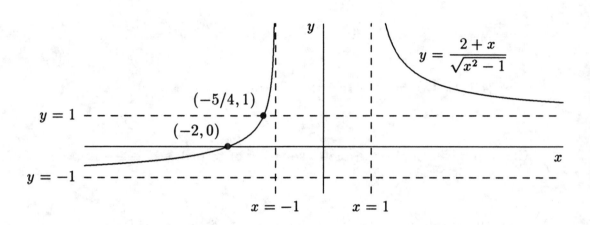

Note!

(N1) First, plot the point corresponding to the zero of f (i.e., $(2, 0)$) and then the point of intersection with the horizontal asymptote $y = 1$ (i.e., $(-5/4, 1)$).

(N2) Then use the behavior of $y = f(x)$ at the vertical asymptotes.

Example 6: Consider $f(x) = \dfrac{\sqrt{1 - x^2}}{x}$.

(A1) Find the domain of f.

$\sqrt{1 - x^2}$ requires $1 - x^2 \geq 0 \implies (1 - x)(1 + x) \geq 0$;

$(1 - x)(1 + x) = 0 \implies x = 1, -1$

$$+ \quad 0 \quad - \quad 0 \quad + \qquad (1-x)(1+x)$$

$$\underset{\substack{\big| \qquad \big| \\ -1 \qquad 1}}{\rule{8cm}{0.4pt}} \quad x$$

$(1-x)(1+x) \geq 0$ if

$-1 \leq x \leq 1$

$\Longrightarrow$ domain of f: $[-1,0) \cup (0,1]$ $\qquad\qquad$ (x in denominator $\Longrightarrow$ $x \neq 0$)

Note! Since the domain of f does not contain intervals of the form $(-\infty, a)$ nor (a, ∞), $\lim\limits_{x \to -\infty} f(x)$ and $\lim\limits_{x \to \infty} f(x)$ can not be considered; hence the graph of $y = f(x)$ has no horizontal asymptotes.

(A2) Find the vertical asymptotes of the graph of $y = f(x)$.

f is continuous on its domain $\Longrightarrow$ $x = 0$: only possible vertical asymptote

$$\lim_{x \to 0} \frac{\sqrt{1-x^2}}{x} : \frac{1}{0} \Longrightarrow \quad \text{check sides}$$

$$\lim_{x \to 0^-} \frac{\sqrt{1-x^2}}{x} = -\infty \Longrightarrow \quad x = 0 \text{: vertical asymptote}$$

$$\uparrow$$

$$\left(x < 0, \ x \text{ near } 0: \ \frac{+}{-} = - \right)$$

$$\lim_{x \to 0^+} \frac{\sqrt{1-x^2}}{x} = +\infty \qquad\qquad\qquad \left(x > 0, \ x \text{ near } 0: \ \frac{+}{+} = + \right)$$

(A3) Find the interval(s) in which f is increasing or f is decreasing and find the local extreme value(s) of f.

$$f(x) = \frac{\sqrt{1-x^2}}{x} = \frac{(1-x^2)^{1/2}}{x}$$

$$f'(x) = \frac{x \left(\frac{1}{2} \right) (1-x^2)^{-1/2}(-2x) - (1-x^2)^{1/2}}{x^2} = \frac{-1}{x^2(1-x^2)^{1/2}}$$

Critical numbers:

$f'(x) = 0$: none

$f'(x)$ DNE: $x^2(1 - x^2)^{1/2} = 0 \implies x = 0, 1, -1$

$$\left\{ \begin{array}{l} x = 0 : \text{ not in domain of } f \\ x = 1, -1 : \text{ endpoints of domain of } f \end{array} \right\} \implies \text{none}$$

$f'(x)$ $f \uparrow$: nowhere

$$\begin{array}{c} - \qquad\qquad - \\ \text{XXXX} \underset{-1}{\rule{0pt}{0pt}} \quad \underset{0}{\text{X}} \quad \underset{1}{\text{XXXX}} \\ \downarrow \qquad \downarrow \end{array}$$

$f \downarrow$: $[-1, 0)$, $(0, 1]$

lx/ln: none

Note! One-sided derivatives

(N1) $f'_+(-1) = \lim\limits_{h \to 0^+} \dfrac{f(-1 + h) - f(-1)}{h} = \lim\limits_{h \to 0^+} \dfrac{\dfrac{\sqrt{1 - (-1 + h)^2}}{-1 + h} - 0}{h}$

$$= \lim\limits_{h \to 0^+} \frac{\sqrt{2h - h^2}}{h(-1 + h)} \underset{\uparrow}{=} \lim\limits_{h \to 0^+} \frac{\sqrt{h}\sqrt{2 - h}}{h(-1 + h)} = \lim\limits_{h \to 0^+} \frac{\sqrt{2 - h}}{\sqrt{h}(-1 + h)} \underset{\uparrow}{=} -\infty$$

$$\left(\frac{0}{0} : \text{ do} \right) \qquad\qquad\qquad \left(\frac{\sqrt{2}}{0} : \frac{+}{(+)(-)} = - \right)$$

(N2) $f'_-(1) = \lim\limits_{h \to 0^-} \dfrac{f(1 + h) - f(1)}{h} = \lim\limits_{h \to 0^-} \dfrac{\dfrac{\sqrt{1 - (1 + h)^2}}{1 + h} - 0}{h}$

$$= \lim\limits_{h \to 0^-} \frac{\sqrt{-2h - h^2}}{h(1 + h)} \underset{\uparrow}{=} \lim\limits_{h \to 0^-} \frac{\sqrt{-h}\sqrt{2 + h}}{h(1 + h)} \underset{\uparrow}{=} \lim\limits_{h \to 0^-} \frac{\sqrt{2 + h}}{-\sqrt{-h}(1 + h)} \underset{\uparrow}{=} -\infty$$

$$\left(\frac{0}{0} : \text{ do; } h < 0 \right) \quad \left(-h = \sqrt{-h}\sqrt{-h} \right) \quad \left(\frac{\sqrt{2}}{0} : \frac{+}{(-)(+)} = - \right)$$

(A4) Use the results of (A1)-(A3) to sketch the graph of $y = f(x)$.

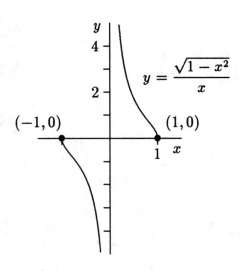

$$y = \frac{\sqrt{1-x^2}}{x}$$

Note: Use $f''(x) = \dfrac{-3x^2 + 2}{x^3(1-x^2)^{3/2}}$ to show

that points of inflection are

$$\left(-\sqrt{\frac{2}{3}}, \frac{-1}{\sqrt{2}}\right) \quad \text{and} \quad \left(\sqrt{\frac{2}{3}}, \frac{1}{\sqrt{2}}\right).$$

Example 7: $\displaystyle\lim_{x\to 3\pi/2} \sec x = ?$

$$\lim_{x\to 3\pi/2} \sec x = \lim_{x\to 3\pi/2} \frac{1}{\cos x} : \quad \frac{1}{0} \quad \Longrightarrow \quad \text{check sides}$$

$$\lim_{x\to(3\pi/2)-} \sec x = \lim_{x\to(3\pi/2)-} \frac{1}{\cos x} = -\infty \qquad \left(x < \frac{3\pi}{2}, \ x \text{ near } \frac{3\pi}{2} : \ \frac{+}{-} = -\right)$$

$$\lim_{x\to(3\pi/2)+} \sec x = \lim_{x\to(3\pi/2)+} \frac{1}{\cos x} = +\infty \qquad \left(x > \frac{3\pi}{2}, \ x \text{ near } \frac{3\pi}{2} : \ \frac{+}{+} = +\right)$$

Therefore: $\displaystyle\lim_{x\to 3\pi/2} \sec x$ DNE

Note!

(N1) $x = \dfrac{3\pi}{2}$: vertical asymptote of the graph $y = \sec x$.

(N2) Recall: the graph of $y = \sec x$.

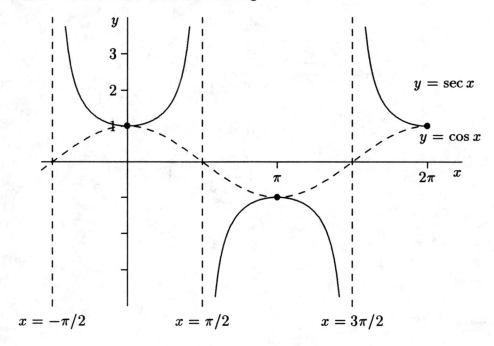

$y = \sec x$

$y = \cos x$

$x = -\pi/2$ $x = \pi/2$ $x = 3\pi/2$

Infinite limits as x increases/decreases without bound can also be defined.

Illustrations:

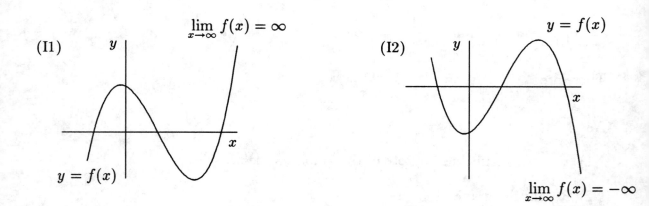

(I1) $\lim_{x \to \infty} f(x) = \infty$

$y = f(x)$

(I2) $y = f(x)$

$\lim_{x \to \infty} f(x) = -\infty$

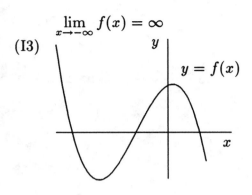

$$\lim_{x \to -\infty} f(x) = \infty$$

(I3) $y = f(x)$

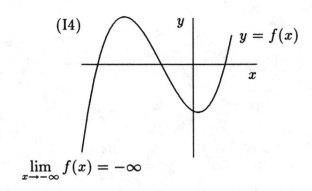

(I4) $y = f(x)$

$$\lim_{x \to -\infty} f(x) = -\infty$$

Extend a result from Section 3.5: let P and Q be polynomials.

$$\lim_{x \to \pm\infty} \frac{P(x)}{Q(x)} = \begin{cases} 0 \text{ if degree of } P < \text{degree of } Q \\ L \ (\neq 0) \text{ if degree of } P = \text{degree of } Q \\ \pm\infty \text{ if degree of } P > \text{degree of } Q \end{cases}$$

Remarks: Consider the case degree of $P >$ degree of Q.

(R1) The sign of the limit depends upon the particular P and Q.

(R2) $\displaystyle\lim_{x \to \infty} \frac{P(x)}{Q(x)}$ can differ in sign from $\displaystyle\lim_{x \to -\infty} \frac{P(x)}{Q(x)}$.

(R3) The graph of $y = \dfrac{P(x)}{Q(x)}$ does not have a horizontal asymptote.

Example 8: Consider $f(x) = \dfrac{1 - x^2}{2 + x}$.

Note! degree of $(1 - x^2) = 2 > 1 = $ degree of $(2 + x)$

(A1) $\displaystyle\lim_{x \to \infty} f(x) = ?$

$$\lim_{x \to \infty} f(x) = -\infty \qquad \left(x \uparrow: \ \frac{1 - x^2}{2 + x} \text{ ``behaves like''} \ \underbrace{\frac{-x^2}{x}} = -x \to -\infty \text{ as } x \to \infty \right)$$

$$\left(\frac{-x^2}{x} = \frac{\text{leading term of } P(x)}{\text{leading term of } Q(x)} \right)$$

(A2) $\lim\limits_{x \to -\infty} f(x) = ?$

$\lim\limits_{x \to -\infty} f(x) = \infty$ $\left(x \downarrow: \dfrac{1-x^2}{2+x} \text{ behaves like } \dfrac{-x^2}{x} = -x \to \infty \text{ as } x \to -\infty \right)$

Note! Consider graph of $y = f(x)$.

Domain of f: $(-\infty, -2) \cup (-2, \infty)$

f: rational function $\implies$ f is continuous on its domain

zeros of f: $x = \pm 1$

Vertical asymptote: $x = -2$ with $\begin{cases} f(x) \to \infty & \text{as } x \to -2^- \\ f(x) \to -\infty & \text{as } x \to -2^+ \end{cases}$

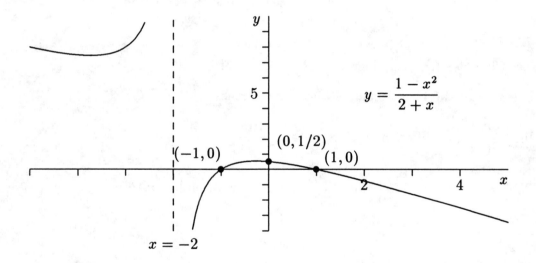

$\lim\limits_{x \to -\infty} f(x) = \infty$ $\lim\limits_{x \to \infty} f(x) = -\infty$

Need to use $f'(x)$ to determine exact location of lx/ln.

Example 9: Consider $f(x) = (x-2)(4-x)^3$.

(A1) $\lim\limits_{x \to \infty} f(x) = ?$

$$\lim_{x \to \infty} (x-2)(4-x)^3 = -\infty$$
$$\uparrow$$
$$\left(\begin{array}{l} x \uparrow: (x-2)(4-x)^3 \text{ behaves like} \\ x(-x)^3 = -x^4 \to -\infty \text{ as } x \to \infty \end{array} \right)$$

(A2) $\lim_{x \to -\infty} f(x) = ?$

$$\lim_{x \to -\infty} (x-2)(4-x)^3 = -\infty$$
$$\uparrow$$
$$\left(\begin{array}{l} x \downarrow: (x-2)(4-x)^3 \text{ behaves like} \\ x(-x)^3 = -x^4 \to -\infty \text{ as } x \to \infty \end{array} \right)$$

Definition (3.39): Let f be a function defined on some open interval that contains the number a, except possibly at a itself. Then

$$\lim_{x \to a} f(x) = \infty$$

means that for every positive number M there is a corresponding number $\delta > 0$ such that

$$f(x) > M \qquad \text{whenever} \qquad 0 < |x-a| < \delta$$

Remarks:

(R1) In words: $f(x)$ can be made as large as we please by taking x close enough to a, but not equal to a.

(R2) Geometrically: choose arbitrary $M > 0$; can one find a $\delta > 0$ such that $f(x) > M$ whenever $a - \delta < x < a + \delta$, $x \neq a$?

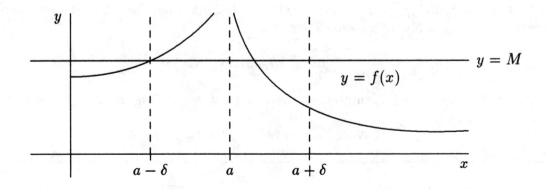

Note!

(N1) The graph of $y = f(x)$ lies above the line $y = M$ for all x satisfying $a - \delta < x < a + \delta$, $x \neq a$.

(N2) Any smaller, non-zero δ will suffice for the chosen M.

(N3) Usually $\delta = g(M)$ (δ depends upon M) in a way that δ must get smaller as M gets bigger.

Example 10: Using the M, δ definition, show $\lim\limits_{x \to 0} \dfrac{1}{x^2} = \infty$.

Note! $a = 0$; $f(x) = \dfrac{1}{x^2}$

Let $M > 0$ be arbitrary.

$$f(x) > M \implies \frac{1}{x^2} > M \implies x^2 < \frac{1}{M} \implies |x| < \frac{1}{\sqrt{M}} \implies |x - 0| < \frac{1}{\sqrt{M}};$$

a choice: $\delta = \dfrac{1}{\sqrt{M}}$ *Note!* $M \uparrow \implies \delta \downarrow$.

Show $\delta = \dfrac{1}{\sqrt{M}}$ works:

$$0 < |x - 0| < \frac{1}{\sqrt{M}} \implies |x| < \frac{1}{\sqrt{M}} \implies x^2 < \frac{1}{M} \implies \underset{\uparrow}{\frac{1}{x^2}} > M$$
$$(x \neq 0)$$

Definition (3.40): Let f be a function defined on some open interval that contains the number a, except possibly at a itself. Then

$$\lim_{x \to a} f(x) = -\infty$$

means that for every negative number N there is a corresponding number $\delta > 0$ such that

$$f(x) < N \quad \text{whenever} \quad 0 < |x - a| < \delta$$

Remarks:

(R1) Geometrically: choose arbitrary $N < 0$; can one find a $\delta > 0$ such that $f(x) < N$ whenever $a - \delta < x < a + \delta$, $x \neq a$?

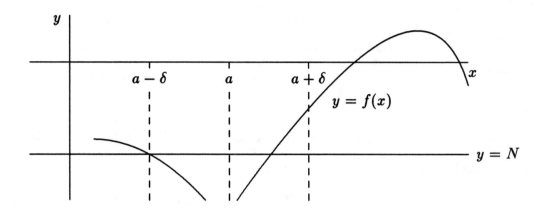

(R2) In words: $f(x)$ can be made as large enough negative as we please by taking x close enough to a, but not equal to a.

(R3) Usually $\delta = g(N)$ (δ depends upon N) in a way that δ must get smaller as N becomes more negative.

(R4) The one-sided infinite limits are correspondingly defined.

Definition (3.41): Let f be a function defined on some interval (a, ∞). Then

$$\lim_{x \to \infty} f(x) = \infty$$

means that for every positive number M there is a corresponding number $N > 0$ such that

$$f(x) > M \qquad \text{whenever} \qquad x > N$$

Remarks:

(R1) Geometrically: choose arbitrary $M > 0$; can one find a number $N > 0$ such that $f(x) > M$ whenever $x > N$?

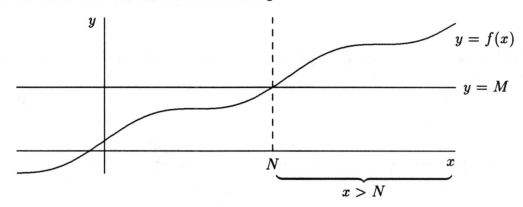

(R2) In words: $f(x)$ can be made as large as we please by taking x large enough.

(R3) The infinite limits

$$\lim_{x \to \infty} f(x) = -\infty, \qquad \lim_{x \to -\infty} f(x) = \infty, \qquad \text{and} \qquad \lim_{x \to -\infty} f(x) = -\infty$$

are correspondingly defined.

3.7   Curve Sketching

Problem: Sketch the graph of $y = f(x)$ with reasonable accuracy.

A Procedure:

✻ (S1) Determine domain of f.   *Note!* Observe continuity of f.

✳(S2) Determine intercepts:
y intercept (set $x = 0$): $f(0)$
x intercepts (set $y = 0$): solve for x such that $f(x) = 0$.

(S3) Establish symmetry:
Even function: $f(-x) = f(x) \implies$ symmetric with respect to the y axis.

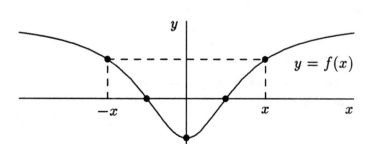

Note! Reflection about the y axis

Odd function: $f(-x) = -f(x) \implies$ symmetric with respect to the origin.

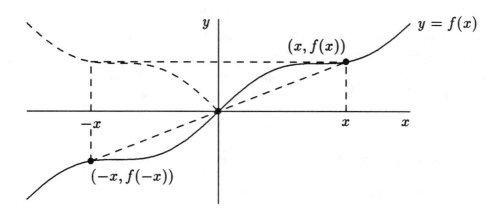

Note!

(N1) To obtain a reflection through the origin: reflect about the y axis and then reflect about the x axis.

(N2) Observation: an odd function continuous at $x = 0$ must go through the origin.

(S4) Determine if f is periodic: if there is a positive number p such that $f(x + p) = f(x)$ for all x in the domain of f, then f has period p.

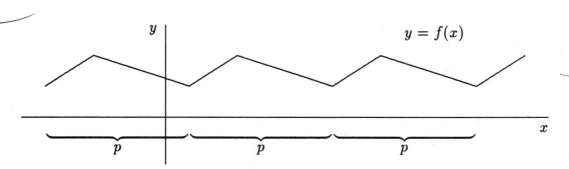

(S5) Determine asymptotes of the graph of f.
  → Horizontal asymptotes: limits as x increases/decreases without bound.
  → Vertical asymptotes: infinite limits as $x \to a$, where a is a finite number.
  Slant asymptotes: If m and b are constants such that $m \neq 0$, then the line $y = mx + b$ is a slant asymptote of the graph of $y = f(x)$ if either

$$\lim_{x \to \infty} [f(x) - (mx + b)] = 0 \qquad \text{or} \qquad \lim_{x \to -\infty} [f(x) - (mx + b)] = 0$$

$\implies$ the graph of $y = f(x)$ approaches (behaves like) the line $y = mx + b$ as $x \uparrow$ or as $x \downarrow$.

Note!

(N1) $m = 0 \implies y = b$: a horizontal asymptote.

(N2) Let P and Q be polynomials. The rational function $f(x) = P(x)/Q(x)$, where degree of $P = 1+$ degree of Q (degree of numerator is one larger than degree of denominator), has a slant asymptote $y = mx + b$, where $mx + b$ is the quotient when long division $P(x)$ divided by $Q(x)$ is carried out.

✗ (S6) Determine intervals of increase and decrease: use f'.

✗ (S7) Determine local extreme values. *Note!* Can use results of (S6).

✗ (S8) Determine intervals of concave upward and concave downward: use f''.

✗ (S9) Determine points of inflection. *Note!* Use results of (S8).

✗ (S10) Sketch the graph of $y = f(x)$ via the results of the applicable steps (S1)-(S9).

Example 1: Sketch the graph of $y = \dfrac{x^2 - 1}{x}$.

(S1) Domain: $(-\infty, 0) \cup (0, \infty)$

Note! $f(x) = \dfrac{x^2 - 1}{x}$ is a rational function $\implies$ f is continuous on its domain.

(S2) y intercept: none ($x = 0$ not in domain)

x intercepts: $f(x) = 0 \implies \dfrac{x^2 - 1}{x} = 0 \implies x^2 - 1 = 0 \implies x = \pm 1$

(S3) $f(-x) = \dfrac{(-x)^2 - 1}{-x} = \dfrac{x^2 - 1}{-x} = -\dfrac{x^2 - 1}{x} = -f(x)$: odd

(S4) f is not periodic

(S5) Horizontal asymptotes:

$$\lim_{x \to \infty} \frac{x^2 - 1}{x} = \infty$$

(behaves like $x^2/x = x$)

$$\lim_{x \to -\infty} \frac{x^2 - 1}{x} = -\infty$$

$\Longrightarrow$ none

Note! Establishes behavior of $y = f(x)$ as $x \uparrow$ and as $x \downarrow$.

Vertical asymptotes: $x = 0$ is only possibility:

$$\lim_{x \to 0^-} \frac{x^2 - 1}{x} = +\infty \quad \Longrightarrow \quad x = 0: \text{ vertical asymptote}$$

$$\left(\frac{-1}{0}; x < 0, x \text{ near } 0 : \frac{-}{-} = + \right)$$

$$\lim_{x \to 0^+} \frac{x^2 - 1}{x} = -\infty \qquad\qquad \left(\frac{-1}{0}; x > 0, x \text{ near } 0 : \frac{-}{+} = - \right)$$

Slant asymptotes: $f(x) = \dfrac{x^2 - 1}{x}$:   rational function with degree of numerator $= 1+$ degree of denominator

$$\Longrightarrow \quad \text{divide: } \frac{x^2 - 1}{x} = x - \frac{1}{x} \quad \Longrightarrow \quad \text{try } y = x.$$

$$\lim_{x \to \infty} \left(\frac{x^2 - 1}{x} - x \right) = \lim_{x \to \infty} \left(-\frac{1}{x} \right) = 0$$

$$\lim_{x \to -\infty} \left(\frac{x^2 - 1}{x} - x \right) = \lim_{x \to -\infty} \left(-\frac{1}{x} \right) = 0$$

$\Longrightarrow \quad y = x$ is a slant asymptote.

Note! The graph of $y = f(x)$ approaches $y = x$ both as $x \uparrow$ and as $x \downarrow$.

(S6) $f'(x) = \dfrac{x(2x) - (x^2 - 1)}{x^2} = \dfrac{x^2 + 1}{x^2}$

Critical numbers:

$f'(x) = 0 : \ x^2 + 1 = 0: \text{ never} \quad \Longrightarrow \quad \text{none}$

$f'(x)$ DNE: $x^2 = 0 \implies x = 0$: not in domain of $f \implies$ none

$$+ \qquad\qquad + \qquad f'(x)$$

$f \uparrow: (-\infty, 0), \ (0, \infty)$

$f \downarrow:$ nowhere

(S7) From (S6): no lx/ln.

(S8) $f''(x) = \dfrac{x^2(2x) - (x^2 + 1)(2x)}{x^4} = -\dfrac{2}{x^3}$

Nominees:

$f''(x) = 0$: none

$f''(x)$ DNE: $x = 0$: not is domain of $f \implies$ none

$$+ \qquad\qquad - \qquad f''(x)$$

CU   0   CD

f CU: $(-\infty, 0)$

f CD: $(0, \infty)$

(S9) From (S8): no points of inflection.

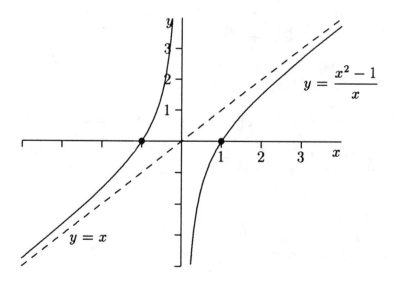

$$y = \frac{x^2 - 1}{x}$$

$$y = x$$

Example 2: Sketch the graph of $y = \dfrac{x}{x^2 - 1}$.

(S1) Domain: $x^2 - 1 = 0 \implies x = \pm 1 \implies (-\infty, -1) \cup (-1, 1) \cup (1, \infty)$

   Note! $f(x) = \dfrac{x}{x^2 - 1}$ is a rational function $\implies$ f is continuous on its domain.

(S2) y intercept: $f(0) = 0$; x intercepts: $f(x) = 0 \implies x = 0$

(S3) $f(-x) = \dfrac{-x}{(-x)^2 - 1} = -\dfrac{x}{x^2 - 1} = -f(x)$: odd

(S4) f is not periodic.

(S5) Horizontal asymptotes: $\displaystyle\lim_{x \to \infty} \frac{x}{x^2 - 1} = 0 = \lim_{x \to -\infty} \frac{x}{x^2 - 1}$

   (degree of numerator < degree of denominator)

   $\implies$ $y = 0$: horizontal asymptote as $x \to \infty$ and as $x \to -\infty$

   Vertical asymptotes:

   $\displaystyle\lim_{x \to -1^-} \frac{x}{(x - 1)(x + 1)} = -\infty \implies x = -1$: vertical asymptote

   $$\left(\frac{-1}{0}; \frac{-}{(-)(-)} = - \right)$$

   $\displaystyle\lim_{x \to -1^+} \frac{x}{(x - 1)(x + 1)} = +\infty$ $\left(\dfrac{-1}{0}; \dfrac{-}{(-)(+)} = + \right)$

   $\displaystyle\lim_{x \to 1^-} \frac{x}{(x - 1)(x + 1)} = -\infty \implies x = 1$: vertical asymptote

   $$\left(\frac{-}{0}; \frac{+}{(-)(+)} = - \right)$$

   $\displaystyle\lim_{x \to 1^+} \frac{x}{(x - 1)(x + 1)} = +\infty$ $\left(\dfrac{1}{0}; \dfrac{+}{(+)(+)} = + \right)$

Slant asymptotes: none, since $\underbrace{y = 0}$ is a horizontal asymptote as $x \to \pm\infty$.
$(y = \text{constant})$

(S6) $f'(x) = \dfrac{(x^2 - 1)(1) - x(2x)}{(x^2 - 1)^2} = -\dfrac{x^2 + 1}{(x^2 - 1)^2}$

Critical numbers:

$f'(x) = 0$: none

$f'(x)$ DNE: $x^2 - 1 = 0 \implies x = \pm 1$: not is domain of $f \implies$ none

$f \uparrow$: nowhere

$f \downarrow$: $(-\infty, -1), \; (-1, 1), (1, \infty)$

(S7) From (S6): no lx/ln

(S8) $f''(x) = -\dfrac{(x^2 - 1)^2(2x) - (x^2 + 1)(2)(x^2 - 1)(2x)}{(x^2 + 1)^4} = \dfrac{2x(x^2 + 3)}{(x^2 - 1)^3}$

Nominees:

$f''(x) = 0: \; 2x(x^2 + 3) = 0 \implies x = 0$

$f''(x)$ DNE: $x^2 - 1 = 0 \implies x = \pm 1$: not in domain of $f \implies$ none

f CU: $(-1, 0), \; (1, \infty)$

f CD: $(-\infty, -1), \; (0, 1)$

(S9) From (S8): point of inflection: $(0, f(0)) = (0, 0)$

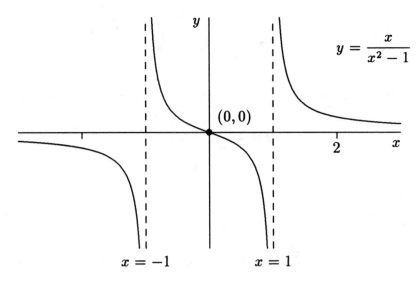

$$y = \frac{x}{x^2 - 1}$$

$(0,0)$

2

$x = -1$ $x = 1$

Example 3: Sketch the graph of $y = \dfrac{\sqrt{x+2}}{x^2}$.

(S1) Domain: $\left\{ \begin{array}{l} x + 2 \geq 0 \implies x \geq -2 \\[2mm] x^2 \neq 0 \implies x \neq 0 \end{array} \right\} \implies$ domain of $f : [-2, 0) \cup (0, \infty)$

Note! $f(x) = \dfrac{\sqrt{x+2}}{x^2} \implies f$ is continuous on its domain

(S2) y intercept: none $(x = 0$ not in domain$)$

x intercepts: $f(x) = 0 \implies x + 2 = 0 \implies x = -2$

(S3) $f(-x) = \dfrac{\sqrt{-x+2}}{(-x)^2} = \dfrac{\sqrt{-x+2}}{x^2} \implies f(-x) \neq \left\{ \begin{array}{ll} f(x) \implies & \text{not even} \\[2mm] -f(x) \implies & \text{not odd} \end{array} \right.$

(S4) f is not periodic

(S5) Horizontal asymptotes:

$$\lim_{x \to \infty} \frac{\sqrt{x+2}}{x^2} = \lim_{x \to \infty} \sqrt{\frac{x+2}{x^4}} = \lim_{x \to \infty} \sqrt{\frac{\dfrac{1}{x^3} + \dfrac{2}{x^4}}{1}} = 0$$

$$\uparrow$$
$$\left(\sqrt{x^4} = |x^2| = x^2 \right)$$

$\implies y = 0$: horizontal asymptote as $x \to \infty$

Note! Domain does not permit discussion of $\displaystyle\lim_{x\to-\infty}\frac{\sqrt{x+2}}{x^2}$

Vertical asymptotes: $\displaystyle\lim_{x\to 0^-}\frac{\sqrt{x+2}}{x^2} = +\infty \implies x = 0$: vertical asymptote

$$\uparrow$$
$$\left(\frac{\sqrt{2}}{0};\ \frac{+}{+} = +\right)$$

$$\lim_{x\to 0^+}\frac{\sqrt{x+2}}{x^2} = +\infty \qquad\qquad \left(\frac{\sqrt{2}}{0};\ \frac{+}{+} = +\right)$$

Slant asymptotes: none, since $y = 0$ is a horizontal asymptote as $x \to \infty$.

(S6) $\displaystyle f'(x) = \frac{x^2\left(\frac{1}{2}\right)(x+2)^{-1/2} - (x+2)^{1/2}(2x)}{x^4} = -\frac{3x+8}{2x^3\sqrt{x+2}}$

Critical numbers:

$f'(x) = 0$: $3x + 8 = 0 \implies x = -\dfrac{8}{3}$: not in domain $\implies$ none

$f'(x)$ DNE: $x^3\sqrt{x+2} = 0 \implies x = 0, -2$

$$\left\{\begin{array}{ll} x = 0 : & \text{not in domain} \\ x = -2 : & \text{endpoint of domain} \end{array}\right\} \implies \text{none}$$

$f \uparrow: [-2, 0)$

$f \downarrow: (0, \infty)$

Note! $\displaystyle f'_+(-2) = \lim_{h\to 0^+}\frac{\dfrac{\sqrt{(-2+h)+2}}{(-2+h)^2} - 0}{h} = \lim_{h\to 0^+}\frac{1}{\sqrt{h}(-2+h)^2} = +\infty$

(S7) From (S6): no lx/ln.

(S8) $f''(x) = -\dfrac{2x^3\sqrt{x+2}(3) - (3x+8)\left[2x^3\left(\dfrac{1}{2}\right)(x+2)^{-1/2} + \sqrt{x+2}(6x^2)\right]}{4x^6(x+2)}$

$$= \frac{15x^2 + 80x + 96}{4x^4(x+2)^{3/2}}$$

Nominees:

$f''(x) = 0:\ 15x^2 + 80x + 96 = 0$

$$\implies\quad x = \frac{-80 \pm \sqrt{(80)^2 - 4(15)(96)}}{2(15)} = \frac{-80 \pm \sqrt{640}}{30} = \frac{-80 \pm 8\sqrt{10}}{30}$$

$$= \frac{-40 \pm 4\sqrt{10}}{15} \approx \frac{-40 \pm 4(3.16)}{15} = -1.82,\ -3.51$$

$\implies\quad x = -1.82$ (since $x = -3.51$ is not in domain)

$f'(x)$ DNE: $x^4(x+2)^{3/2} = 0 \implies x = 0,\ -2$

$\left\{ \begin{array}{l} x = 0:\quad \text{not in domain} \\ x = -2:\quad \text{endpoint of domain} \end{array} \right\} \implies \text{none}$

f CU: $(-1.82, 0),\ (0, \infty)$

f CD: $(-2, -1.82)$

(S9) From (S8): point of inflection: $(-1.82, f(-1.82)) = (-1.82, 0.13)$

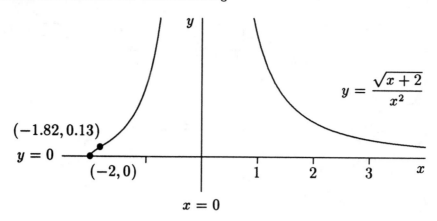

$$y = \frac{\sqrt{x+2}}{x^2}$$

$(-1.82, 0.13)$

$y = 0$

$(-2, 0)$

$x = 0$

Example 4: Sketch the graph of $y = x + \sin x, \quad -2\pi \le x \le 3\pi$

(S1) Domain: $[-2\pi, 3\pi]$ *Note!* $f(x) = x + \sin x \implies f$ is continuous on its domain

(S2) y intercept: $f(0) = 0$

x intercepts: $f(x) = 0 \implies x + \sin x = 0 \implies \sin x = -x$

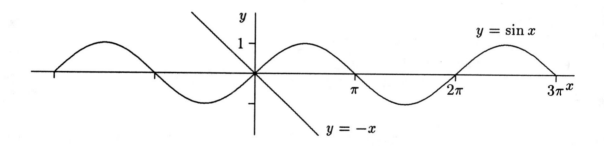

$\implies$ intersect at $x = 0$ only

(S3) $f(-x) = -x + \sin(-x) = -x - \sin x$

$\uparrow$

(sin x: odd function)

$$\implies f(-x) \ne \begin{cases} f(x) & \implies \quad \text{not even} \\ -f(x) & \implies \quad \text{not odd} \end{cases}$$

(S4) f is not periodic

(S5) Horizontal asymptotes: domain does not permit discussion of $\lim\limits_{x \to \infty} f(x)$ and $\lim\limits_{x \to -\infty} f(x)$

$\implies$ no horizontal asymptotes

Note! If domain were $(-\infty, \infty)$:

$|\sin x| \leq 1 \implies -1 \leq \sin x \leq 1 \implies x - 1 \leq x + \sin x \leq x + 1$

$$\left.\begin{array}{ll} \lim\limits_{x \to \infty} (x + \sin x) = +\infty & (x - 1 \to \infty \text{ as } x \to \infty) \\[2mm] \lim\limits_{x \to -\infty} (x + \sin x) = -\infty & (x + 1 \to -\infty \text{ as } x \to -\infty) \end{array}\right\}$$

$\implies$ no horizontal asymptote, but establishes behavior of $y = f(x)$ as $x \uparrow$ and as $x \downarrow$

Vertical asymptotes:

$\lim\limits_{x \to a} (x + \sin x) = a + \sin a$: exists for all $a \in (-\infty, \infty) \implies$ no vertical asymptotes
$\qquad\qquad \uparrow \qquad\qquad\qquad\qquad\qquad\qquad\qquad\qquad\quad \uparrow$
$\qquad$ (continuity of $x + \sin x$) $\qquad\qquad\qquad\qquad ([-2\pi, 3\pi] \subset (-\infty, \infty))$

Slant asymptotes: none, since domain is restricted to $[-2\pi, 3\pi]$

Note! From (S5) above: $x - 1 \leq x + \sin x \leq x + 1$

$$\implies \quad y = x + \sin x \text{ lies } \left\{\begin{array}{l} \text{on or above the line } y = x - 1 \\[2mm] \text{on or below the line } y = x + 1 \end{array}\right\}$$

(S6) $f'(x) = 1 + \cos x$

Critical numbers:

$f'(x) = 0 : 1 + \cos x = 0 \implies \cos x = -1 \implies x = -\pi, \pi, 3\pi \implies x = -\pi, \pi$
$\qquad\qquad\qquad\qquad\qquad\qquad\qquad\qquad\qquad\qquad\qquad\qquad\qquad\qquad\quad \uparrow$
$\qquad\qquad\qquad\qquad\qquad\qquad\qquad\qquad\qquad\qquad\qquad (x = 3\pi: \text{endpoint of domain})$

$f'(x)$ DNE: none

$$+ \quad 0 \quad + \quad 0 \quad + \qquad\qquad f'(x)$$

XXX—————————————————XXX

$$-2\pi \uparrow -\pi \quad \uparrow \quad \pi \quad \uparrow \quad 3\pi \quad x \quad f$$

$f \uparrow: [-2\pi, 3\pi]$

$f \downarrow:$ nowhere

(S7) From (S6): no lx/ln

(S8) $f''(x) = -\sin x$

Nominees:

$f''(x) = 0:$ $\sin x = 0$ $\implies$ $x = -2\pi, -\pi, 0, \pi, 2\pi, 3\pi$ $\implies$ $x = -\pi, 0, \pi, 2\pi$

$\uparrow$

$(x = -2\pi,\ 3\pi:$ endpoints of domain)

$f''(x)$ DNE: none

$$- \quad 0 \quad + \quad 0 \quad - \quad 0 \quad + \quad 0 \quad - \qquad f''(x)$$

XXX—————————————————XXX

$$-2\pi \quad -\pi \quad 0 \quad \pi \quad 2\pi \quad 3\pi \quad x$$

$$\text{CD} \quad \text{CU} \quad \text{CD} \quad \text{CU} \quad \text{CD} \qquad f$$

f CU: $(-\pi, 0)$, $(\pi, 2\pi)$

f CD: $(-2\pi, -\pi)$, $(0, \pi)$,

$(2\pi, 3\pi)$

(S9) From (S8): points of inflection are $(-\pi, f(-\pi)) = (-\pi, -\pi)$, $(0, f(0)) = (0, 0)$,

$(\pi, f(\pi)) = (\pi, \pi)$, $(2\pi, f(2\pi)) = (2\pi, 2\pi)$

Note! All points of inflection lie on the line $y = x$.

Points on graph at endpoints of domain are $(-2\pi, f(-2\pi)) = (-2\pi, -2\pi)$

and $(3\pi, f(3\pi)) = (3\pi, 3\pi)$

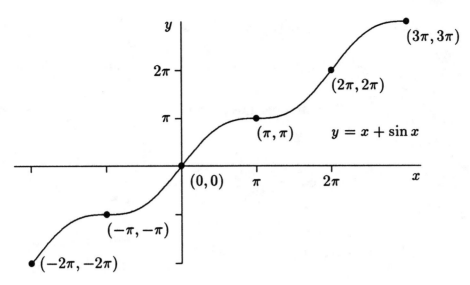

Note!

(N1) From (S6): the slope of the tangent line is zero at the critical numbers $x = -\pi,\ \pi$

(N2) $f'_-(3\pi) = 0$ (N3) $f'_+(-2\pi) = 2$

Example 5: Sketch the graph of $y = 2\cos x - \cos^2 x, \quad -2\pi \le x \le 3\pi$.

(S1) Domain: $[-2\pi, 3\pi]$

> *Note!* $f(x) = 2\cos x - \cos^2 x \quad \Longrightarrow \quad f$ is continuous on its domain

(S2) y intercept: $f(0) = 1$

x intercepts: $f(x) = 0 \quad \Longrightarrow \quad \cos x(2 - \cos x) = 0$

$$\Longrightarrow \quad \begin{cases} \cos x = 0 \quad \Longrightarrow \quad x = -\dfrac{3\pi}{2},\ -\dfrac{\pi}{2},\ \dfrac{\pi}{2},\ \dfrac{3\pi}{2},\ \dfrac{5\pi}{2} \\[2mm] \cos x = 2: \text{ never} \quad \Longrightarrow \quad \text{none} \end{cases}$$

(S3) $f(-x) = 2\cos(-x) - \cos^2(-x) = 2\cos x - \cos^2 x = f(x)$: even

$\uparrow$

($\cos x$: even function)

(S4) $f(x + 2\pi) = 2\cos(x + 2\pi) - \cos^2(x + 2\pi) = 2\cos x - \cos^2 x = f(x)$: period 2π

$$\uparrow$$

$$(\cos(x + 2\pi) = \cos x: \text{ period } 2\pi)$$

(S5) Horizontal asymptotes: Domain does not permit discussion of $\lim\limits_{x \to \infty} f(x)$ and $\lim\limits_{x \to -\infty} f(x)$

$\implies$ no horizontal asymptotes

Note! If domain were $(-\infty, \infty)$: $\lim\limits_{x \to \pm\infty} (2\cos x - \cos^2 x)$ DNE, since

for $n = 0, \pm 1, \pm 2, \ldots,$ $\left\{ \begin{array}{l} f(2n\pi) = 1 \\ f((2n+1)\pi) = -3 \end{array} \right\}$ $\implies$ no horizontal asymptotes

Vertical asymptotes: f continuous on $[-2\pi, 3\pi]$ $\implies$ no vertical asymptotes

Slant asymptotes: none, since domain is restricted to $[-2\pi, 3\pi]$

(S6) $f'(x) = -2\sin x - 2\cos x(-\sin x) = -2\sin x(1 - \cos x)$

Critical numbers:

$f'(x) = 0 :$ $\sin x(1 - \cos x) = 0$

$$\implies \left\{ \begin{array}{l} \sin x = 0 \implies x = -2\pi, -\pi, 0, \pi, 2\pi, 3\pi \implies x = -\pi, 0, \pi, 2\pi \\ \qquad\qquad\qquad\qquad\qquad\qquad\quad \uparrow \\ \qquad\qquad (x = -2\pi, 3\pi: \text{endpoints of domain}) \\ \\ \cos x = 1 \implies x = -2\pi, 0, 2\pi \implies x = 0, 2\pi \\ \qquad\qquad\qquad\qquad\quad \uparrow \\ \qquad\quad (x = -2\pi: \text{endpoint of domain }) \end{array} \right.$$

$f'(x)$ DNE: none

$f \uparrow: [-\pi, 0], [\pi, 2\pi]$

$f \downarrow: [-2\pi, -\pi], [0, \pi],$

$[2\pi, 3\pi]$

(S7) From (S6): lx: $f(0) = 1$, $f(2\pi) = 1$;   ln: $f(-\pi) = -3$, $f(\pi) = -3$

(S8) $f''(x) = -2\cos x + 2(\cos^2 x - \sin^2 x) = -2\cos x + 2\cos^2 x - 2(1 - \cos^2 x)$

$$= 4\cos^2 x - 2\cos x - 2 = 2(2\cos^2 x - \cos x - 1) = 2(2\cos x + 1)(\cos x - 1)$$

Nominees:

$f''(x) = 0$: $(2\cos x + 1)(\cos x - 1) = 0$

$$\implies \begin{cases} \cos x = -\dfrac{1}{2} \implies x = -\dfrac{4\pi}{3}, -\dfrac{2\pi}{3}, \dfrac{2\pi}{3}, \dfrac{4\pi}{3}, \dfrac{8\pi}{3} \\[2mm] \cos x = 1 \implies x = -2\pi, 0, 2\pi \implies x = 0, 2\pi \\[1mm] \qquad\qquad\qquad\qquad\qquad\quad \uparrow \\ \qquad\qquad (x = -2\pi\text{: endpoint of domain}) \end{cases}$$

$f''(x)$ DNE: none

| | $-$ | 0 | $+$ | 0 | $-$ | 0 | $-$ | 0 | $+$ | 0 | $-$ | 0 | $-$ | 0 $+$ | $f''(x)$ |

XX————————————————————————————————XX

-2π $-\dfrac{4\pi}{3}$ $-\dfrac{2\pi}{3}$ 0 $\dfrac{2\pi}{3}$ $+\dfrac{4\pi}{3}$ 2π $\dfrac{8\pi}{3}$ 3π x

CD   CU   CD   CD   CU   CD   CD   CU f

f CU: $\left(-\dfrac{4\pi}{3}, -\dfrac{2\pi}{3}\right)$, $\left(\dfrac{2\pi}{3}, \dfrac{4\pi}{3}\right)$, $\left(\dfrac{8\pi}{3}, 3\pi\right)$

f CD: $\left(-2\pi, -\dfrac{4\pi}{3}\right)$, $\left(-\dfrac{2\pi}{3}, \dfrac{2\pi}{3}\right)$, $\left(\dfrac{4\pi}{3}, \dfrac{8\pi}{3}\right)$

(S9) From (S8): points of inflection are

$$\left(-\frac{4\pi}{3}, f\left(-\frac{4\pi}{3}\right)\right) = \left(-\frac{4\pi}{3}, -\frac{5}{4}\right), \quad \left(-\frac{2\pi}{3}, f\left(-\frac{2\pi}{3}\right)\right) = \left(-\frac{2\pi}{3}, -\frac{5}{4}\right),$$

$$\left(\frac{2\pi}{3}, f\left(\frac{2\pi}{3}\right)\right) = \left(\frac{2\pi}{3}, -\frac{5}{4}\right), \quad \left(\frac{4\pi}{3}, f\left(\frac{4\pi}{3}\right)\right) = \left(\frac{4\pi}{3}, -\frac{5}{4}\right),$$

$$\left(\frac{8\pi}{3}, f\left(\frac{8\pi}{3}\right)\right) = \left(\frac{8\pi}{3}, -\frac{5}{4}\right)$$

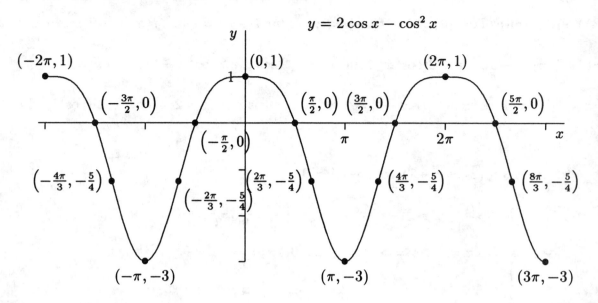

$$y = 2\cos x - \cos^2 x$$

Note!

(N1) Endpoints of domain: $(-2\pi, 1)$, $(3\pi, -3)$

(N2) $f'_+(-2\pi) = 0$ (N3) $f'_-(3\pi) = 0$

(N4) Periodic with period 2π (N5) Even function

3.8 Applied Maximum and Minimum Problems

Problem: Find the absolute extrema of an applied maximum/minimum problem.

A Procedure:

(S1) Draw a diagram and identify the given and required quantities on the diagram.

(S2) Determine equation(s) relating variables.

(S3) Determine quantity to be maximized/minimized.

(S4) Use equation(s) from (S2) to express quantity to be maximized/minimized in terms of one variable.

    Note! Keep the domain in mind.

(S5) Find the appropriate absolute extrema.

Example 1: Two nonnegative integers sum to 120. Find these integers if the product of one integer with the square of the other is to be a maximum.

Solution:

x, y: real numbers such that $x \geq 0$ and $y \geq 0$ with $x + y = 120$

$\uparrow \qquad \uparrow$

(nonnegative)

Maximize $P = x^2 y \implies P(x) = x^2(120 - x), \quad 0 \leq x \leq 120$

$\uparrow$

$(y = 120 - x; \ y \geq 0)$

Note! P: polynomial $\implies$ P is continuous on the closed interval $[0, 120]$

$P'(x) = x^2(-1) + (120 - x)(2x) = 3x(80 - x)$

Critical numbers:

$P'(x) = 0 : \ 3x(80 - x) = 0 \implies x = 0, \ 80 \implies x = 80$

$\uparrow$

($x = 0$: endpoint of domain)

$P'(x)$ DNE: none

Find ax on closed interval:

	(endpoints)		(critical #)
x	0	120	80
$P(x)$	0	0	256000

$\implies x = 80, \ y = 120 - 80 = 40$

ax

Example 2: A sheet of paper for a poster contains 18 ft². The margins at the top and bottom are 9 inches and at the sides are 6 inches. What are the dimensions of the poster if the printed area is to be a maximum?

Solution:

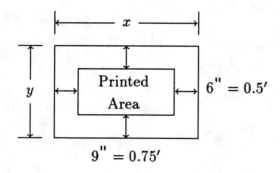

Note! Convert all units to feet (a choice)

$$xy = 18 \text{ where } \begin{cases} x \geq .5 + .5 = 1 \\ y \geq .75 + .75 = 1.5 \end{cases}$$

Maximize $A = (x - 1)(y - 1.5)$

$$\implies A(x) = (x - 1)\left(\frac{18}{x} - 1.5\right), \quad 1 \leq x \leq 12$$

$\uparrow$
$\left(y = \dfrac{18}{x}\right)$

$\uparrow$
$\left(x = \dfrac{18}{y} \leq \dfrac{18}{1.5} = 12\right)$

$\uparrow$
$(y \geq 1.5)$

Note! A: rational function such that A is continuous on $[1, 12]$

$$A'(x) = (x - 1)\left(-\frac{18}{x^2}\right) + \left(\frac{18}{x} - 1.5\right) = \frac{3}{2}\frac{12 - x^2}{x^2}$$

Critical numbers:

$A'(x) = 0: \ 12 - x^2 = 0 \implies x = \pm 2\sqrt{3} \implies x = 2\sqrt{3}$

$\uparrow$
$\left(x = -2\sqrt{3}: \text{ not in domain}\right)$

$A'(x)$ DNE: $x^2 = 0 \implies x = 0 \implies$ none

$\uparrow$
$(x = 0: \text{ not in domain})$

Find ax on closed interval:

x	1	12	$2\sqrt{3}$
$P(x)$	0	0	9.10

$$\begin{cases} x = 2\sqrt{3} \text{ ft (width)} \\[2mm] y = \dfrac{18}{2\sqrt{3}} = 3\sqrt{3} \text{ ft (height)} \\[2mm] \left(y = \dfrac{18}{x} \right) \end{cases}$$

(endpoints)   (critical #)

$\underbrace{\qquad\qquad}_{\text{ax}}$

$\Longrightarrow$

Example 3: Find the point on the graph of $y = x^2$ that is closest to the point $(4, -1/2)$.

Solution:

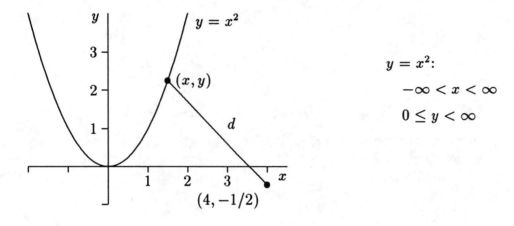

$y = x^2$:

$$-\infty < x < \infty$$

$$0 \le y < \infty$$

Minimize $d = \sqrt{(x-4)^2 + \left(y + \dfrac{1}{2} \right)^2}$ (≥ 0)

Note! $0 \le d_1 < d_2 \implies d_1^2 < d_2^2 \implies d$: minimum, then d^2: minimum

Minimize: $D = d^2 = (x-4)^2 + \left(y + \dfrac{1}{2} \right)^2$ (easier to work with)

$\implies D(x) = (x-4)^2 + \left(x^2 + \dfrac{1}{2} \right)^2, \qquad -\infty < x < \infty$

$\uparrow$

$\left(y = x^2 \right)$

Note! D: polynomial $\implies$ D is continuous on $(-\infty, \infty)$

$$D'(x) = 2(x - 4) + 2\left(x^2 + \frac{1}{2}\right)(2x) = 4(x^3 + x - 2) = 4(x - 1)(x^2 + x + 2)$$
$$\uparrow$$

$$\left(\begin{array}{l} \text{let } f(x) = x^3 + x - 2; \;\; f(1) = 0 \implies x - 1 \text{ is a factor of } f(x); \\ \text{using long division or synthetic division: } x^3 + x - 2 = (x - 1)(x^2 + x + 2) \end{array}\right)$$

Critical numbers:

$$D'(x) = 0 : \;\; (x - 1)(x^2 + x + 2) = 0 \implies \begin{cases} x - 1 = 0 \implies x = 1 \\ x^2 + x + 2 = 0 \implies \text{none} \\ \qquad\qquad\qquad\uparrow \\ \left(b^2 - 4ac = -7 < 0\right) \end{cases}$$

$D'(x)$ DNE: none

Find an on open interval:

$$
\begin{array}{ccccc}
- & 0 & + & D'(x) & \\
\hline
\downarrow & 1 & \uparrow & \quad x & D \\
& \text{ln, an} & & &
\end{array}
$$

$\implies x = 1, \;\; y = 1 \quad$ at an
$$\uparrow$$
$$\left(y = x^2\right)$$

$$\implies d = \sqrt{(1 - 4)^2 + \left(1 + \frac{1}{2}\right)^2} = \frac{1}{2}\sqrt{45} \;\; \text{at the point } (1, 1)$$

Note! The absolute minimum in Example 3 above is established as a result of the following theorem.

First Derivative Test for Absolute Extrema (3.43): Suppose that c is a critical number of a continuous function f defined on an interval.

(a) If $f'(x) > 0$ for all $x < c$ and $f'(x) < 0$ for all $x > c$, then $f(c)$ is the absolute maximum value of f.

(b) If $f'(x) < 0$ for all $x < c$ and $f'(x) > 0$ for all $x > c$, then $f(c)$ is the absolute minimum value of f.

Remarks:

(R1) If f is continuous on a closed interval, then the endpoints/critical number(s) test for ax/an can be used (see (3.8) of Section 3.1).

    Note! Used this test in Examples 1 and 2 above.

(R2) If c is the only critical number of a function f continuous on an interval I, then the First Derivative Test for Absolute Extrema can be used.

    Illustration: Consider $I = (a, b)$.

(I1) Part (a):

$$
\begin{array}{ccccccc}
 & + & & 0,\ \text{DNE} & & - & f'(x) \\
\hline
a & \uparrow & & c & & b \quad x & f \\
 & & & \text{lx, ax} & & &
\end{array}
$$

(I2) Part (b):

$$
\begin{array}{ccccccc}
 & - & & 0,\ \text{DNE} & & + & f'(x) \\
\hline
a & \downarrow & & c & & b \quad x & f \\
 & & & \text{ln, an} & & &
\end{array}
$$

Example 4: An open-topped box is to be constructed by cutting equal squares from each corner of a square sheet of aluminum and folding up the sides. If the sheet has side length 3 feet, find the largest volume of such a box.

Solution:

Maximize volume: $V = lwh \implies V(x) = (3 - 2x)^2 x, \quad 0 \le x \le \dfrac{3}{2}$

$$\uparrow$$

$$(l = 3 - 2x = w; \quad h = x)$$

Note! V: polynomial $\implies$ V is continuous on the closed interval $[0, 3/2]$.

$$V'(x) = (3 - 2x)^2 + x(2)(3 - 2x)(-2) = 3(3 - 2x)(1 - 2x)$$

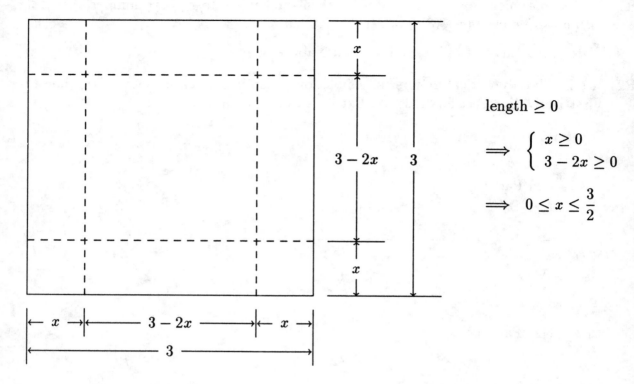

length ≥ 0

$$\implies \begin{cases} x \geq 0 \\ 3 - 2x \geq 0 \end{cases}$$

$$\implies 0 \leq x \leq \frac{3}{2}$$

Critical numbers:

$$V'(x) = 0: \ (3 - 2x)(1 - 2x) = 0 \implies \begin{cases} 3 - 2x = 0 \implies x = 3/2 \implies \text{none} \\ \qquad\qquad\qquad\qquad\quad \uparrow \\ \qquad\qquad\quad (x = 3/2: \text{endpoint of doma} \\ 1 - 2x = 0 \implies x = 1/2 \end{cases}$$

$V'(x)$ DNE: none

Find ax on closed interval:

x	0	3/2	1/2
$V(x)$	0	0	2

(endpoints) — 0, 3/2; (critical #) — 1/2; ax

$\implies \quad x = 1/2$ at ax

$$\implies \quad V = 2 \text{ ft}^3 \text{ with } l = w = 2 \text{ ft and } h = 1/2 \text{ ft}$$

Example 5: A wire of length 36 cm is cut into two pieces such that the shorter piece has a length of at least 1 cm. One piece is bent into a circle while the other piece forms a square. How should the wire be cut in order to maximize (minimize) the enclosed area?

Solution:

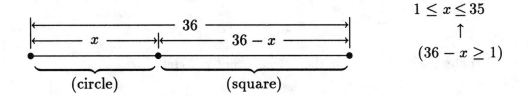

$$1 \leq x \leq 35$$
$$\uparrow$$
$$(36 - x \geq 1)$$

Area of circle: $A_1 = \pi r^2$ Circumference of circle: $2\pi r = x$

$$r = \frac{x}{2\pi} \quad \Longrightarrow \quad A_1(x) = \pi \left(\frac{x}{2\pi} \right)^2 = \frac{x^2}{4\pi}$$

Area of square: $A_2 = s^2$

Perimeter of square: $4s = 36 - x$

$$s = \frac{36 - x}{4} \quad \Longrightarrow \quad A_2(x) = \left(\frac{36 - x)}{4} \right)^2 = \frac{1}{16}(36 - x)^2$$

Maximize/minimize: $A = A_1 + A_2 \quad \Longrightarrow \quad A(x) = \frac{x^2}{4\pi} + \frac{1}{16}(36 - x)^2, \quad 1 \leq x \leq 35$

Note! A: polynomial $\Longrightarrow$ A is continuous on the closed interval $[1, 35]$

$$A'(x) = \frac{x}{2\pi} + \frac{1}{8}(36 - x)(-1) = \frac{(4 + \pi)x - 36\pi}{8\pi}$$

Critical numbers:

$$A'(x) = 0 : \ (4 + \pi)x - 36\pi = 0 \quad \Longrightarrow \quad x = \frac{36\pi}{4 + \pi} \approx 15.83$$

$A'(x)$ DNE: none

Find ax/an on closed interval:

	(endpoints)		(critical #)
x	1	35	15.83
$A(x)$	76.64	97.59	45.38

ax (under 97.59)   an (under 45.38)

$$\begin{cases} \text{ax:} & A = 97.59 \text{ cm}^2 \text{ at } x = 35 \text{ cm} \\ \text{an:} & A = 45.38 \text{ cm}^2 \text{ at } x = 15.83 \text{ cm} \end{cases}$$

Note! If the length of the smaller piece is not restricted, then

$$A(x) = \frac{x^2}{4\pi} + \frac{1}{16}(36 - x)^2, \quad 0 \le x \le 36$$

	(endpoints)		(critical #)
x	0	36	15.83
$A(x)$	81	103.18	45.38

ax (under 103.18)   an (under 45.38)

$$\begin{cases} \text{ax:} & A = 103.18 \text{ cm}^2 \text{ at } x = 36 \text{ cm} \\ & \text{(only the circle would be formed:} \\ & \text{do not cut the wire)} \\ \text{an:} & A = 45.38 \text{ cm}^2 \text{ at } x = 15.83 \text{ cm} \end{cases}$$

Example 6: An island is located at point A which is 6 miles from the nearest point B on a straight shore line. A store is located at point C which is 7 miles down the shore from point B. A person can row a boat at 4 miles per hour and walk at 5 miles per hour. Where should the boat be landed in order to reach the store in the least possible time?

Solution:

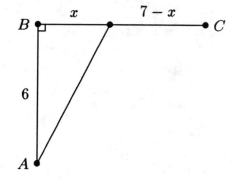

x: distance from B to where boat is beached

$\implies \quad 0 \le x \le 7$

Note!

(N1) $x = 0$: row to point B.

(N2) $x = 7$: row to point C.

Distance rowed: $d_1 = \sqrt{x^2 + 6^2} = \sqrt{x^2 + 36}$

Distance walked: $d_2 = 7 - x$

Minimize: $t = $ total time from island to store

$$\text{time} = \frac{\text{distance}}{\text{constant velocity}} \implies \begin{cases} \text{Time rowed: } t_1 = \dfrac{d_1}{4} = \dfrac{\sqrt{x^2 + 36}}{4} \\[4mm] \text{Time walked: } t_2 = \dfrac{d_2}{5} = \dfrac{7 - x}{5} \end{cases}$$

Minimize: $t = t_1 + t_2 \implies t(x) = \dfrac{\sqrt{x^2 + 36}}{4} + \dfrac{7 - x}{5}, \quad 0 \le x \le 7$

Note! t is continuous on the closed interval $[0, 7]$

$$t'(x) = \frac{1}{4}\left(\frac{1}{2}\right)(x^2 + 36)^{-1/2}(2x) + \frac{1}{5}(-1) = \frac{5x - 4\sqrt{x^2 + 36}}{20\sqrt{x^2 + 36}}$$

Critical numbers:

$t'(x) = 0: \ 5x - 4\sqrt{x^2 + 36} = 0 \implies 5x = 4\sqrt{x^2 + 36} \implies 25x^2 = 16(x^2 + 36)$

$\implies x^2 = 64 \implies x = \pm 8 \implies$ none

$\uparrow$

$(x = \pm 8: \text{ not in domain})$

$t'(x)$ DNE: none

Find an on closed interval:

$$\overbrace{\phantom{\begin{array}{cc} 0 & 7 \end{array}}}^{\text{(endpoints)}}$$

x	0	7
$t(x)$	2.9	2.32

$\underbrace{\phantom{\begin{array}{cc} & \end{array}}}_{\text{an}}$

$\implies \begin{cases} x = 7 \text{ miles at an:} \\ \text{row directly to the store} \end{cases}$

Note! If length x is assigned only the restriction $x \ge 0$, then $x = 8$ is a critical number of $t(x)$:

```
               −        0        +           t'(x)
  XXX ─────────────────────────────────────────
       0          ↓       8       ↑        x
                        ln, an                t
```

$\Longrightarrow$ person rowed 1 mile beyond the store and then walked back to the store: of course, the time involved is greater than rowing directly to the store; note that $t_2 \geq 0$ for any x, but
$t_2(8) = -0.2 < 0.$

$\Longrightarrow$ *Moral*: Carefully state the domain.

3.9 Applications to Economics

Recall: From Section 2.3: Let $C(x)$, the **cost function**, denote the cost of producing x units ($x \geq 0$) of a certain commodity. Then the **marginal cost function** is $C'(x)$.

Remark: The marginal cost function is the slope of the tangent line to the curve $y = C(x)$ at the point $(x, C(x))$

The **average cost function**

$$c(x) = \frac{C(x)}{x} \left(= \frac{\text{cost of a number of units}}{\text{number of units}} \right),   x > 0,$$

represents the cost per unit when x units are produced.

Question: What is the absolute minimum value of the average cost function c ?

$$c'(x) = \frac{d}{dx} \left(\frac{C(x)}{x} \right) = \frac{xC'(x) - C(x)}{x^2}$$

Critical numbers:

$$c'(x) = 0 : xC'(x) - C(x) = 0 \Longrightarrow C'(x) = \frac{C(x)}{x} = c(x)$$

$\Longrightarrow$ a critical number of c occurs when marginal cost $=$ average cost.

$c'(x)$ DNE: $x^2 = 0 \Longrightarrow x = 0 \Longrightarrow$ none
$$\uparrow$$
$$(x = 0: \text{ not in domain})$$

Therefore, if the average cost function has an absolute minimum, it must occur when marginal cost $=$ average cost.

Example 1: A company estimates the cost (in dollars) of producing x components is $C(x) = 400 + 0.1x + 0.0002x^2$.

(A1) Find the cost, average cost, and marginal cost of producing 1000 components, 1500 components, and 2000 components.

average cost: $c(x) = \dfrac{C(x)}{x} = \dfrac{400}{x} + .1 + .0002x$

marginal cost: $C'(x) = .1 + .0004x$

x	$C(x)$	$c(x)$	$C'(x)$
1000	700.00	0.70	0.50
1500	1,000.00	0.67	0.70
2000	1,400.00	0.70	0.90

Note!

(N1) The value $c(x)$ can be determined by substituting the value of x into the expression $c(x)$ or by dividing the value of $C(x)$ in column 2 by its corresponding value of x in column 1.

(N2) The value of $c(x)$ decreases and, then, increases as x changes from $x = 1000$ to $x = 2000$, so the absolute minimum value of c may occur between these values of x.

(A2) Find the production level that will minimize average cost.

$$C'(x) = c(x) \implies .1 + .0004x = \frac{400}{x} + .1 + .0002x$$

$$\implies x^2 = \frac{400}{.0002} = 2,000,000 \implies x \approx 1414$$

Note!

(N1) $1000 < 1414 < 2000$: satisfies observation in (N2) above.

(N2) Find the absolute minimum value of $c(x)$ by using the expression $c(x)$ directly:

$$c'(x) = -\frac{400}{x^2} + .0002$$

Critical numbers:

$$c'(x) = 0: \quad -\frac{400}{x^2} + .0002 = 0 \implies x^2 = 2{,}000{,}000 \implies x \approx 1414$$

$$c'(x) \text{ DNE: } x^2 = 0 \implies x = 0 \implies \text{none}$$
$$\uparrow$$
$$(x = 0: \text{not in domain})$$

```
            -        0        +        c'(x)
 XXXX ───────────────┬──────────────
      0         ↓   1414    ↑      x    c
                          
               ln, an
```

an: $c(1414) \approx 0.67$

or 67 cents per component

(N3) $c''(x) = \dfrac{800}{x^3} > 0$ for $x > 0$

$\implies$ the graph of $y = c(x)$ is concave up on $(0, \infty)$.

(A3) Compare the marginal cost of producing 1500 components with the cost of producing the 1501st component.

Marginal cost: $C'(x) \implies$ for $x = 1500: \ C'(1500) = 0.70$

Cost of producing next item: $C(x+1) - C(x)$

$\implies$ for $x = 1500: \ C(1501) - C(1500) \approx 1000.70 - 1000.00 = 0.70$

Hence, $C'(1500) \approx C(1501) - C(1500)$

Definition: A company produces a certain commodity,

(D1) Let $p(x)$, the **demand function**, denote the price per unit if x units ($x \geq 0$) of the commodity are sold.
Remark: Usually expect p to be a decreasing function of x.

(D2) If $R(x)$ denotes the **revenue function** then $R(x) = xp(x)$, where x is the number of units sold.
Remark: $R(x)$ represents the total revenue when x units are sold at the price $p(x)$.

(D3) The **marginal revenue function** is $R'(x)$, the rate of change of the revenue with respect to the number of units sold.

(D4) If $P(x)$ denotes the **profit function**, then $P(x) = R(x) - C(x)$, where x is the number of units sold and $C(x)$ represents the cost function for producing those x units.

(D5) The **marginal profit function** is $P'(x)$, the rate of change of the profit with respect to the number of units sold.

Question: What is the absolute maximum value of the profit function P ?

$$P'(x) = R'(x) - C'(x)$$

Critical numbers:

$$P'(x) = 0 : \quad R'(x) - C'(x) = 0 \quad \Longrightarrow \quad R'(x) = C'(x)$$

$\Longrightarrow$ a critical number of P occurs when marginal revenue = marginal cost.

$P'(x)$ DNE: none (under the assumption R' and C' exist for $x > 0$)

Therefore, if the profit function has an absolute maximum value, it must occur when marginal revenue = marginal cost.

Remark: Let a (> 0) be a critical number of P $\Longrightarrow$ $P'(a) = 0$.

If $P''(a) = R''(a) - C''(a) < 0$, then $\underbrace{P(x) \text{ is a lx}}$

(Second Derivative test)

$\underbrace{\text{with } R''(a) < C''(a)}$

(rate of change of marginal revenue is less than the rate of change of marginal cost)

Additionally, if a is the only critical point of P, then $P(a)$ is an ax.

Example 2: Find the production level that will maximize profits of a commodity whose cost function is $C(x) = 125 + 0.42x + 0.0001x^2$ and whose demand function is $p(x) = 5 - 0.025x$.

Note! $p'(x) = -.025 < 0 \quad \Longrightarrow \quad p \downarrow$

$$R(x) = xp(x) = x(5 - .025x) = 5x - .025x^2 \quad \Longrightarrow \quad R'(x) = 5 - .05x$$

$$C'(x) = .42 + .0002x$$

$$R'(x) = C'(x) : \quad 5 - .05x = .42 + .0002x$$

$$\implies \quad .0502x = 4.58 \quad \implies \quad x = \frac{4.58}{.0502} \approx 91 \quad \text{(unique critical number of } P\text{)}$$

$$\left\{ \begin{array}{l} R''(x) = -.05 \quad \implies \quad R''(91) = -.05 \\ C''(x) = .0002 \quad \implies \quad C''(91) = .0002 \end{array} \right\} \implies \left\{ \begin{array}{l} R''(91) < C''(91) : \\ \text{maximum profit at 91 units.} \end{array} \right.$$

Note! Maximize profits by considering $P(x)$ directly:

$$P(x) = R(x) - C(x) = 5x - .025x^2 - (125 + .42x + .0001x^2) = -125 + 4.58x - .0251x^2$$

$$\implies \quad P'(x) = 4.58 - .0502x$$

Critical numbers:

$$P'(x) = 0 : \quad 4.58 - .0502x = 0 \quad \implies \quad x = \frac{4.58}{.0502} \approx 91$$

$P'(x)$ DNE: none

ax: $P(91) = -125 + 4.58(91) - .0251(91)^2$

≈ 83.93

Example 3: A travel agency, in order to sell x package-deal vacations, assigns a price per vacation of $1000 - 14x$ (in dollars) for $1 \le x \le 50$. The cost to the agency for the x vacations is estimated to be $1200 + 2x + .01x^2$ (in dollars). Find the number of vacations that will maximize profits and state the maximum profit.

Note! $p(x) = 1000 - 14x \quad \implies \quad p'(x) = -14 < 0 \quad \implies \quad p \downarrow \text{ on } [1, 50]$

$$R(x) = xp(x) = x(1000 - 14x) = 1000x - 14x^2 \quad \implies \quad R'(x) = 1000 - 28x$$

$$C(x) = 1200 + 2x + .01x^2 \quad \implies \quad C'(x) = 2 + .02x$$

$$R'(x) = C'(x) : \quad 1000 - 28x = 2 + .02x$$

$$28.02x = 998 \quad \implies \quad x = \frac{998}{28.02} \approx 36 \quad \text{(unique critical numberof } P\text{)}$$

$$\left\{ \begin{array}{l} R''(x) = -28 \implies R''(36) = -28 \\ C''(x) = .02 \implies C''(36) = .02 \end{array} \right\} \implies \left\{ \begin{array}{l} R''(36) < C''(36): \\ \text{maximum profit at } x = 36. \end{array} \right.$$

$$P(36) = R(36) - C(36) = 17856 - 1084.96 = \$16{,}771.04$$

Example 4: A public lecture is being given in an auditorium with a seating capacity of 1400. A previous lecture, with a ticket price of \$7, had an attendance of 1100. It is estimated that an increase to \$8 per ticket will decrease the attendance to 900. Assuming the demand function is linear, what price per ticket should be set in order to maximize revenues?

$$\text{Let} \left\{ \begin{array}{l} x: \text{ number of tickets sold} \\ y: \text{ price per ticket (demand function)} \end{array} \right.$$

Know y is a linear function of x which passes through the points $(1100, 7)$ and $(900, 8)$.

$$\text{Use point-slope form:} \left\{ \begin{array}{l} \text{point: } (1100, 7) \\ \text{slope: } \dfrac{8-7}{900-1100} = -\dfrac{1}{200} \end{array} \right\}$$

$$\implies \quad y - 7 = -\frac{1}{200}(x - 1100) \implies y = 12.5 - \frac{x}{200}$$

$$\implies \quad p(x) = 12.5 - \frac{x}{200}, \quad 0 \le x \le 1400$$

Note! $p'(x) = -\dfrac{1}{200} < 0 \implies p \downarrow$ on $[0, 1400]$

$$R(x) = xp(x) = x\left(12.5 - \frac{x}{200}\right) = 12.5x - \frac{x^2}{200} \implies R'(x) = 12.5 - \frac{x}{100}$$

Critical numbers:

$$R'(x) = 0: \quad 12.5 - \frac{x}{100} = 0 \implies x = 1250$$

$R'(x)$ DNE: none

$$
\begin{array}{ccccc}
 & + & 0 & - & R'(x) \\
\text{XXX} & & & \text{XXX} & \\
0 & \uparrow & 1250 \quad \downarrow & 1400 \quad x & R
\end{array}
$$

ax at $x = 1250$:

$p(1250) = \$6.25$

lx, ax

3.10   Antiderivatives

Definition (3.45): A function F is called an **antiderivative** of f on an interval I if $F'(x) = f(x)$ for all x in I.

Example 1: $f(x) = x + 2$

(A1) $F(x) = \dfrac{x^2}{2} + 2x \implies F'(x) = x + 2 = f(x) \implies F$ is an antiderivative of f.

(A2) $F(x) = \dfrac{(x+2)^2}{2} \implies F'(x) = x + 2 = f(x) \implies F$ is an antiderivative of f.

Note! Different $F(x)$ in (A1) and (A2), yet each F is an antiderivative of the same f.

Theorem (3.46): If F is an antiderivative of f on an interval I, then the most general antiderivative of f on I is
$$F(x) + C$$
where C is an arbitrary constant.

Proof:

Given: F antiderivative of f on $I \implies F'(x) = f(x)$ for all $x \in I$

Let G be any antiderivative of f on I

$\implies G'(x) = f(x)$ for all $x \in I \implies G'(x) = F'(x)$ for all $x \in I$.

$\implies G(x) = F(x) + C$ for all $x \in I$, where C an arbitrary constant

$\uparrow$

(Corollary (3.17))

∗ there are an ˉinfinite # of antiderivatives ∗

Remarks:

(R1) $F(x)$: an antiderivative

(R2) $F(x) + C$: the general antiderivative; contains all possible antiderivatives

Example 1 (continued from above): $f(x) = x + 2$

(A1) $F(x) = \dfrac{x^2}{2} + 2x$: an antiderivative

$$\implies \quad G(x) = \frac{x^2}{2} + 2x + C: \text{the general antiderivative } (C: \text{arbitrary constant})$$

(A2) $F(x) = \dfrac{(x+2)^2}{2} = \dfrac{x^2 + 4x + 4}{2} = \dfrac{x^2}{2} + 2x + 2$:

an antiderivative such that $C = 2$ is a particular choice in $G(x)$ of (A1)

$$\implies \quad G(x) = \frac{x^2}{2} + 2x + 2 + C = \frac{x^2}{2} + 2x + C_1: \text{the general antiderivative}$$

$$\uparrow$$

$(C: \text{arbitrary} \quad \implies \quad C_1 = C + 2: \text{arbitrary})$

Note! $G(x)$ has same expression in (A1) and (A2)

Remark: For a given $f(x)$, the set of its antiderivatives $F(x) + C$, for assigned values to C, have graphs which are vertical translations of one another but each curve has the same slope $y' = F'(x) = f(x)$ at x.

antiderivative of a sum is the sum of an antiderivative

same rule for difference

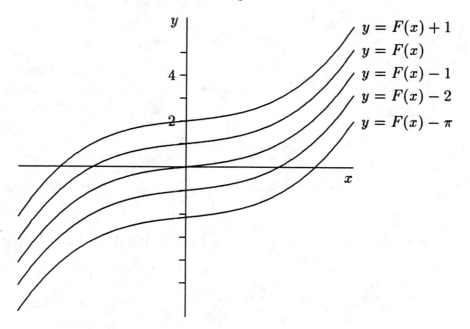

Formulas For Antiderivatives:

(F1) $f(x) = 1 \implies F(x) = x$

   Proof: $F(x) = x \implies F'(x) = 1 = f(x)$

(F2) $f(x) = x^n$, $n \neq -1 \implies F(x) = \dfrac{x^{n+1}}{n+1}$

can't be neg one because would have zero in the denomenator

   Proof: $F(x) = \dfrac{x^{n+1}}{n+1} \implies F'(x) = x^n = f(x)$

Note!

(N1) In words: Antiderivative of a variable to a fixed power, provided the power is different from negative one, is the variable to the power plus one, all divided by the power plus one.

(N2) $n = -1 \implies n + 1 = 0$, so $F(x) = \dfrac{x^{n+1}}{n+1}$ is not defined. The case for $n = -1$ is treated in Section 6.5.

(N3) The case $n = 0$ corresponds to $f(x) = 1$ treated in (F1) above.

(F3) $f(x) = cg(x)$, where c is a constant

$\implies$ $F(x) = cG(x)$, where G is an antiderivative of g.

Note! In words: Antiderivative of a constant times a function is the constant times an antiderivative of the function.

Proof: $F(x) = cG(x)$ $\implies$ $F'(x) = cG'(x) = cg(x) = f(x)$

(F4) $f(x) = g(x) + h(x)$ $\implies$ $F(x) = G(x) + H(x)$, where G and H are antiderivatives of g and h, respectively.

Note! In words: Antiderivative of sum is sum of antiderivatives.

Proof: $F(x) = G(x) + H(x)$ $\implies$ $F'(x) = G'(x) + H'(x) = g(x) + h(x) = f(x)$

(F5) $f(x) = g(x) - h(x)$ $\implies$ $F(x) = G(x) - H(x)$, where G and H are antiderivatives of g and h, respectively.

Note! In words: Antiderivative of difference is the corresponding difference of antiderivatives.

Proof: $F(x) = G(x) - H(x)$ $\implies$ $F'(x) = G'(x) - H'(x) = g(x) - h(x) = f(x)$

(F6) $f(x) = \cos x$ $\implies$ $F(x) = \sin x$

Proof: $F(x) = \sin x$ $\implies$ $F'(x) = \cos x = f(x)$

(F7) $f(x) = \sin x$ $\implies$ $F(x) = -\cos x$

Proof: $F(x) = -\cos x$ $\implies$ $F'(x) = -(-\sin x) = \sin x = f(x)$

(F8) $f(x) = \sec^2 x$ $\implies$ $F(x) = \tan x$

(F9) $f(x) = \sec x \tan x$ $\implies$ $F(x) = \sec x$

Remark: The general antiderivative is obtained from the above formulas by adding an arbitrary constant.

Comments:

(C1) No formulas for antiderivative of products or quotients are given.

(C2) The process of calculating an antiderivative $F(x)$ from a given $f(x)$ is called **antidifferentiation**.

Example 2: Antidifferentiate $f(x) = 3x^2 - 2x^{-2}$

$$F(x) = 3\left(\frac{x^3}{3}\right) - 2\left(\frac{x^{-1}}{-1}\right) + C = x^3 + 2x^{-1} + C$$
$$\uparrow$$

((F5), (F3), (F2))

Check: $F'(x) = 3x^2 + 2(-x^{-2}) = f(x)$

Example 3: Antidifferentiate $f(x) = \dfrac{3x^4 - 2x^2 + 1}{x^2}$

$\quad f(x) = 3x^2 - 2 + x^{-2}$ (rewrite $f(x)$: no rule for antiderivative of quotient)

$$\implies F(x) = 3\left(\frac{x^3}{3}\right) - 2x + \left(\frac{x^{-1}}{-1}\right) + C = x^3 - 2x - \frac{1}{x} + C$$
$$\uparrow$$

((F5), (F4), (F3), (F2), (F1))

Check: $F'(x) = 3x^2 - 2 - (-x^{-2}) = f(x)$

Example 4: Antidifferentiate $f(x) = \sin x + 3\sec x \tan x$

$\quad F(x) = -\cos x + 3\sec x + C$ ((F4), (F7), (F3), (F9))

Check: $F'(x) = -(-\sin x) + 3\sec x \tan x = f(x)$

Quiz on 3.5 - Friday
Test   - Tuesday

Example 5: Antidifferentiate $f(x) = \dfrac{3 + \sin x}{\cos^2 x}$

$$f(x) = \underset{\uparrow}{\frac{3}{\cos^2 x}} + \frac{\sin x}{\cos^2 x} = 3\sec^2 x + \sin x \sec^2 x = \underset{\uparrow}{3\sec^2 x + \tan x \sec x}$$

$$\left(\begin{array}{c}\text{rewrite } f(x)\text{: no rule for}\\ \text{antiderivative of quotient}\end{array}\right) \qquad \left((\sin x)(\sec x) = \frac{\sin x}{\cos x} = \tan x\right)$$

$$\implies \quad F(x) = 3\tan x + \sec x + C \qquad\qquad ((F4), (F3), (F8), (F9))$$

Check: $f'(x) = 3\sec^2 x + \sec x \tan x = f(x)$

Definition: A **differential equation** is an equation involving derivatives of a function.

Problem: Given a differential equation find a function whose derivatives satisfy the differential equation.

Remark: A differential equation may have given conditions associated with it in order to specify a particular function which satisfies the differential equation. These conditions enable the values of the arbitrary constants introduced during antidifferentiation to be determined.

Example 6: Determine f satisfying the differential equation $f''(x) = 36x^2 + 10$ subject to the conditions $f(0) = -7$ and $f'(0) = -3$.

$$f'(x) = 36\left(\frac{x^3}{3}\right) + 10x + C = 12x^3 + 10x + C \qquad\qquad (f'\text{: antiderivative of } f'')$$

$$f'(0) = -3 \implies 12(0)^3 + 10(0) + C = -3 \implies C = -3 \implies f'(x) = 12x^3 + 10x - 3$$

$$f(x) = 12\left(\frac{x^4}{4}\right) + 10\left(\frac{x^2}{2}\right) - 3x + C = 3x^4 + 5x^2 - 3x + C \quad (f\text{: antiderivative of } f')$$

$$f(0) = -7 \implies 3(0)^4 + 5(0)^2 - 3(0) + C = -7 \implies C = -7$$

$$\implies \quad f(x) = 3x^4 + 5x^2 - 3x - 7$$

Check: $f(0) = -7$; $f'(x) = 12x^3 + 10x - 3 \implies f'(0) = -3$; $f''(x) = 36x^2 + 10$

Example 7: Find f if $f''(x) = 6x - 2$, $f(0) = 5$ and $f(1) = 11$.

$$f'(x) = 6\left(\frac{x^2}{2}\right) - 2x + \underbrace{C_1}_{} = 3x^2 - 2x + C_1$$
$$(C_1: \text{no condition for } f')$$

$$\Longrightarrow \quad f(x) = 3\left(\frac{x^3}{3}\right) - 2\left(\frac{x^2}{2}\right) + C_1 x + C_2 = x^3 - x^2 + C_1 x + C_2$$

$$f(0) = 5 \implies 0 - 0 + C_1(0) + C_2 = 5 \implies C_2 = 5 \implies f(x) = x^3 - x^2 + C_1 x + 5$$

$$f(1) = 11 \implies 1 - 1 + c_1(1) + 5 = 11 \implies C_1 = 6 \implies f(x) = x^3 - x^2 + 6x + 5$$

Check: $f(0) = 5$; $f(1) = 1 - 1 + 6 + 5 = 11$; $f'(x) = 3x^2 - 2x + 6$; $f''(x) = 6x - 2$

Recall: (from Section 2.7): An object in rectilinear motion:

$s(t)$: position of object

$v(t) = s'(t)$: velocity of object

$a(t) = v'(t) = s''(t)$: acceleration of object

Example 8: Find $s(t)$ if $a(t) = 12t - 10$, $s(0) = 4$ and $v(0) = 3$.

$$v(t) = 12\left(\frac{t^2}{2}\right) - 10t + C = 6t^2 - 10t + C \qquad\qquad (v: \text{antiderivative of } a)$$

$$v(0) = 3 \implies C = 3 \implies v(t) = 6t^2 - 10t + 3$$

$$s(t) = 6\left(\frac{t^3}{3}\right) - 10\left(\frac{t^2}{2}\right) + 3t + C = 2t^3 - 5t^2 + 3t + C \qquad\qquad (s: \text{antiderivative of } v)$$

$$s(0) = 4 \implies C = 4 \implies s(t) = 2t^3 - 5t^2 + 3t + 4$$

Check: $s(0) = 4$; $v(t) = s'(t) = 6t^2 - 10t + 3 \implies v(0) = 3$; $a(t) = v'(t) = 12t - 10$

Example 9: A ball is released with an initial velocity v_0 from a position of s_0 feet from ground level. Find $s(t)$, the position function, at time t (≥ 0).

$$a(t) = -32 \ \ (\text{ft/s}^2)$$

$$\left\{ \begin{array}{l} \text{gravity acts downwards} \\ \text{measuring positive height upwards (a choice)} \end{array} \right\} \implies a < 0$$

Note!

(N1) Constant gravitational attraction

(N2) $s(t)$ in feet: acceleration is 32 ft/s^2

(N3) $s(t)$ in meters: acceleration is 9.8 m/s^2

Time of release: $t = 0$

$$v(0) = v_0 \ (\text{constant}) \ \left\{ \begin{array}{c} > \\ < \\ = \end{array} \right\} 0 \ \text{if} \ \left\{ \begin{array}{l} \text{thrown upward} \\ \text{thrown downward} \\ \text{released} \end{array} \right\}$$

$$s(0) = s_0 \ (\text{constant}) \ \left\{ \begin{array}{c} > \\ < \\ = \end{array} \right\} 0 \ \text{if} \ \left\{ \begin{array}{l} \text{above} \\ \text{below} \\ \text{at} \end{array} \right\} \text{ground level}$$

(A1) A ball is thrown upward with an initial velocity of 16 ft/s from a point 96 ft above the ground. Find the time required for the ball to strike the ground and the velocity as the ball strikes the ground.

Note! Must find the expression $s(t)$ in order to set $s(t) = 0$ (ground level)

Given: $a(t) = -32, \ \underbrace{s(0) = 96}_{(\text{above ground})}, \ \underbrace{v(0) = 16}_{(\text{thrown upward})}$

$$a(t) = -32 \implies v(t) = -32t + C$$

$$v(0) = 16 \implies C = 16 \implies v(t) = -32t + 16$$

$$s(t) = -32 \left(\frac{t^2}{2} \right) + 16t + C = -16t^2 + 16t + C$$

$$s(0) = 96 \implies C = 96 \implies s(t) = -16t^2 + 16t + 96$$

$$s(t) = 0: \quad -16(t^2 - t - 6) = -16(t - 3)(t + 2) = 0 \implies t = 3, \ -2$$

$$\implies t = 3 \text{ s} \quad \text{when ball strikes ground}$$
$$\uparrow$$
$$(t \geq 0)$$

velocity $v(3) = -32(3) + 16 = -80$ ft/s

Note! $v(3) = -80 < 0 \implies$ ball is falling

(A2) A ball is dropped from 144 ft above the ground. Find the time required for the ball to strike the ground.

Note! Must set $s(t) = 0$

Given: $a(t) = -32, \quad \underbrace{s(0) = 144}_{\text{(above ground)}}, \quad \underbrace{v(0) = 0}_{\text{(released)}}$

$$a(t) = -32 \implies v(t) = -32t + C$$

$$v(0) = 0 \implies C = 0 \implies v(t) = -32t$$

$$s(t) = -32 \left(\frac{t^2}{2} \right) + C = -16t^2 + C$$

$$s(0) = 144 \implies C = 144 \implies s(t) = -16t^2 + 144$$

$$s(t) = 0: \quad -16(t^2 - 9) = -16(t - 3)(t + 3) = 0 \implies t = 3, \ -3$$

$$\implies t = 3 \text{ s} \quad \text{when ball strikes the ground}$$
$$\uparrow$$
$$(t \geq 0)$$

(A3) A ball, which is thrown upward with a speed of 20 ft/s, strikes the ground 5 seconds later. From what height was the ball thrown?

Note! Must find the expression $s(t)$ in order to determine the value $s(0)$.

Given: $a(t) = -32,$ $\underbrace{s(5) = 0,}_{\text{(ground level)}}$ $\underbrace{v(0) = 20}_{\text{(thrown upward)}}$

$a(t) = -32 \implies v(t) = -32t + C$

$v(0) = 20 \implies C = 20 \implies v(t) = -32t + 20$

$s(t) = -32 \left(\dfrac{t^2}{2} \right) + 20t + C = -16t^2 + 20t + C$

$s(5) = 0 \implies -16(5)^2 + 20(5) + C = 0 \implies C = 300$

$\implies s(t) = -16t^2 + 20t + 300$

At release: $t = 0$ and $s(0) = 300$ ft